Climate Change Modelling, Planning and Policy for Agriculture

Anil Kumar Singh • Jagdish Chander Dagar •
Ayyanadar Arunachalam • Gopichandran R •
Kirit Nanubhai Shelat
Editors

Climate Change Modelling, Planning and Policy for Agriculture

Springer

Editors
Anil Kumar Singh
RVS Krishi Vishwa Vidayalaya
Gwalior, Madhya Pradesh, India

Jagdish Chander Dagar
Central Soil Salinity Research Institute
Karnal, Haryana, India

Ayyanadar Arunachalam
Indian Council of Agricultural Research
New Delhi, India

Gopichandran R
Vigyan Prasar
Noida, Uttar Pradesh, India

Kirit Nanubhai Shelat
NCCSD
Ahmedabad, Gujarat, India

ISBN 978-81-322-2959-9 ISBN 978-81-322-2157-9 (eBook)
DOI 10.1007/978-81-322-2157-9

Springer New Delhi Heidelberg New York Dordrecht London

Softcover reprint of the hardcover 1st edition 2015

Printed on acid-free paper

Springer (India) Pvt. Ltd. is part of Springer Science+Business Media (www.springer.com)

Preface

As one of the world's largest agrarian economies, agriculture sector in India accounts for about 14 % of the GDP and over 10 % of country's exports. Agriculture remains the core sector, providing employment to over 50 % of the work force, food security as well as inclusive growth and development of the Indian economy. Development and infusion of appropriate technologies have enabled annual production of over 250 mt of foodgrains and 248 mt of horticultural produce. India is among 15 leading exporters of agricultural products.

A major challenge at this stage, however, is to ensure nutritional security for over a billion people, tackle widespread hunger and child malnutrition. Related issues as the growing impacts of climate change on agriculture, biodiversity losses, erosion of natural resources base, competing demands, abiotic stresses, emerging pests and diseases, losses in harvest and post-harvest interventions compound the context. Youth from rural areas are moving away from agriculture. This enhances the vulnerability of 'stress agriculture' calling for changes in multiple dimensions of inputs, infrastructure, policy, governance and regulation, with science as the starting point.

In order to discuss the above stated and related issues pertaining to climate change impact on agriculture per se, the Indian Council of Agricultural Research and the National Council for Climate Change, Sustainable Development and Public Leadership joined hands to organize the International Conference on Climate Change, Sustainable Agriculture and Public Leadership in New Delhi on 7–9 February 2012. Over 400 national and international experts from different disciplines participated and deliberated on the pros and cons of climate change and the best options to mitigate and/or adapt in response to the challenges that emerge duly integrating agroforestry, livestock and fisheries along with agriculture.

In this synthesis volume, we have attempted to compile 22 research papers with analytical and policy planning perspectives. These insights along with the others fed into the 'New Delhi Charter 2012'. These insights are important given the fact that the country is implementing the National Mission on

Sustainable Agriculture under the Prime Minister's National Action Plan on Climate Change and that the country is committed to secure food security as well.

Gwalior, India A. K. Singh
Karnal, India J. C. Dagar
New Delhi, India A. Arunachalam
Noida, India R. Gopichandran
Ahmedabad, India K. N. Shelat

Prologue

The objective of the present volume is to highlight some salient work reported in India at the international conference organized by the Indian Council for Agricultural Research and the National Council for Climate Change, Sustainable Development and Public Leadership. While a large number of papers on several themes were presented at the conference, the present initiative is to focus on papers related to modeling of parameters relevant for interpretation of climate change impacts on crops. Such related aspects as sustainability conjectures and an overarching framework of policies to promote climate resilient agriculture are also included in the collection of papers.

It is well known that the impacts of climate change are tangible and hence there can be no debate about the need for appropriate adaptation measures, on a priority basis. However, it is equally important to recognize the fact that adaptation measures actually represent a dynamic synthesis of interventions pertaining to multiple systems. These are particularly of water, soil characteristics, genotypic and phenotypic variations and their expressions, age-correlated biochemical changes aligned with planting schedules and favorable weather/climate conditions. Nutrients, occurrence and distribution of associated vegetation including crop mixes also influence productivity. The overarching aspect of farming practice wields significant influence on the outcome and hence it is important to be clear about the particular focus of the investigations being carried out and reported in a suitable manner.

It is essential to recognize that scientific research in agriculture in India has always produced valuable results of direct relevance to her people. Importantly, preparedness to tackle disasters due to inclement weather system has prominently featured on the agenda. The recent focus on climate change and impacts has provided the necessary impetus to reorganize the framework of investigation to capture the specifics of such impacts. In this context, the importance of micro-climate variations too viz-a-viz the larger scales of impacts cannot be overemphasized. It will also be useful to help characterize natural variations versus artificially induced variations, helping us understand the complexities of individual and synergistic impacts too. Obviously, the limits and limitations of models could determine the spread and depth of the outcomes of investigations. Empirical evidences to reinforce assumptions have also to be documented with utmost care, guided by an understanding of the limits of tolerance, limiting factors, and the precautionary principle especially in the public policy interface.

The National Mission on Sustainable Agriculture has created the context to consolidate existing and emerging insights from India to set the roadmap for value added investigations for the future. Research infrastructure in India is also being embellished to rise up to the challenges. The present volume is, therefore, a useful compendium of insights at a time when these initiatives are emerging and are set to grow substantially very soon. Some of the most important questions that have designed the guiding principles for the present volume are:

1. What are the recent interpretations of the dynamics of crop productivity across several states of India? This is relevant because the soil structures, agricultural practices, and crop mixes are distinctly different across the country, further modulated by location specific agricultural practices, including alternative systems.
2. What are the sets of assumptions and models used by scientists, and is it possible to capture some initial and emerging insights especially in the Indian context? While it might be a bit too early to ask for significantly greater depth in investigation, the nature of findings reported could become useful inspiration for the way forward.
3. What is the nature of policy interventions proposed on the basis of comparable initiatives from other parts of the world, so that agriculture is mainstreamed as an integrated adaptation and mitigation option to tackle challenges posed by climate change?

The present volume is an attempt to present developments, indicative at best, from India in response to the questions raised above.

The conference also took note of some interesting segments of information reported from India and other parts of the world, centered on the aspects stated. These include the following:

- Dev 2011 (Dev MS 2011. Climate Change, Rural Livelihood and Agriculture "Focus on Food Security" in Asia-Pacific region. WP-2011-014; IGIDR) highlighted vulnerabilities associated with agriculture with special reference to livelihoods. The roles of such parameters as exposure, sensitivity and adaptive capacities have been discussed. A wide range of adaptation options and supportive policies have also been presented.
- The Food and Agriculture Organization in its report of the roundtable on organic agriculture and climate change (FAO 2011. Organic Agriculture and Climate Change Mitigation. A report of the roundtable on organic agriculture and climate change. December 2011, Rome, Italy) highlighted the dynamics of soil carbon sequestration of organic crops and importantly gaps in data for assessing mitigation potential of organic agriculture. Lifecycle assessments and related methodological challenges are also stated. This sets the context for understanding emerging trends in interpretation and their relationship with such market mechanisms as carbon credits to quantify mitigation and adaptation benefits.
- The CGIAR Research Programme on Climate Change, Agriculture and Food Security (CCAFS) in March 2012 presented its final report on sustainable agriculture and climate change (CCAFS 2012. Achieving

Food Security in the Face of Climate Change. Final report from the Commission on Sustainable Agriculture on Climate Change 63P) focused on the need for location-specific interventions duly recognizing the wide variety of options for adaptation and such preventive practices as emission reduction through suitably designed agricultural practices. The need to understand current capacities to meet technical challenges and hence the design and implementation of appropriate capacity building programmes has also been emphasized.

- The Meridian Institute (2011 Agriculture and Climate Change: A Scoping Report ISBN:978-0-615-49585-9; 116P) indicated that the options for early action on climate smart agriculture have to be shaped only by the specific circumstances and capacities within countries determined by the periodicity of productivity deficit, further modulated by food price volatility. It is therefore essential to establish and validate evidences for countries to design their respective portfolios of mitigation and adaptation options.
- Smith et al 2007 as part of the fourth assessment report of the IPCC present an excellent overview of the assessment of mitigation technologies, practices, options, potential costs with respect to sustainability and links with policies that will foster responses. (Smith P, D Martino, Z CAI, D Gwary, H Janzen, P Kumar, B McCarl, S Ogle, F O'Mara, C Rice, D Scholes and O. Sirotenko 2007: Agriculture. In Climate Change 2007: Mitigation. Contribution of Working Group-III to the Fourth Assessment Report of the Intergovernmental Panel on Climate Change (B Metz, O R Davidson, P R Bosch, R Dev, L A Meyer (eds). Cambridge University Press, UK and New York)).
- The report of the Working Group-I of the IPCC in the form of the summary for policy makers (IPCC, 2007: Summary for Policymakers. In: Climate Change 2007: The Physical Science Basis. Contribution of Working Group I to the Fourth Assessment Report of the Intergovernmental Panel on Climate Change [Solomon, S., D. Qin, M. Manning, Z. Chen, M. Marquis, K.B. Averyt, m. Tignor and H.L. Miller (eds.)]. Cambridge University Press, Cambridge, United Kingdom and New York, NY (USA)) provides a comprehensive overview of the scenarios considered for interpreting the dynamics of the phenomenon and its impacts.
- The US Department of Agriculture (Malcolm S, Marshall E, Aillery M, Heisey P, Livingston M, and Day-Rubenstein K, Agricultural Adaptation to a Changing Climate. Economic and Environmental implications vary by US region. ERR-137, USDA, Economic Research Service July 2012) indicates the region specific economic and environmental mitigation options for adaptation to climate change. The influence of crop rotations, tillage and other land use practices are also highlighted.
- The GEF and UNEP (Clemants, R., J. Haggar, A. Quezada, and J. Tomes (2011) Technologies for Climate Change Adaptation – Agriculture Sector. X. Zhu (Ed.). UNEP Rise Centre, Roskilde, 2011) discussed technologies for climate change adaptation in agriculture with a special emphasis on vulnerability assessments and criteria to prioritize related technologies. Some of the important sectors addressed by them include

water use and management, soil, crop, livestock management and sustainable farming systems.

- The IFPRI (Fofana I. 2011. Simulating the impact of Climate Change and Adaptation Strategies on Farm Productivity and Income – A Bio-economic Analysis. IFPRI discussion paper 01095, June 2011) presents an interesting case of assessing variations in land productivity as a function of temperature and precipitation patterns, with implications for farm income.
- The IFAD (2011. Climate – Smart Small Holder Agriculture: What's Different) argues for increasing access to an efficient use of water especially for the small holders followed by institutional capacities for adaptation.

These essentially represent some predominant strands of thinking and interventions. India too is witnessing several of these in various stages of development. The chapters presented reflect this emerging status. The present volume, therefore, showcases these strands with the fond hope that they will stimulate further thinking and enable appropriate action. The scale of action and its timeliness is equally important. The sources of information presented below are of the references cited in this prologue and of some others in addition to them. The objective of the listing (compiled on 09-02-2013) is also to further help readers access information cited.

Sources

1. http://www.igidr.ac.in/pdf/publication/WP-2011-014.pdf
2. http://www.fao.org/docrep/015/i2537e/i2537e00.pdf
3. http://ccafs.cgiar.org/sites/default/files/assets/docs/climate_food_commission-final-mar2012.pdf
4. http://www.ifw-members.ifw-kiel.de/publications/climate-change-agricultural-production-and-food-security-evidence-from-yemen/KAP_1747.pdf
5. http://www.gscsa2011.org/LinkClick.aspx?fileticket=dAJxg2imVx8%3D&tabid=3251
6. http://www.ipcc.ch/pdf/assessment-report/ar4/wg3/ar4-wg3-chapter8.pdf
7. http://www.ipcc.ch/pdf/assessment-report/ar4/wg1/ar4-wg1-spm.pdf
8. http://www.ers.usda.gov/media/848748/err136.pdf
9. http://tech-action.org/Guidebooks/TNA_Guidebook_AdaptationAgriculture.pdf
10. http://www.oecd.org/tad/agriculturalpoliciesandsupport/50544691.pdf
11. http://www.ifpri.org/sites/default/files/publications/ifpridp01095.pdf
12. http://www.eclac.org/portofspain/noticias/paginas/0/44160/Guyana_lCARl323.pdf
13. http://www.ecologyandsociety.org/vol16/iss3/art3/
14. http://www.ifad.org/pub/op/3.pdf
15. http://www.bis.gov.uk/assets/foresight/docs/food-and-farming/11-546-future-of-food-and-farming-report.pdf

Contents

About the Editors

Anil Kumar Singh is currently the Vice-Chancellor of the Rajmata Vijayaraje Scindia Krishi VishwaVidyalaya, Gwalior. Earlier he was the Deputy Director General (NRM) in ICAR—2007–2012; Acting Director, IARI-2005–2006; Project Director, Water Technology Centre-1999–2007. Dr. Singh joined the Agricultural Research Service in 1976 and grew up to Principal Scientist until 1999. His areas of research interest have been soil-water-nutrient-plant-atmosphere interactions, crop growth simulation, irrigation water management, and integrated watershed management. Dr. Singh has been known as a DAAD Fellow at University of Hohenheim, Germany, and has served in International Rice Research Institute between 1993 and 1995.

Jagdish Chander Dagar has been well recognized for his research in the area of agroforestry, both nationally and internationally. He is presently working as Scientist Emeritus in Central Soil Salinity Research Institute, Karnal; previously Dr. Dagar was the Assistant Director General (Agronomy/Agroforestry) at the headquarters of ICAR (2010–2012). His research interest has been in the areas of biosaline agriculture, agroforestry, rehabilitation of degraded lands, management of natural resources, bio-drainage, climate change, and sustainable agriculture and policy. He has written several books and has published more than 200 research papers of high repute in international and national level journals.

A. Arunachalam is a trained ecologist from the North-Eastern Hill University, Shillong, and has grown professionally through strong pursuit in natural resource management ecological research that is evident from his 150 research papers, 5 synthesis volumes, 100 book chapters and 25 popular articles. He started his service as a Lecturer/Assistant Professor in Forestry, North Eastern Regional Institute of Science and Technology (Arunachal Pradesh) and currently he is working as a Principal Scientist in the headquarters of Indian Council of Agricultural Research, Krishi Bhawan, New Delhi. His areas of research have been: restoration ecology and environmental management; ecosystems; forest ecology and management; agriculture and soil nutrient management; and conservation biology.

R. Gopichandran serves as the Director of Vigyan Prasar, an autonomous institution of the Department of Science and Technology, Government of India. His two doctoral degrees are in the areas of microbial ecology and chemical ecology. He holds a degree in law. A significant part of his career that has spanned nearly 25 years till now was spent on chemicals management for ozone layer protection. His insights on substitution, phase out and integrated management of ozone depletion and climate protection including inclusive mitigation and adaptation approaches were derived through strong working links with the Compliance Assistance Programme of the OzonAction programme, United Nations Environment Programme, at the regional levels. This was predominantly as part of his output at the Centre for Environment Education, Ahmedabad, and the Gujarat Energy Research and Management Institute, Gandhinagar.

Kirit Nanubhai Shelat was a public administrator for the Government of Gujarat, India. He is currently associated with NGOs and Trusts. Dr. Kirit Nanubhai Shelat is Ph.D. in Public Administration. He recently retired from the Indian Administrative Service. He started his career in public administration by joining the Gujarat Administrative Service in the year 1967. During his career of 40 years, he has worked at the grass-root level and had his hand in the formulation and implementation of policies for agricultural, rural and industrial development. He has worked as

Commissioner of Rural Development, Industries Commissioner, Commissioner for Employment and Training, Commissioner for disabled persons and Secretary Energy Department. He has also worked as Chairman of State Level Public undertakings like Gujarat Agro Industries Corporation and Land Development Corporation. He also worked in Afghanistan as Land Settlement Advisor to Government of Afghanistan. He has designed and implemented large-scale projects for poor families, farmers, and micro-entrepreneurs. He developed guidelines for micro-level planning with focus on individual poor family and village development plan. He was responsible for "cluster development approach" for small industries and "step up project" for rural micro-level entrepreneurs. He developed micro-level production plan module for individual farmer and has his hand in restructuring the Gujarat agriculture sector. He introduced new extension approach of meeting with farmers at their door step prior to monsoon by team led by agriculture scientist. He introduced scientific agriculture based on soil health moisture analysis giving a soil health card to every farmer of Gujarat. This effort went in a long way in developing sustainable agriculture in Gujarat.

Climate Change Adaptation and Mitigation Strategies in Rainfed Agriculture

B. Venkateswarlu and Anil Kumar Singh

Abstract

Climate change impacts on agriculture have been dealt at several national and international fora wherein it has always been indicated as a vulnerable ecosystem to climate change and reports do indicate that these ecosystems to contribute to the growing CO_2 level in the atmosphere, whilst a few studies do establish negative impact on the productivity of a few crops and also positive impact on crop movement along altitudinal gradient. With projected increase in water requirements, sustaining production in the rainfed areas is a challenge and in a country like India where a major junk of agricultural practices are monsoon-dependent, and has a strong socio-cultural and socio-economic bondages with farms and farming communities. Within the paradox of climate resilience in agriculture, opportunities for adaptation and mitigation strategies have been discussed in this paper with specific reference to rainfed agriculture.

Keywords

Climatic change adaption • Mitigation • Rainfed agriculture • Carbon sequestration • Conservation agriculture • Biochar

Introduction

Climate change impacts on agriculture are being witnessed all over the world. However, countries like India are more vulnerable in view of the high population depending on agriculture and excessive pressure on natural resources. The warming trend in India over the past 100 years (1901–2007) was observed to be 0.51 °C with accelerated warming of 0.21 °C for every 10 years since 1970 (Krishna Kumar 2009). The projected impacts are likely to further aggravate yield fluctuations of many crops with implications for food security and prices. Cereal productivity is projected to decrease by 10–40 % by 2100 and greater loss is expected in *rabi*. There are already evidences of negative impacts on yield of wheat and paddy in parts of India due to increased temperature, increasing

B. Venkateswarlu (✉)
Vasantrao Naik Marathwada Krishi Vidyapeth, Krishinagar, Parbhani 431402, India
e-mail: vbandi_1953@yahoo.com

A.K. Singh
Rajmata Vijayaraje Scindia Krishi Vishwavidyalaya, Gwalior 475005, India
e-mail: aksingh.icar@gmail.com

A.K. Singh et al. (eds.), *Climate Change Modelling, Planning and Policy for Agriculture*,
DOI 10.1007/978-81-322-2157-9_1, © Springer India 2015

water stress, and reduction in number of rainy days. Modeling studies project a significant decrease in cereal production by the end of this century (Mujumdar 2008). Climate change impacts are likely to vary in different parts of the country. Parts of western Rajasthan, Southern Gujarat, Madhya Pradesh, Maharashtra, Northern Karnataka, Northern Andhra Pradesh, and Southern Bihar are likely to be more vulnerable in terms of extreme events (Mall et al. 2006). For every 1° increase in temperature, yields of wheat, soybean, mustard, groundnut, and potato are expected to decline by 3–7 % (Agarwal 2009a). Similarly, rice yields may decline by 6 % for every 1° increase in temperature (Saseendran et al. 2000). Water requirement of crops is also likely to go up with projected warming, and extreme events are likely to increase.

Greater Vulnerability of Rainfed Agriculture

While climate change impacts agriculture sector in general, rainfed agriculture is likely to be more vulnerable in view of its high dependency on monsoon, the likelihood of increased extreme weather events due to aberrant behavior of south west monsoon. Nearly 85 m ha of India's 141 m ha net sown area is rainfed. Rainfed farming area falls mainly in arid, semiarid, and dry subhumid zones. About 74 % of annual rainfall occurs during southwest monsoon (June to September). This rainfall exhibits high coefficient of variation particularly in arid and dry semiarid regions. Skewed distribution has now become more common with reduction in number of rainy days. Aberrations in southwest monsoon which include delay in onset, long dry spells, and early withdrawal, all of which affect the crops, strongly influence productivity levels (Lal 2001). These aberrations are likely to further increase in the future. The risk of crop failure and poor yields always influence farmers' decision on investing on new technologies and level of input use (Pandey et al. 2000). Numerous technological (e.g., cropping patterns, crop diversification, and shifts to drought-/salt-resistant varieties) and socioeconomic (e.g., ownership of assets, access to services, and infrastructural support) factors will enhance or constrain the current capacity of rainfed farmers to cope with climate change.

Trends in Key Weather Parameters and Crop Impacts

Rainfall is the key variable influencing crop productivity in rainfed farming. Intermittent and prolonged droughts are a major cause of yield reduction in most crops. Long-term data for India indicates that rainfed areas witness 3–4 drought years in every 10-year period. Of these, 2–3 are moderate and one may be of severe intensity. However, so far no definite trend is seen on the frequency of droughts as a result of climate change. For any R&D and policy initiatives, it is important to know the spatial distribution of drought events in the country.

A long-term analysis of rainfall trends in India (1901–2004) using Mann Kendall test of significance by AICRPAM, CRIDA indicates significant increase in rainfall trends in West Bengal, Central India, coastal regions, southwestern Andhra Pradesh, and central Tamil Nadu. A significant decreasing trend was observed with respect to the central part of Jammu Kashmir, Northern MP, central and western part of UP, and northern and central part of Chattisgarh (Fig. 1). Analysis of number of rainy days based on the IMD grid data from 1957 to 2007 showed declining trends in Chattisgarh, Madhya Pradesh, and Jammu Kashmir. In Chattisgarh and eastern Madhya Pradesh, both rainfall and number of rainy days are declining. This is a cause for concern as this is a rainfed rice production system that supports a large tribal population with poor coping abilities.

Temperature is another important variable influencing crop production particularly during *rabi* season. A general warming trend has been predicted for India. It is however important to know the temporal and spatial distribution of the trend. An analysis carried out by AICPRAM, CRIDA using maximum and minimum

Fig. 1 Rainfall trends over India from 1901 to 2004 (*Source*: NPCC Annual Report, CRIDA 2009–2010)

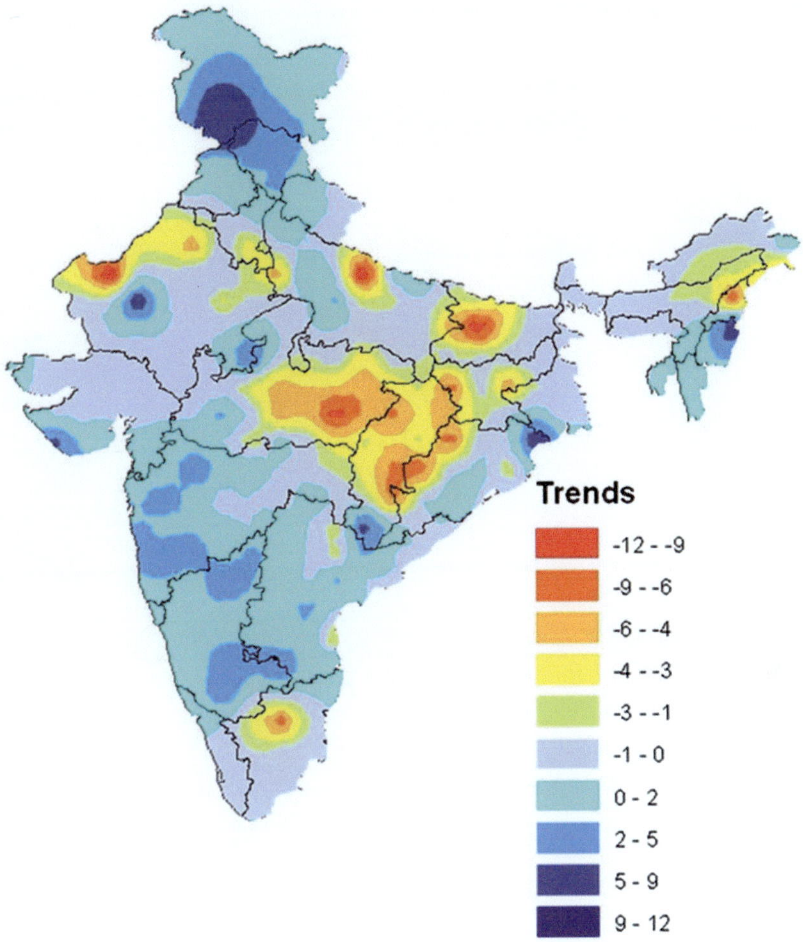

temperature data for 47 stations across India (DARE 2009) showed 9 of 12 locations in south zone with an increasing trend for maximum temperature, whereas the north, only 20 % locations showed increasing trend (Fig. 2). With respect to minimum temperature, most of the stations in India are showing an increasing trend. This is a cause of concern for agriculture as increased night temperatures accelerate respiration, hasten crop maturity, and reduce yields. The increasing trend is more evident in central and eastern zones where rainfall is also showing a declining trend. This is an area of concern and requires high attention for adaptation research.

Besides hastening crop maturity and reducing crop yields, increased temperatures will also increase the crop water requirement. A study carried out by CRIDA (unpublished) on the major crop-growing districts in the country for four crops, viz., groundnut, mustard, wheat, and maize, indicated a 3 % increase in crop water requirement by 2020 and 7 % by 2050 across all the crops/locations. The climate scenarios for 2020 and 2050 were obtained from HadCM3 model outputs using 1960–1990 as base line weather data (Table 1).

Adaptation and Mitigation Strategies

Successful adaptation to climate change requires long-term investments in strategic research and new policy initiatives that mainstream climate change adaptation into development planning.

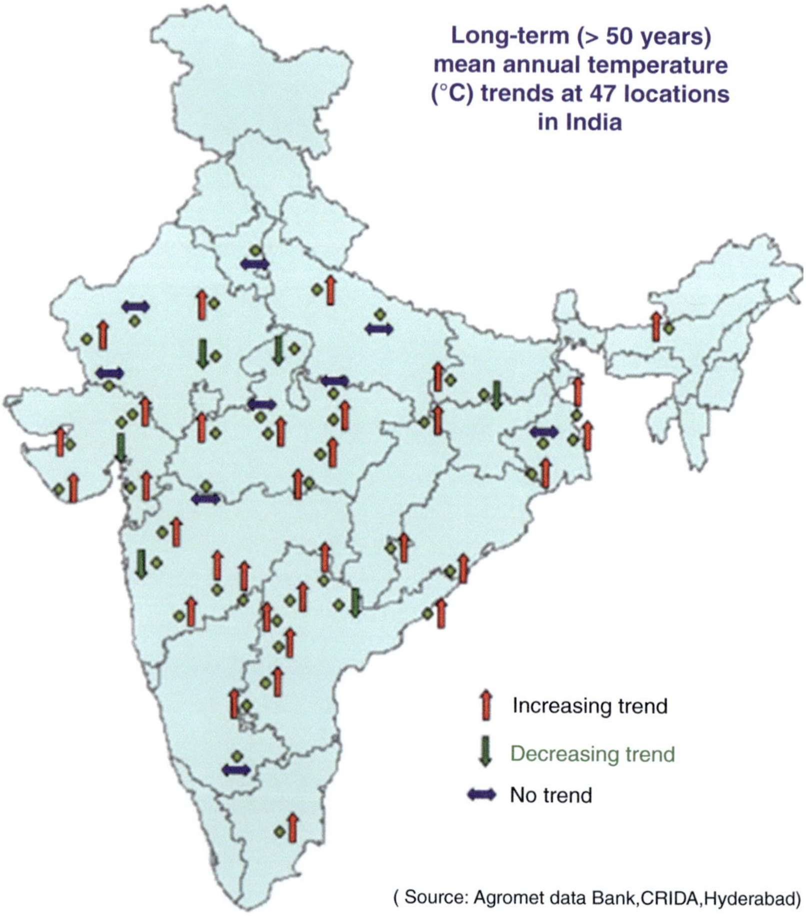

Fig. 2 Trends in mean temperature over different parts of India

As a first step, we need to document all the indigenous practices farmers have been following over time for coping with climate change. Secondly, we need to quantify the adaptation and mitigation potential of the existing best bet practices for different crop and livestock production systems in different agroecological regions of the country. Thirdly, a long-term strategic research planning is required to evolve new tools and techniques including crop varieties and management practices that help in adaptation.

Initiative of the ICAR

The Indian Council of Agricultural Research (ICAR) has initiated a Network Project on Climate Change (NPCC) in the X Five-Year Plan with 15 centers. This has been expanded in the XI Plan covering 23 centers. The initial results of the project through crop modeling have helped to understand the impacts of changes in rainfall and temperature regimes on important crops and livestock (Agarwal 2009b). Currently, the focus is on evolving cost-effective adaptation strategies. More recently during 2010, ICAR has launched the National Initiative on Climate Resilient Agriculture (NICRA) as a comprehensive project covering strategic research, technology demonstration, and capacity building. Targeted research on adaptation and mitigation is at a nascent stage in India However, some options for adaptation to climate variability can be suggested based on the knowledge already generated. These can be with respect to such induced effects as droughts, high temperatures,

Table 1 Estimated crop water requirement (mm) of four crops in major growing districts of the country under climate change scenario

District (state)	1990	2020	2050	% change over 1990 in 2020	2050
Groundnut					
Tiruvannamalai (TN)	506.0	515.2	544.0	1.8	7.5
Rajkot (Gujarat)	559.3	562.3	582.2	0.5	4.1
Junagadh (Gujarat)	522.6	528.5	550.2	1.1	5.3
Belgaum (Karnataka)	354.9	366.3	386.0	3.2	8.7
Anantapur (AP)	517.6	567.1	650.3	9.6	25.6
Bangalore (Karnataka)	490.9	510.9	559.3	4.1	13.9
Mustard					
Agra (UP)	276.4	284.0	295.5	2.7	6.9
Bharatpur (Raj)	276.2	283.8	295.1	2.8	6.8
Hisar (Haryana)	357.0	369.6	380.8	3.5	6.7
Nadia (WB)	483.2	491.5	508.8	1.7	5.3
Morena (MP)	263.4	269.0	282.5	2.1	7.2
Wheat					
Sirsa (Haryana)	281.8	293.1	301.4	4.0	7.0
Ahmedabad (Gujarat)	523.0	536.8	551.0	2.6	5.4
Ahemedanagar (Mah)	485.8	496.1	509.5	2.1	4.9
Ganganagar (Raj)	278.9	290.3	298.2	4.1	6.9
Hardoi (UP)	475.0	488.2	502.2	2.8	5.7
Kangra (HP)	367.7	380.7	391.2	3.5	6.4
Vidisha (MP)	437.1	446.9	460.4	2.3	5.3
Sangrur (Punjab)	391.1	405.4	416.3	3.7	6.4
Maize					
Udaipur (Raj)	388.8	392.4	400.9	0.9	3.1
Karimnagar (AP)	424.7	433.4	440.0	2.0	3.6
Jhabua (MP)	424.5	430.6	441.9	1.4	4.1
Begusarai (Bihar)	370.0	374.7	388.9	1.3	5.1
Bahraich (UP)	407.4	412.1	426.5	1.1	4.7
Godhra (Gujarat)	426.3	432.3	444.0	1.4	4.2
Khargaon (MP)	354.3	365.0	381.0	3.0	7.6
Aurangabad (Mah)	413.4	423.1	435.7	2.3	5.4

floods, and sea water inundation. These could be crop-based and/or resource management-based strategies.

Crop-Based Strategies

Crop-based approaches include:

- Growing crops and varieties that fit into the changed rainfall and seasons
- Development of varieties:
  - With changed duration that can over winter the transient effects of change
  - For heat stress, drought, and submergence tolerance
- Evolving varieties which respond positively in terms of growth and yield under high CO_2

In addition to the above, varieties with high fertilizer and radiation use efficiency and novel crops/varieties that can tolerate coastal salinity and sea water inundation are needed. Intercropping is a time tested practice to cope with climate variability and climate change. If a crop fails due to floods or droughts, the alternative crop could give some minimum assured returns for livelihood security. Germplasm of wild relatives and local land races could provide valuable sources of climate-ready traits. We need to revisit the germplasm collected so far to examine tolerance for heat and cold stresses and consider them even if they have been relegated earlier due to low yield considerations. A detailed account of crop-based approaches is beyond the scope of this paper. Susheel Kumar (2006) provides a succinct account of breeding objectives under the climate change context in India.

Strategies Based on Resource Conservation and Management

There are several options in soil, water, and nutrient management technologies that contribute to both adaptation and mitigation. Much of the research done in rainfed agriculture in India relates to conservation of soil and rain water and drought proofing which is an ideal strategy for adaptation to climate change (Venkateswarlu et al. 2009). Important technologies include in situ moisture conservation, rainwater harvesting and recycling, efficient use of irrigation water, conservation agriculture, energy efficiency in agriculture, and use of poor quality water. Watershed management is now considered an accepted strategy for the development of rainfed agriculture. Watershed approach has many elements which help both adaptation and mitigation. For example, soil and water conservation works, farm ponds, check dams, etc., moderate runoff and minimize floods during high intensity rainfall. The plantation of multipurpose trees in degraded lands helps in carbon sequestration. The crop and soil management practices can be tailored for both adaptation and mitigation at the landscape level. Some of the most important adaptation and mitigation approaches with high potential are described below:

Rainwater conservation and harvesting interventions are based on in situ and ex situ conservation of rainwater for recycling to rainfed crops. The arresting of soil loss contributes to reduced carbon losses. Lal (2004) estimates that if water and wind erosion are arrested, it can contribute to 3–4.6 Tg year^{-1} of carbon in India. Increased groundwater utilization and pumping water from deep tube wells is the largest contributor to GHG emissions in agriculture. If surface storage of rainwater in dugout ponds is encouraged and low lift pumps are used to lift that water for supplemental irrigation, it can reduce dependence on ground water. Sharma et al. (2010) estimated that about 28 m ha of rainfed area in eastern and central states has the maximum potential to generate runoff of 114 billion cubic meters which can be used to provide one supplemental irrigation in about 25 m ha of rainfed area. For storing such quantum of rainwater, about 50 million farm ponds are required. This is one of the most important strategies to control runoff and soil loss and contribute to climate change mitigation. Conjunctive use of surface and ground water is an important strategy to mitigate climate change. Innovative approaches in groundwater sharing can also help equitable distribution of water and reduce energy use in pumping.

Soil Carbon Sequestration

Soil carbon sequestration is yet another strategy towards mitigation of climate change. Although tropical regions have limitation of sequestering carbon in soil due to high temperatures, adoption of appropriate management practices helps in sequestering reasonable quantities of carbon in some cropping systems particularly in high rainfall regions. The potential of cropping systems can be optimized through soil carbon sequestration and sequestration into vegetation. Tree-based systems can sequester substantial quantities of carbon into biomass in a short period.

The total potential of soil C sequestration in India is 39–49 Tg year^{-1} (Lal 2004). This is inclusive of the potential of the restoration of degraded soils and ecosystems, estimated at 7–10 TgC year^{-1} (Table 2). The potential of adoption of recommended package of practices on agricultural soils is 6–7 Tg year^{-1}. This is in addition to the potential of soil inorganic carbon sequestration estimated at 21.8–25.6 TgC year^{-1}. Long-term manurial trials conducted in arid

Table 2 Soil organic carbon sequestration potential through restoration of degraded soils (Source: Lal, 2004)

Degradation process	Area (Mha)	SOC sequestration rate (kg/ha/y)	Total SOC sequestration potential (Tg C/y)
Water erosion	32.8	80–120	2.62–3.94
Wind erosion	10.8	40–60	0.43–0.65
Soil fertility decline	29.4	120–150	3.53–4.41
Waterlogging	3.1	40–60	0.12–0.19
Salinization	4.1	120–150	0.49–0.62
Lowering of water table	0.2	40–60	0.01–0.012
Total			7.20–9.82

regions of Andhra Pradesh (at Anantapur) under rainfed conditions indicate that the rate of carbon sequestration in groundnut production system varied from 0.08 to 0.45 t ha^{-1} $year^{-1}$ with different nutrient management systems (Srinivasarao et al. 2009). Under semiarid conditions in alfisol region of Karnataka, the rate of carbon sequestration was 0.04–0.38 t ha^{-1} $year^{-1}$ in finger millet system under diverse management practices. Under rabi sorghum production system in vertisol region of Maharashtra (semiarid), the sequestration rate ranged from 0.1 to 0.29 t ha^{-1} $year^{-1}$ with different integrated management options. In soybean production system in black soils of Madhya Pradesh (semiarid), the potential rate of carbon sequestration is up to 0.33 t ha^{-1} $year^{-1}$ in top 20 cm soil depth.

Site-Specific Nutrient Management

Integrated Nutrient Management and Site-Specific Nutrient Management (SSNM) is another approach with potential to mitigate effects of climate change. Demonstrated benefits of these technologies are increased rice yields and thereby increased CO_2 net assimilation and 30–40 % increase in nitrogen use efficiency. This offers important prospect for decreasing GHG emissions linked with N fertilizer use in rice systems. It is critical to note here that higher CO_2 concentrations in the future will result in temperature stress for many rice production systems, but will also offer a chance to obtain higher yield levels in environments where temperatures are not reaching critical levels. This effect can only be tapped under integrated and site directed nutrient supply, particularly N. Phosphorus (P) deficiency, for example, not only decreases yields but also triggers high root exudation and increases CH_4 emissions. Judicious fertilizer application, a principal component of SSNM approach, thus has twofold benefit, i.e., reducing greenhouse gas emissions, simultaneously improving yields under high CO_2 levels. The application of a urease inhibitor, hydroquinone (HQ), and a nitrification inhibitor, dicyandiamide (DCD), together with urea also is an effective technology for reducing N_2O and CH_4 from paddy fields. Very little information is available on the potential of SSNM in reducing GHG emissions in rainfed crops.

Conservation Agriculture (CA)

In irrigated areas, zero tillage (ZT) in particular has effectively reduced the demand for water in rice-wheat cropping system of Indo-Gangetic plains and is now considered a viable option to combat climate change. ZT has some mitigation effect in terms of enhancing soil carbon, reducing energy requirement, and improving water and nutrient use efficiency, but its actual potential has to be quantified from long-term experiments. The scope of CA in rainfed agriculture has been reviewed by Singh and Venkateswarlu (2009). While reduced tillage is possible in few production systems in high rainfall regions in eastern and northern India, non-availability of crop residue for surface application is a major constraint, particularly in peninsular and western India where it is mainly used as fodder.

Biomass Energy and Waste Recycling

A large amount of energy is used in cultivation and processing of such crops as sugarcane, food grains, vegetables, and fruits, which can be recovered by utilizing residues for energy production. This can be a major strategy in climate change mitigation by avoiding burning of fossil fuels and recycling crop residues. The integration of biomass-fuelled gasifiers and coal-fired energy generation would be advantageous in terms of improved flexibility in response to fluctuations in biomass availability with lower investment costs. Waste-to-energy plants offer twin benefits of environmentally sound waste management and disposal, as well as the generation of clean energy.

Livestock production has been an integral part of agriculture in India. Livestock provides an excellent recycling arrangement for most of the crop residue. Most by-products of cereals, pulses, and oilseeds are useful as feed and fodder for livestock, while that of cotton, maize, pigeon pea, castor and sunflower, and sugarcane are used as low-calorie fuel or burnt to ashes or left in the open to decompose over time. Ideally such residue is incorporated into soil to enhance physical properties of the soil and its water holding capacity. Lack of proper chipping and soil incorporation equipment is one of the major reasons for the colossal wastage of agricultural biomass in India. Increased cost of labor and transport is another reason for lack of interest in utilizing the biomass. This is an area where little or no effort has gone in despite availability of opportunities for reasons such as aggregation, transport, and investment in residue processing facilities.

Many technologies as briquetting, anaerobic digestion vermin-composting, biochar, etc., exist, but they have not been commercially exploited. This area is gradually receiving attention now as a means to producing clean energy by substituting forest biomass for domestic needs. Modest investments in decentralized facilities for anaerobic digestion of agricultural residue through vermin-composting and biogas generation can meet the needs of energy-deficit rural areas and simultaneously contribute to climate change mitigation.

Biomass-Based Biogas Production

There is renewed interest in the use of anaerobic digestion processes for efficient management and conversion of cattle dung and other agro-industrial wastes (livestock, paper and pulp, brewery, and distillery) into clean renewable energy and organic fertilizer source. The biogas captured could not only mitigate the potential local and global pollution but could either be combusted for electricity generation using combined heat and power generator in large to medium enterprises or used for cooking and lighting for small households. A 2 m^3 digester can generate up to 4.93 t CO_2e year-1 of certified emission reduction (CER). Animal wastes are generally used as feedstock in biogas plants. But the availability of these substrates is one of the major problems hindering the successful operation of biogas digesters. Khandelwal (1990) reported that the availability of cattle waste could support only 12–30 million family-size biogas plants against the requirement of 100 million plants. A significant portion of 70–88 million biogas plants can be run with fresh/dry biomass residues. Of the available 1,150 billion tons of biomass, a fifth would be sufficient to meet this demand.

Biochar

When biomass is exposed to moderate temperatures, between about 400 and 500 °C (a kind of low-temperature pyrolysis), under complete or partial exclusion of oxygen, it goes through exothermic processes and releases many gases in addition to heat along with biochar (Czernik and Bridgewater 2004). Pyrolysis produces biochar, a carbon-rich, fine-grained, porous substance and solid by-product, similar in its appearance to charcoal. This can be returned to the soil and help several environmental benefits, including enhanced soil carbon sequestration and soil fertility improvement (Lehmann 2007). Both heat and gases can be captured to produce energy carriers such as electricity, hydrogen, or bio-oil which can be used as a fuel for various purposes in the process of manufacturing biochar. In addition to energy, certain valuable coproducts, including wood preservative, food flavoring substances, adhesives, etc., can be obtained (Czernik and Bridgewater 2004).

In India, it has been projected that about 309 m t of biochar could be produced annually, the application of which might offset about 50 % of carbon emission (292 TgC year^{-1}) from fossil fuel (Lal 2005). Rice-wheat cropping system in the Indo-Gangetic plains of India produces substantial quantities of crop residues, and if these residues can be pyrolyzed, 50 % of the carbon in biomass can be returned to the soil as biochar, increasing soil fertility and crop yields while sequestering carbon. At CRIDA, research on biochar use in rainfed crops has been initiated. Biochar from castor, cotton, and maize stalks was produced by using a portable kiln and used as an amendment for pigeon pea during *kharif* 2010. The crop growth was significantly superior in biochar-applied plots from all three sources (Venkatesh 2010).

Agroforestry

Agroforestry systems like agri-silviculture, silvopasture, and agri-horticulture offer adaptation and mitigation opportunities. They buffer farmers against climate variability and reduce atmospheric loads of greenhouse gases much as they can sequester carbon and produce a range of economic, environmental, and socioeconomic benefits. The extent of sequestration can be up to 10 t ha^{-1} year^{-1} in short rotation Eucalyptus, leucaena plantations (Table 3). Agri-silviculture systems with moderate tree density with intercrops however have lower potential.

Policy Issues

Apart from the use of technological advances to combat climate change, there has to be a sound and supportive policy framework. The framework should address issues of redesigning the social sector with focus on vulnerable areas/ populations, introduction of new credit instruments with deferred repayment liabilities during extreme weather events and weather insurance as a major vehicle of risk transfer. Governmental initiatives should be undertaken to identify and prioritize adaptation options in key sectors (storm warning systems, water storage and diversion, health planning, and infrastructure needs). Focus on integrating national development policies into a sustainable development framework that complements adaptation should accompany technological adaptation methods.

In addition, the role of local institutions in strengthening capacities, e.g., SHGs, banks, and agricultural credit societies, should be promoted. The role of community institutions and private sector in relation to agriculture should be a matter of policy concern. Political will is essential to enable economic diversification in terms of risk spreading, diverse livelihood strategies, migrations, and financial mechanisms. Policy initiatives in relation to access to banking, microcredit/insurance services before, during, and after a disaster event, access to communication and information services is imperative in the envisaged climate change scenario. Some of the key policy initiatives that are to be considered are:

- Mainstream adaptations by considering impacts in all major development initiatives.
- Facilitate greater adoption of scientific and economic pricing policies, especially for water, land, energy, and other natural resources.
- Consider financial incentives and package for improved land management and explore CDM benefits for mitigation strategies.
- Establish a "Green Research Fund" to strengthen research on adaptation, mitigation, and impact assessment (Venkateswarlu and Shankar 2009).

Table 3 Carbon storage (Mg/ha/year) in different Agri-silvicultural systems

Location	System	Carbon sequestration (Mg ha^{-1} $year^{-1}$)	References
Raipur	Gmelina-based system	2.96[a]	Swamy and Puri (2005)
Chandigarh	Leucaena-based system	0.87	Mittal and Singh (1989)
Jhansi	Anogeissus-based system	1.36	Rai et al. (2002)
Coimbatore	Casuarina-based system	1.45	Viswanath et al. (2004)

[a]Includes soil carbon storage of 0.42 Mg ha^{-1} $year^{-1}$ (up to 60 cm depth)

Conclusions

Even though climate change in India is now a reality, a more certain assessment of the impacts and vulnerabilities of rainfed agriculture sector and a comprehensive understanding of adaptation options across the full range of warming scenarios and regions is needed and will would go a long way in preparing the nation to tackle climate change challenges. A multipronged strategy of using indigenous coping mechanisms, wider adoption of the existing technologies, and/or concerted R&D efforts for evolving new technologies are needed for adaptation and mitigation. Policy incentives will play a crucial role in adoption of climate-ready technologies in rainfed agriculture too as in other sectors. The state agricultural universities and regional research centers will have to play major roles in adaptation research on region- and location-specific systems, while national level efforts are required to come up with cost-effective mitigation options, new policy initiatives, and global cooperation.

References

Agarwal PK (2009a) Global climate change and Indian agriculture; case studies from ICAR network project. Indian Council of Agricultural Research, p 148

Agarwal PK (2009b) Vulnerability of Indian agriculture to climate change; current stage of knowledge. Paper presented at the national workshop – review of implementation of work program towards Indian Network of Climate Change Assessment, Ministry of Environment and Forests, New Delhi, 14 October

CRIDA (2009) Annual report Central Research Institute for Dryland Agriculture. Indian Council of Agricultural Research, New Delhi

Czernik S, Bridgewater AV (2004) Overview of applications of biomass fast pyrolysis oil. Energy Fuels 18:590–598

DARE (2009) Annual report of Department of Agricultural Research and Education, 2008–2009, pp 1–137

Khandelwal KC (1990) Biogas technology development implementation strategies: Indian experience. In: Proceedings of the international conference on biogas technology, implementation strategies, Pune, India, 10–15 January 1990 (Bremen, Ed. Borda)

Kumar S (2006) Climate change and crop breeding objectives in the twenty first century. Curr Sci 90: 1053–1054

Kumar K (2009) Impact of climate change on India's monsoon climate and development of high resolution climate change scenarios for India. Presented at MoEF, New Delhi on October 14, 2009. http://moef.nic.in

Lal R (2001) Future climate change: implications for India Summer monsoon and its variability. Curr Sci 81:1205–1207

Lal R (2004) Soil carbon sequestration in India. Clim Change 65:277–296

Lal R (2005) Carbon sequestration and climate change with special reference to India. In: Proceedings international conference on soil, water and environmental quality – issues and strategies, Indian Society of Soil Science, Division of SS & Ac, IARI, New Delhi, India, pp 295–302

Lehmann J (2007) A handful of carbon. Nature 447: 143–144

Mall R, Singh R, Gupta A, Srinivasan G, Rathore L (2006) Impact of climate change on Indian agriculture: a review. Clim Change 78:445–478

Mittal SP, Singh P (1989) Intercropping field crops between rows of *Leucaena leucocephala* under rainfed conditions in northern India. Agrofor Syst 8: 165–172

Mujumdar PP (2008) Implications of climate change for sustainable water resources management in India. Phys Chem Earth Pt A/B/C 33:354–358

Pandey S, Behura D, Villano R, Naik D (2000) Economic cost of drought and farmers coping mechanisms: a study of rainfed rice systems in eastern India. Discussion paper series, IRRI Laguna, Philippines

Rai AK, Solanki KR, Rai P (2002) Performance of *Anogeissus pendula* genotypes under agrisilvicultural system. Indian J Agrofor 4(1):71–74

Saseendran ASK, Singh KK, Rathore LS, Singh SV, Sinha SK (2000) Effects of climate change on rice production in the tropical humid climate of Kerala, India. Clim Change 44:495–514

Sharma BR, Rao KV, Vittal KPR, Ramakrishna YS, Amarasinghe U (2010) Estimating the potential of rainfed agriculture in India: prospects for water productivity improvements. Agric Water Manag 97(1):23–30

Singh AK, Venkateswarlu B (2009) Conservation agriculture in arid and semi regions: opportunities and constraints. Indian Farm (Special issue) 58:3–8

Srinivasarao C, Ravindra Chary G, Venkateswarlu B, Vittal KPR, Prasad JVNS, Kundu S, Singh SR, Gajanan GN, Sharma RA, Deshpande AN, Patel JJ, Balaguravaiah G (2009) Carbon sequestration strategies under rainfed production systems of India. Central Research Institute for Dryland Agriculture, Hyderabad, p 102 (ICAR)

Swamy SL, Puri S (2005) Biomass production and C-sequestration of *Gmelina arborea* in plantation and agroforestry system in India. Agrofor Syst 64(3): 181–195

Venkatesh G (2010) Biochar for carbon sequestration. News letter. Central Research Institute for Dryland Agriculture, Hyderabad, January–June 2010, p 12

Venkateswarlu B, Shankar AK (2009) Climate change and agriculture adaptation and mitigation strategies. Indian J Agron 54:226–230

Venkateswarlu B, Arun Kumar S, Gogoi AK (2009) Climate change adaptation and mitigation in Indian agriculture. In: Proceedings national seminar on climate change adaptation strategies in agriculture and allied sectors, KAU, Trissur, December 2008, pp 109–121

Viswanath S, Peddappaiah RS, Subramoniam V, Manivachakam P, George M (2004) Management of *Casuarina equisetifolia* in wide-row intercropping systems for enhanced productivity. Indian J Agrofor 6(2):19–25

Diversification of Agriculture in India: Challenges Ahead

J.P. Mishra and Ayyanadar Arunachalam

Abstract

Agriculture in its present form face challenges to meet the growing needs of the Indian population and therefore warrants convergence of policies and technologies to handhold the farmer for increased production, productivity and income. Therefore, diversification has been dealt as a way out to address the above challenge and to provide a kind of safety-net for the Indian farms and farming communities. The changes over the last few decades with respect to crop diversification has been addressed to indicate the potential in the scenario of reduced land availability and also the need for diversifying Indian agricultural systems to make it more remunerative while balancing the ecological principles of conservation and diversity.

Keywords

Crop diversification • Agricultural production • Contract farming • Land use policy

Introduction

Agriculture is the primary source of livelihood of a significant segment of India's population; about 70 % population lives in rural areas where the main occupation is agriculture. Indian agriculture is dominated by small and marginal farm holdings with an average farm size of only 1.23 ha. Around 94 % of farmers have land holdings smaller than 4 ha and they cultivate nearly 65 % of the arable land. On the other hand, only one per cent of the total farmers have operational land holdings above 10 ha, and they utilize 11.8 % of the total cultivated land.

This has to be viewed in the context of the fact that the country is bestowed with diverse agro-climatic conditions favourable for the cultivation of diverse crops like food grain crops, commercial crops, horticulture, plantation, spices, trees, etc. 'Self reliance' in food grains has been the cornerstone of India's policies in the last five decades. Around 65 % of the total cultivated area is under food grain crops. The cultivation of commercial crops especially cotton, sugarcane,

J.P. Mishra (✉)
Planning Commission, Government of India, Yojna Bhawan, New Delhi 110001, India
e-mail: mishrajaip@gmail.com

A. Arunachalam
Indian Council of Agricultural Research, Krishi Bhawan, New Delhi 110001, India

A.K. Singh et al. (eds.), *Climate Change Modelling, Planning and Policy for Agriculture*,
DOI 10.1007/978-81-322-2157-9_2, © Springer India 2015

spices, etc. continue to be developed to meet the growing demand of the increasing population. However, surplus of the commercial crops has been earning foreign exchange also from the exports. There was renewed focus on food grains in the eleventh plan in view of their stagnating production and increasing demand.

A number of initiatives such as the National Food Security Mission and Rashtriya Krishi Vikas Yojana were taken to improve the production. As a result of increased attention to agriculture in recent times, there is marked improvement in production of food grains, fruits and vegetables and other related commodities. A significant public policy challenge is to twin the benefits bestowed by diverse systems and a plan milieu duly recognizing the impacts of inclement weather systems. Landscapes are important elements that can determine the structure and function of the diverse systems stated with implications for the occurrence and distribution of crop systems that can mutually reinforce productivity through physiological and agronomic traits that are resilient. The present paper consummates important aspects of these elements in the context of challenges to sustainable agriculture in the country.

Current Production Scenario

The Indian Agriculture has changed emphatically from a food shortage stage to self-reliance and surplus owing to technological breakthrough achieved during the Green Revolution. This was well supported with appropriate fiscal and policy initiatives of the government. The present food grain production is ca. 259 million tonnes. There is significant increase in pulses production with a renewed focus during the eleventh five-year plan. From a stagnating production level of 14 million tons during the last two decades, a production of more than 18 million tons was achieved during 2010–2011 (Ministry of Agriculture 2012). Oilseeds too have seen a jump in production to about 32.5 million tons. Increased pulses and oilseeds production is encouraging; however, a large quantity of these commodities are still being imported to bridge the demand supply gap.

The area under food grains has increased by 30.2 %, while production exhibited a 397 % signifying the quantum jumps in productivity over a period of 61 years. This major breakthrough was achieved through increase in the irrigated area, intensive fertilizer use and adoption of HYV. The area under irrigation is 58.7 % in rice, 91.3 % in wheat and 93.7 % in sugarcane. Some horticultural crops as potato and banana also registered significant increase. However, more efforts are required to popularize the horticultural/plantation crops in view of changing dietary pattern and growing fruit/vegetable demand. Similarly efforts under the National Horticulture Mission and Technology Mission on Horticulture for North Eastern and Hill States have resulted in impressive gains during the last five-year plan with the production of horticultural crops reaching a level of about 350 million tons. Importantly more and sustained efforts are required to popularize the horticultural/plantation crops in view of changing dietary pattern and ever growing fruit/vegetable demand.

The gross cropped area has increased by about 63.21 million ha over the period of the last 58 years. However, there has been a differential dynamic in this increase. Of the total increase, rice and wheat alone account for 33.33 million ha, while oilseeds account for 15.75 million ha and pulses for 7.13 million ha, alongside coarse cereals that have reduced by 11.05 million ha. The trend in area coverage under various crops since 1950–1951 is given in Fig. 1.

An overall analysis of percentage increase in area under different crops indicates that the share of rice area has been almost constant (22–24 %) over the years. However, the share of wheat area has increased from 7.4 % in 1950–1951 to 14.9 % in 2010–2011 the majority of which came from coarse cereals. The area under other crops that include horticulture/plantation crops, spices, fodder crops, etc., has increased from 16.01 m ha, in 1950–1951, to 23.10 million ha, an increase of 8.40 million ha in 2010–2011 in terms of gross cropped area. This indicates a process of diversification though not properly designed. With adequate reserve of rice and wheat, it is possible to design a plan of

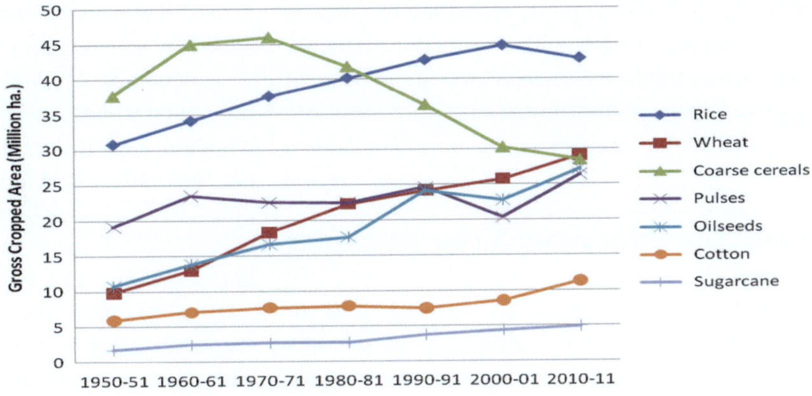

Fig. 1 Temporal change in gross cropped area (*Source*: Agriculture Statistics at a Glance (2012) Directorate of Economics and Statistics, Ministry of Agriculture, Government of India)

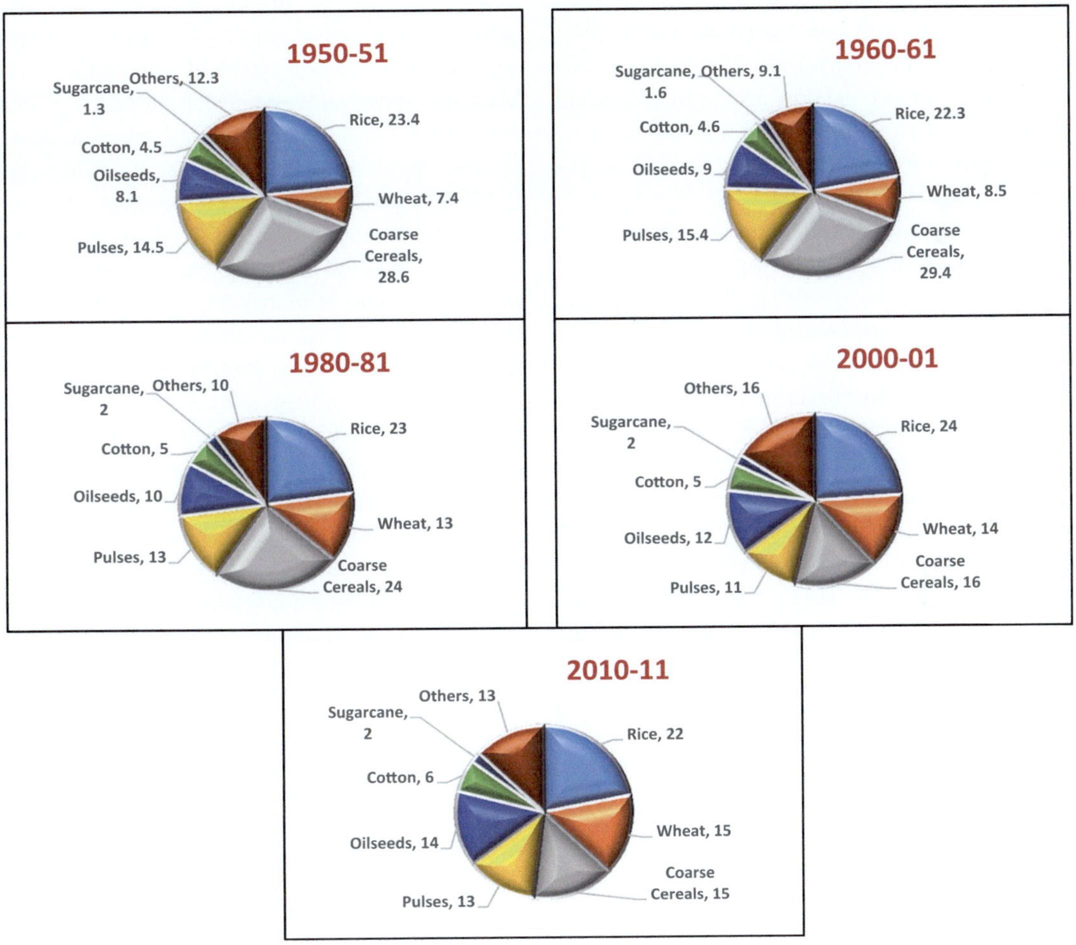

Fig. 2 Per cent share of crops in the gross cropped area (*Source*: Agriculture Statistics at a Glance (2012). Directorate of Economics & Statistics, Ministry of Agriculture, Government of India)

diversification to divert some of the area under staple food grains to other crops particularly oilseeds, pulses, horticultural crops, medicinal and aromatic plants, floriculture and silvi-pasture systems. The integrated farming system approach including crops, animal husbandry, piggery/poultry/fish culture and allied business is the proper way to deal with the situation. The per cent share of different crops in the gross cropped area has been given in Fig. 2.

Silent Diversification: A Few Examples

Diversification in agriculture implies shift in cropping patterns from traditionally grown less remunerative crops to more remunerative oilseeds, pulses, fodder crops, horticulture, medicinal and aromatic plants, floriculture, etc. This is aligned with land-based activities, livestock and fishery enterprises. The interventions for crop production and postharvest management are also technology driven, thus offering employment opportunities in the rural as well as urban areas.

Crop diversification enhances choice and reduces risk Thornton and Lipper (2013). These transitions are further enabled by the government's policies and thrust on some crops over a given time. This includes support mechanism on production; marketing, etc., induce the crop shift. The Technology Mission on Oilseeds (TMO) started during 1986 to give thrust on oilseeds production as a national need for the country's requirement and less dependency on imports. The coverage of the oilseeds increased from 18.63 m ha in 1986–1987 to 26.48 m ha in 2011–2012. The specific favourable cropping situations also encourage crop shift, e.g. cultivation of rice in the command areas due to plenty of water, wheat in well irrigated areas due to availability of irrigation water, etc.

Liberal financial assistance for improved planting material, high-tech inventions like micropropagation, drip irrigation, greenhouse cultivation, etc., has encouraged farmers to grow horticultural crops. Production of horticultural crops which was only 96.56 million tonnes during 1991–1992 has increased to 233.14 million tonnes during 2010–2011, and the area under horticultural crops increased by 8.97 million ha as in Table 1. Thus, there has been a gradual shift towards horticultural crops in crop diversification.

Development of market infrastructure also helps in the introduction of alternate crops, e.g. introduction of soybean in vertisols of Madhya Pradesh and Maharashtra. Often low-volume high-value crops as spices also aid in crop diversification. Higher profitability and stability in production also induce crop diversification, for example, sugarcane replacing rice and wheat. Diversified cropping systems are adopted in rainfed areas primarily to reduce the risk of crop failures owing to drought or inadequate rains. There has been a significant growth in the area under rice, wheat, oilseeds and sugarcane during the period 1984–1994, whereas the pace of growth reduced considerably during the last decade. The negative growth in the area of coarse cereals has been noticed during 1984–2004. However, the renewed efforts through various crop development programmes created enabling environment for increase in growth of production of all the crops during XI Plan.

Table 1 Area (A, m ha) and production (P, m ha) of horticultural crops

	1991–1992		1996–1997		2001–2002		2010–2011	
Crops	A	P	A	P	A	P	A	P
Fruit	2.87	28.63	3.58	40.46	3.89	45.37	6.62	75.83
Vegetables	5.14	58.56	5.51	75.10	6.25	93.92	8.22	137.69
Spices	2.05	1.90	2.38	2.80	2.50	3.02	2.46	4.02
Coconut	1.53	6.93	1.89	8.98	1.84	8.67	1.90	10.82
Cashew	0.53	0.30	0.66	0.43	0.72	0.45	0.46	0.13
Others	0.21	0.24	0.26	0.30	0.51	1.07	1.64	4.65
Total	**12.33**	**96.56**	**14.28**	**128.07**	**15.71**	**152.5**	**21.30**	**233.14**

Source: Agricultural Statistics at a Glance 2012, Directorate of Economics and Statistics, Ministry of Agriculture, Government of India

Need for Diversification

As indicated, diversification in agriculture is required to address several issues, viz. environmental, food security, nutritional security, food security sustainability, risk coverage, employment opportunity, etc. Continuous cultivation of a group of crops in the Indo-Gangetic Plain of the country has created many agro-ecological problems including deterioration in soil fertility, soil physical conditions, etc. Excessive exploitation of underground water led to fall in water table, while excessive use of water in canal network area has brought about soil salinity and alkalinity. Depletion of organic carbon content of soil (from 0.8 to 0.2 % or even low) as evident from experimental results of Punjab Agricultural University, Ludhiana, indicates poor soil support system for crop growth. The fall in water table has been observed almost half a metre per year. Continuous and heavy use of pesticides in rice, wheat and cotton has led to environmental hazards including the depletion in ozone layer. The intensive use of fertilizers has increased the nitrate concentration in water. Use of industrial effluents for agriculture especially near cities has created problem of heavy metals in the food item. Emission of methane from low-lying rice areas also causes environmental hazards including depletion of ozone layer. These problems can be reduced to minimum by incorporating horticultural/plantation crops, oilseeds and pulses in the system based on the suitability of land resources and need of the locality/nation.

Although India has the surplus food grain stock, food and nutritional security at household level continue to be an area of concern. The per capita availability of food grains is 438.6 g against the recommended level of 500 g/day. The availability of pulses, a rich source of protein is less than 31.6 g/day/capita against the recommended intake of 55 g/day/capita. The per capita availability of vegetables is below the Recommended Dietary Allowance of 295 g/day by about 40 g/day. The availability of oils and fats is about 20.2 g/day against the requirement of 40 g/day. With increasing standards of living, the need for horticultural produce will further increase. This calls for increasing oilseed, pulse and horticultural crop production. There is a need for diversification from cereal-dominated production system to horticulture, oilseeds, and pulses to increase the area under these crops to enhance the production to meet the increasing demand of diversified food basket.

Monoculture and continuous cropping of rice-wheat system have created various disadvantages besides deteriorating soil fertility. The physical condition of soil like aeration has deteriorated leading to build-up of diseases and pests. The decline in productivity in the most productive regions in the country has become a major concern. All these developments have endangered the basic fabric of sustainability in some of the most productive zones of the country.

Pulses, offer a unique opportunity in building up soil fertility, when included in the cropping systems (PAN et al. 2012). This is by virtue of converting atmospheric nitrogen into a form that could be assimilated through biological nitrogen fixation (BNF). Besides BNF, they have several added advantages including substantial amount of organic matter through leaf litter and roots, break in disease cycle, creation of soil aeration, etc. They fit well in rainfed environment due to their low water demand and capacity to extract water from deeper soil profiles. The inclusion of pulses in cereal-cereal system economizes nitrogen use to the tune of 30–40 kg/ha for succeeding cereal crops. About 60 % of the gross cropped area of the country is rainfed where mono-cropping is prevalent. These areas have higher risk of crop failure due to failure of rains. Diversified agriculture through intercropping, agroforestry, horticulture and plantation crops may help in minimizing the risk of complete crop failure and ensuring income to the farmers.

Under mono-cropping situation, labour requirement is limited to peak crop season leaving the remaining period dry for employment. Moreover, the attraction towards conventional agriculture is reducing in educated rural youth due to their non-remunerative nature. Indeed, the technology-driven interventions in crops as well as horticulture sector (floriculture, poly house

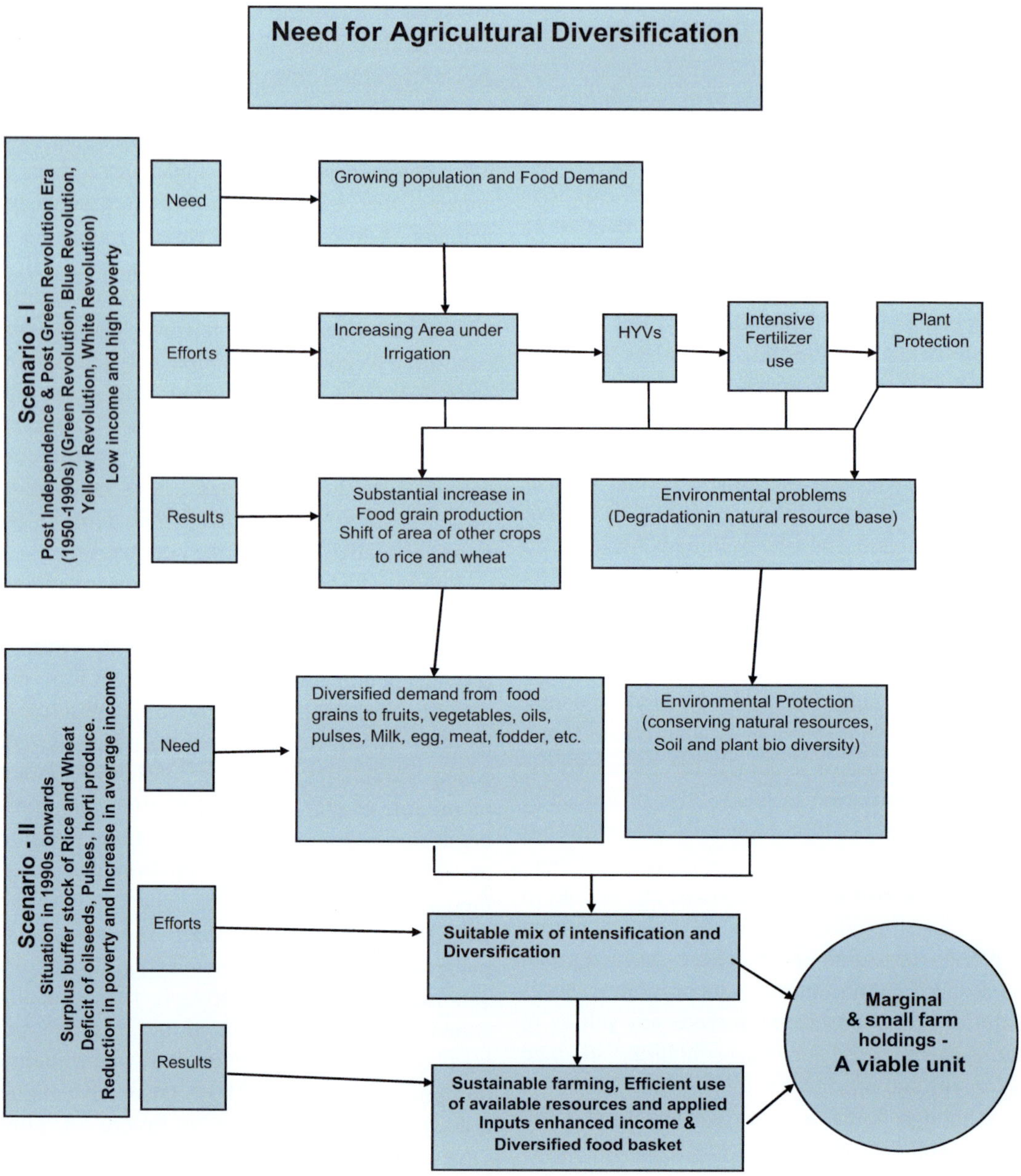

Fig. 3 Agricultural diversification framework for agricultural development

farming, etc.) will provide intellectual satisfaction in attracting the rural youths. Diversified agriculture may help in creating additional employment opportunities. Substantial growth in the area of rice and wheat coupled with productivity gain by virtue of improved varieties and input intensive cropping has resulted in huge food grain stocks; storage and the disposal of which have become a problem. Contrary to this, the nation has to depend increasingly on import of edible oils and pulses to meet the domestic needs. This again needs consideration for agriculture diversification from rice-wheat system. Milk, meat and egg production has to be increased substantially to meet the growing demands. This would be possible only through improved cattle breeds, better feeding and foddering quality. Green fodder plays a vital role in augmenting the milk production. The demand for fodder will increase proportionately to the cattle population. There is an immense need to bring additional area under fodder cultivation through diversification or by utilizing waste and fallow lands. The need for agricultural diversification in different scenarios of agricultural development in the country is depicted in Fig. 3.

Potential Areas for Diversification

Rice-Wheat Cropping System of Indo-Gangetic Plains

About 10.5 million ha area in the Indo-Gangetic plains is under rice-wheat system where continuous cropping of the same species of plants has led to several problems related to soil fertility, factor productivity, alkalinity/salinity and build-up of specific diseases and several environmental hazards. The system offers unique opportunity for diversification through short season oilseeds as catch crop and spring summer pulses like urd bean and mung bean and summer vegetables in sequential cropping. During kharif season, the pulses and oilseeds may be grown in place of upland rice. This will also increase soil fertility and avoid excessive exploitation of ground water. Concerted efforts are required for rational crop planning according to land suitability, need of locality and marketing avenues.

Rice-Rice System of Peninsular States

Rice monoculture is prevalent on a vast area in the peninsular region of the country, i.e. Andhra Pradesh and Tamil Nadu. The system in the long run has created anaerobic conditions thereby leading to reduced zone in the rhizosphere which often restricts the root and plant growth. Incorporation of pulses/oilseeds in the cropping system needs to be promoted to overcome these problems.

Upland Rice of Eastern States

Over a large area in Eastern India, rice is grown under upland rainfed conditions. Productivity of upland rice is low and the crop generally suffers from moisture stress conditions. Complete failure of upland rice crops is not uncommon. Because of high risk involved, the input use in upland rice is also low which further affects on the productivity of the crop. These upland paddy areas, if diversified, could be more productive with cultivation of maize, oilseeds, pulses, millets, vegetables, horticultural and floriculture crops.

Rice Fallows

Of the total 44 million ha area under rice production, about 15 million ha is under rainfed conditions. Majority of this area is left fallow after the harvest of the rice crop. The rice fallows offer an attractive opportunity of crop intensification and diversification by virtue of their potential to grow short-duration pulses, viz. urd bean, mung bean, pea, lentil and oilseeds like rapeseed, linseed, groundnut, vegetables, floriculture, etc., on residual moisture. In the country, rice fallows exists under three distinct agro-climatic conditions.

Coastal Ecosystem

In coastal ecosystem, there is sufficient moisture in soil profile at the time of harvesting allowing sowing of rabi season crop. The winters are mild providing most congenial conditions for the growth of winter season oilseeds, pulses, vegetables, floriculture, etc. The soils of this region are acidic and acid-tolerant crops may be included in the cropping system. The most suitable crops are black gram and green gram in pulses, rapeseed and mustard and linseed in oilseeds and vegetables.

Semiarid Rainfed Ecosystem

This ecosystem spreads over Chhattisgarh, part of Madhya Pradesh and Maharashtra and is characterized by moisture stress at the time of harvesting of rice. Winters are relatively cool with occurrence of frost. Under these conditions, broadcasting of small seed—oilseeds and pulses—in standing crop of rice before 4–7 days of its harvest is the ideal practice. The crops invariably suffer due to moisture stress especially during their terminal growth stages which calls for suitable moisture/water conservations measures to be adopted for obtaining optimum yields and introducing floriculture and oilseed crops.

Humid Ecosystem

This ecosystem consists of rice fallows of Bihar, West Bengal, Orissa and North Eastern Hill States and is characterized by heavy rainfall and sufficient moisture in the soil profile at the time of the harvest of the rice. About 8–10 lakh ha of saturated soils remain least exploited in the rainfed lowland areas of Eastern India. In many cases, the soils remain saturated throughout the year, particularly in inter-hill valleys. Winters are cool and provide opportunity for the introduction of a range of oilseeds, pulses and vegetable crops. Under saturated conditions winter season rice is also a possible option for crop intensification.

Sugarcane-Based Cropping System

Sugarcane is grown under two distinct biophysical conditions existing in Central and Peninsular India and the northern part of the country. Mostly long-duration (18 months) crop is grown in the central and southern parts of the country, while in Northern India, the crop duration is only 12 months or even less. Sugarcane, initially a slow-growing crop and widely planted, offers an attractive opportunity for the introduction of early maturing pulses like mung bean, urd bean, cowpea (spring/summer planted crop), chickpea and oilseeds like rapeseed and mustard and sunflower (autumn-planted crop) and vegetables as intercrop. The system provides a bonus yield of intercrops besides the optimum yield of the main crop, i.e. sugarcane. Pulses when incorporated in the system also enrich the soil through biological nitrogen fixation and add to soil fertility.

Agroforestry Systems

Alley cropping of arable crops with widely planted fodder/shrubs offer an opportunity to grow short-duration less water demanding oilseeds and pulses in the fragile rainfed ecosystem. These systems besides providing additional output in terms of food grain also enrich the soil and help as a natural soil cover against water and wind erosion. The important agroforestry systems may be Casuarina/Leucaena + green gram/black gram/cowpea/sunflower.

Designing Crop Diversification

The appropriate technology suiting to local agroclimatic conditions as well as to the local needs of the people is available for promoting the diversification in agriculture. It may not be practical to adopt a uniform strategy for crop diversification for the whole of the country having varied ecological zones and production systems. Different diversification strategies are required to be considered for different zones and areas. Therefore, the strategies on crop diversification may be a regionally differentiated one based on the geographical and agro-ecological considerations as suggested below:

- Zone I – Northern plains, embracing Punjab, Haryana and western U.P. The major thrust in the region would be to promote diversification towards less water demanding crops such as oilseeds, pulses, horticulture and vegetables from rice-wheat system.
- Zone II – Western arid zone embracing Rajasthan and Gujarat where more emphasis is required to promote oilseeds, pulses and perennial plants preferably medicinal and aromatic plants, quality millets and tree fodders in place of rice, wheat, sugarcane and cotton.
- Zone III – In Central Plateau region embracing Madhya Pradesh, Chhattisgarh and Maharashtra, there is a need to promote

Table 2 Traditional crops and proposed crops to be grown in some states

Sl. no	State	Traditional crops grown	Proposed crops to be undertaken
1.	Madhya Pradesh	Upland paddy, kodon kutki, low-lying paddy, paddy bunds	Tur, til, niger, castor, barley, soybean, maize, urd and mung, fruits
2.	Jharkhand	Marua, niger, wheat, paddy	Til, tur, durum wheat, vegetables, gram, lentil
3.	Chhattisgarh	Upland and midland paddy, kodon-kutki	Niger, soybean, maize, tur, horticulture crops
4.	Bihar	Paddy, lowland paddy, wheat, kharif tur	Fine and scented paddy, pulses, oilseeds, banana, litchi, rabi tur, floriculture
5.	Uttar Pradesh	Paddy, wheat, sugarcane, tobacco	Scented paddy, rabi maize, pulses, groundnut, cotton, soybean and vegetable crops
6.	Gujarat	Upland paddy, coarse cereals	Maize, sesame, castor, date palm, medicinal plants, spices and fodder crops
7.	Andhra Pradesh	Cotton, paddy, groundnut, jowar, rabi paddy	Castor, red gram, soybean, maize, sunflower, sesame, pulses and vegetables, coriander, horticultural crops
8.	Karnataka	Groundnut, cotton, millets, castor, soybean	Horticulture crops, castor, intercropping with sugarcane, coconut and others
9.	West Bengal	Boro rice, upland crops, kharif rice	Wheat, summer groundnut, til, maize, soybean, vegetables, fruits, flowers and spices
10.	Haryana	Rice, wheat	Arhar, gram, mung, maize, cotton, sugarcane, fruits and vegetables
11.	Punjab	Rice, wheat	Basmati rice, organic Basmati, maize, sunflower, pulses, barley, hayola, winter maize
12.	Orissa	Rice, minor millet	Pulses, groundnut, cowpea, kharif vegetables, horticultural crops
13.	Tamil Nadu	Rice, pulses	Banana, onion, sugarcane
14.	Maharashtra	Paddy, coarse cereals, cotton	Oilseeds, soybean, horticultural crops, medicinal plants and floriculture
15.	Rajasthan	Bajra, cotton, pulses	Intermixed cropping of bajra, pulses, oilseeds, green fodder crops, guar, moth, spices and horticulture crops

oilseeds and pulses in place of rice and sugarcane. The rice fallows need to be exploited through oilseeds and pulses. The floriculture and horticulture and quality millets should also get priority.

- Zone IV – The southern peninsula region embracing Tamil Nadu, Karnataka, Andhra Pradesh and Kerala requires emphasis on diversifying the rice-rice system and also sugarcane-based cropping system towards pulses, oilseeds, horticulture, plantation crops, spices and floriculture.
- Zone V – In the eastern region embracing Bihar, Jharkhand, Orissa and West Bengal, crop diversification/intensification in rice fallows should get top priority.
- Zone VI – In hilly areas including NE states, diversification of non-remunerative rice and rice fallows through organic pulses and perennial fruits, floriculture and plantation crops on hilly slopes needs to be promoted. In Sub-Himalayan northern states like Uttaranchal, Himachal Pradesh and Jammu and Kashmir, promotion of horticultural/plantation crops and floriculture on hill slopes and off-season vegetables in valleys is required to be promoted.

The central government has been advising states to promote agricultural diversification keeping in view the availability of natural resources, domestic demand and the potential of exports. The issue has been discussed and deliberated upon with the state governments on various occasions. Based on such interactions with the state governments, state-specific strategy for agricultural diversification is given in Table 2.

Policy Interventions

An environment of favourable policy support particularly in the areas of land policy, land leasing, contract farming, land share company and price support mechanism would be necessary to promote designed agricultural diversification.

Land Use Policy Reforms

It is now widely recognized that agricultural growth can be accelerated through mutually supportive forward and backward integrations leading to better postharvest management and higher value addition in the agricultural sector. This would require substantive reforms in the land use policy of the country to achieve economies of scale in agriculture, boost agro-processing, facilitate development of postharvest and marketing infrastructure in rural areas to promote agricultural diversification and thereby help improve socio-economic conditions of small farmers and landless labourers. In this context, promotion of contract farming, land leasing and land sharing company with some adaptation will lead to the desired vertical integration of all aspects of diversification leading to rural transformation.

Contract Farming

Contract farming can help in promoting demand-driven agricultural diversification in a big way. At present, contract farming is not widespread in India, although cultivation of commercial crops like cotton, sugarcane, tobacco, tea, coffee, rubber and dairy enterprises have had some elements of contract farming. Crops like tomato, groundnut, chilli, barely, potato, cotton, etc., have come under contractual agreements in recent years with some centralized processing and marketing units.

The notable examples are Hindustan Lever Ltd. in tomato; PepsiCo in tomato and basmati rice; United Breweries in barley; Mahindra Shubhlabh Services Ltd. in maize; Sharp Menthol India Ltd. in mentha in Punjab; Maxworth fruits in horticultural crops in A.P., Karnataka and Tamil Nadu; VST National Products Ltd. in cucumber and paprika in A.P.; and Cadbury in cocoa in Karnataka and NDDB in Maharashtra. Similar success stories of contract farming are Amul and NDDB for milk procurement, sugarcane cooperative in Maharashtra and prawn aquaculture in Andhra Pradesh. However, presently most of the contract farming arrangements are informal in nature, and in case of violation of contract, there is no legal remedy. In case of pests' attacks and diseases, contract farmers are often left in lurch by contracting parties. Due to inadequate access to institutional credit, small farmers tend to get discouraged to participate in contract farming. Sometimes, there is lack of effective linkage between the company and agricultural research and extension system in the matter of technology diffusion which is crucial for expansion and sustainability of contract farming. Share croppers who are not generally recognized by law of most states do not enjoy security of tenure to participate in contract farming. Small farmers and marginal farmers are unorganized and have limited bargaining power vis-à-vis the companies. There is a need to have a legal framework which can take care of all these constraints and provide fair and just environment for promotion of the contract farming.

Land Leasing

Land leasing is another instrument which can help in promoting agricultural diversification. If permitted, land leasing can provide economy of scale by attracting potential investors in agriculture. This can also help in creating large scale captive production centres for processing as well as export units. Presently there is an informal land lease market in all the regions of the country despite the legal provisions to the contrary in the states. The main apprehensions regarding liberalization of land leasing that the land leasing in areas with poor infrastructure for nonfarm development and employment may encourage reverse tenancy by alienating the marginal farmers from land without having alternative sources of employment and income. It may also lead to concentration of operational holdings in a few hands. There is a need to have a legal framework which promotes land leasing but also protects the interests of small and marginal farmers by having appropriate safeguards inbuilt. A proper regulatory mechanism may help in allaying some of these fears.

Agriculture has become a capital-intensive and knowledge-based enterprise in the era of globalization. High-tech interventions like micropropagation, micro-irrigation, etc., are required along with intensive use of inputs to make agriculture competitive. The small and marginal farmers who account for more than 80 % of land holdings are unable to manage agriculture production professionally in view of their limited capacity and reach. The concept of land sharing company may provide an answer to this problem. Land sharing companies will not only provide economy of scale but will also help in establishing forward and backward linkages to production systems. The concept of land share in agriculture does not exist in India. However, it is possible to float a land share agro-processing company in which farmers of all categories may have the option to become share holders in proportion to their size of holding. Development of such participating land share companies in agriculture is likely to accelerate the pace of both agricultural and nonagricultural developments in rural areas. The company based on the land share system should be made eligible to receive concessional credit and other investment subsidies allowed for the promotion of agro-processing enterprises.

Price Support Policy Reforms

Though the crop diversification has been a continuous process in agriculture production system in the country, there is no proper mechanism in place to promote designed diversification. The Minimum Support Price (MSP) policy is being used as an instrument to promote crop diversification by announcing higher prices of certain commodities in comparison to others. The Minimum Support Prices of food grains like wheat and rice were increased substantially in comparison to other crops. Therefore, there was substantial increase in production of these commodities.

Of late, the focus of the government has shifted to increasing the production of oilseeds and pulses. Therefore, the increase in MSP of oilseeds and pulses has been much higher in comparison to cereals. This has encouraged farmers to increase area coverage under oilseeds and pulses. Thus, the instrument of MSP has been effective to some extent in increasing the production of a particular crop through diversification. However, the benefit of present MSP regime has gone to cereal growers in certain areas like Punjab, Haryana, U.P., A.P., Tamil Nadu and West Bengal only, whereas the farmers in other areas have not been benefited to the desired extent. Further, the cereal crops like rice and wheat have benefited most by the existing MSP policies, whereas coarse cereal crops have been neglected. The capacity of the nodal agencies like FCI and NAFED in undertaking procurement under MSP throughout the country is also often questioned. The role of the state governments has become important in the recently introduced decentralized procurement system but majority of the states have limitations in undertaking large-scale procurement operations under PSS. There is a need to appropriately strengthen the existing system of implementing MSP in an effective and transparent manner to promote crop diversification in the country.

In view of the inherent limitations of the existing MSP system, there is a need to explore alternate Price Support System to promote agricultural diversification. In view of the surplus production of cereals like rice and wheat, diversification is required to be focused to high-value and more remunerative crops like oilseeds, vegetables, horticulture, floriculture and commercial crops. Cultivation of rice and wheat crops is relatively risk free. The rice-wheat and rice-rice systems have therefore become attractive for the farmers in most productive regions in the country. To manage the production of these crops in proportion to demand, there is a need to have a strategy to promote designed diversification in the areas having rice-rice and rice-wheat systems in the country. Since these systems are confined to certain areas only, it is easier to implement designed diversification in these areas. The payment of cash compensation of the difference between the income from rice or wheat crop and the alternative crop to be promoted can be an attractive proposition for the farmers to go for designed diversification. Based on a long-term assessment, specific areas can be targeted under this strategy for a specific period. There should be flexibility of withdrawing the incentives from these areas as and when there is need to increase production of wheat and rice at any stage.

The Price Support Policies can also be designed to increase production of a particular crop during a particular period. The MSP of a targeted crop may be increased exponentially to increase the production of that crop significantly throughout the country. However, the government should have the option of scaling down the level of MSP of that commodity as and when the production and productivity has reached to the desired level. Such policy may help in designing the diversification for increasing the production of a particular crop in a particular period in the country. However, this policy is required to be supported with proper procurement infrastructure throughout the country. Alternatively, government may make payment to the producers of the difference between fair market price and the MSP announced by the government for that particular commodity. This will avoid any distortion in the market and interests of both producers as well as consumers will be taken care of.

Marketing Infrastructure

Marketing and processing are the two basic prerequisites for promoting crop diversification in a given agro-ecological conditions. Once the alternate produce is ready with the farmers, there must be a chain of retail outlets or regulated markets so that the farmers can have open opportunity to sell their produce at a remunerative price. Similarly, the rural marketing system needs to be strengthened and modernized to provide ample opportunity for the marketing of production of alternate crops. The various options in APMC Act need to be looked into and streamlined for efficient marketing of the agro-produce throughout country.

Storage and Transportation

Storage and transportation are equally important in achieving the goal of diversification. The producer may be provided with enough storage facilities preferably nearest to the production site to minimize the loss of precious commodities especially the one of perishable nature like fruits and vegetables. According to an estimate, about one third of the fruits and vegetables are damaged either in transportation or due to nonavailability of proper storage facilities. The policy decisions are required to promote storage and processing facilities to avoid the loss of agri-horti produce. There is a need to strengthen the cold storage facilities, cold chain and refrigerated transport infrastructure.

References

Ministry of Agriculture (2012) Agricultural statistics at a glance, Directorate of economics of statistics, p 388

PAN, FOE Europe, IFOAM EU Group & APRODEV 2012 Crop rotation, benefitting farmers, the environment and the economy. http://aprodev.eu/files/Trade/crop%20rotation%20briefing_pan_ifoam_aprodev_foee_fina.pdf

Thornton P, Lipper L (2013) How does climate change alter agricultural strategies to support food security? Draft, 11 March 2013. In: Background paper for the conference “Food security futures: research priorities for the 21st century”, 11–12 April 2013, Dublin. http://www.pim.cgiar.org/files/2013/03/ClimateChangeAndFoodSecurity_PrioritiesForPublicResearch.pdf

Rainfall Changes and Possible Linkages with the Synoptic Disturbances in Eastern India

D.K. Panda and A. Kumar

Abstract

In this study, the rainfall of Orissa, a tropical region in eastern India, and the cyclonic disturbances over the Bay of Bengal were analyzed to assess the trends and variations using parametric statistical procedures. This study revealed noticeable changes in the last century (1901–2000) in regard to the rainfall of Orissa and in the cyclonic disturbances over the Bay of Bengal. Test of structural changes and the locally weighted regression technique (LOWESS) indicated that the spatially averaged annual rainfall of Orissa experienced three significant regime shifts: 1,487 mm with standard deviation 167 mm during 1901–1940, 1,412 mm with standard deviation 184 mm during 1940–1969, and 1,354 mm with standard deviation 241 mm during 1970–2000, respectively. The study of structural breaks in rainfall during the period 1901–1940 and 1941–1969 indicated a significant change in both slope and the intercepts. However, structural break analysis during the period 1941–1969 and 1970–2000 indicated a nonsignificant change. The annual depression in the Bay of Bengal also exhibited a structural change with occurrence of an average of 10.7 depressions during 1901–1948 in comparison to 9.6 during 1949–2000.

Keywords

Synoptic disturbances • Cyclones • Rainfall

Introduction

The State of Orissa is most affected by climate change. This is due to frequent occurrence of hydrologic extremes in the recent past (Swiss Re 2002; Mirza 2003). The Center of Environmental Studies (CES 2007) reported that the erratic behavior of the climate in Orissa is primarily due to the combination of anthropogenic factors such as deforestation, extensive construction activities, uncontrolled mining, elimination of water bodies, and extensive coal fossil fuel consumption over a period of time. Further, a minor change in the pressure anomaly of the Bay of Bengal can have profound hydrological

D.K. Panda (✉) • A. Kumar
Directorate of Water Management (ICAR), Bhubaneswar 751023, India
e-mail: dileeppanda@rediffmail.com

A.K. Singh et al. (eds.), *Climate Change Modelling, Planning and Policy for Agriculture*,
DOI 10.1007/978-81-322-2157-9_3, © Springer India 2015

impacts on the land mass of Orissa due to its geographical location at the head of the Bay where the weather forms (CES 2007). Therefore, it is imperative to explore the long-term changes in rainfall in the study area and synoptic disturbances.

Methodology

Test of Structural Changes

The test of structural change is a general form of the Chow test, commonly used in forecasting (Johnston and Dinardo 1997). If the parameters of interest in a model differ from one subset of data to another, then structural breaks or change occurs in that time series. The parametric linear regression of response variable, i.e., rainfall with time (year), can be defined as

$$y = \alpha + \beta X + u$$

where y represents the dependent variable, α is the intercept, β is the slope, X is the independent variable year, and u is the random error. This model can be defined in the matrix form as

$$y = X\beta + u$$

I. $$\begin{bmatrix} y_1 \\ y_2 \end{bmatrix} = \begin{bmatrix} i_1 & X_1^* \\ i_2 & X_2^* \end{bmatrix} \begin{bmatrix} \alpha \\ \beta^* \end{bmatrix} + \mathbf{u} \quad \text{(common parameters : RSS1)}$$

II. $$\begin{bmatrix} y_1 \\ y_2 \end{bmatrix} = \begin{bmatrix} i_1 & 0 & X_1^* \\ 0 & i_2 & X_2^* \end{bmatrix} \begin{bmatrix} \alpha 1 \\ \alpha 2 \\ \beta* \end{bmatrix} + \mathbf{u} \quad \text{(different intercept, common slope : RSS2)}$$

III. $$\begin{bmatrix} y_1 \\ y_2 \end{bmatrix} = \begin{bmatrix} i_1 & 0 & X_1^* & 0 \\ 0 & i_2 & 0 & X_2^* \end{bmatrix} \begin{bmatrix} \alpha 1 \\ \alpha 2 \\ \beta_1^* \\ \beta_2^* \end{bmatrix} + \mathbf{u} \quad \text{(different intercept, common slope : RSS3)}$$

where $\mathbf{y}$ is a vector of n observation representing the dependent variable, $\mathbf{X}$ is a matrix of independent variable, $\boldsymbol{\beta}$ is a column vector of parameters, i.e., α and β, and $\mathbf{u}$ is the column vector of errors which is normally distributed with mean 0 and constant variance $\boldsymbol{\sigma}^2$.

In this study, for the simplification of the procedure, we investigate the structural change in two subsets of data that form the complete time series, although there is a possibility of several subsets with structural breaks. Let the total time series of n observation be divided into two subsets of samples with n_1 and n_2 observation, i.e., $n = n_1 + n_2$. The structural change may occur between the subset of time series due to the change in intercept or change in slope or a combination of both the parameters.

The models representing the subsets of time series can be defined as

$$y_1 = X_1\beta_1 + u_1$$
$$y_2 = X_2\beta_2 + u_2$$

i.e.,

$$\begin{bmatrix} y_1 \\ y_2 \end{bmatrix} = \begin{bmatrix} X_1 & 0 \\ 0 & X_2 \end{bmatrix} \begin{bmatrix} \beta_1 \\ \beta_2 \end{bmatrix} + \mathbf{u}$$

Here, $X_1 = [i_1\ X_1{*}]$ and $X_2 = [i_2\ X_2{*}]$, where i_1 and i_2 are n_1 and n_2 vectors of ones, and $X_1{*}$ and $X_2{*}$ are subset vectors of years. Further, the β vectors can be partitioned to a conformable form as

$$\beta_1' = \left[\alpha_1\ \beta_1{*}'\right] \ \beta_2' = \left[\alpha_2\ \beta_2{*}'\right]$$

Null hypothesis of no structural breaks is H_0: $\beta_1 = \beta_2$.

Under null hypothesis of common parameters

The following hypotheses are evaluated for the structural change test:

$$H_0 : \alpha_1 = \alpha_2 \quad F = \frac{\text{RSS}_1 - \text{RSS}_2}{\text{RSS}_2/(n-k-1)}$$

$$H_0 : \beta_1^* = \beta_2^* \quad F = \frac{(\text{RSS}_2 - \text{RSS}_3)/(k-1)}{\text{RSS}_3/(n-2k)}$$

$$H_0 : \beta_1 = \beta_2 \quad F = \frac{(\text{RSS}_1 - \text{RSS}_3)/k}{\text{RSS}_3/(n-2k)}$$

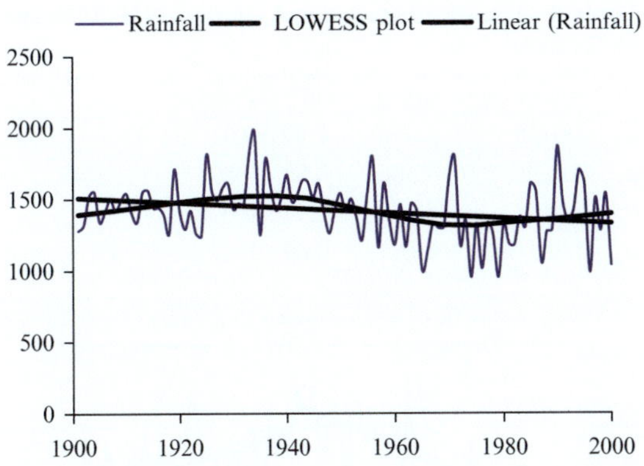

Fig. 1 Structural breaks of annual rainfall (mm) distribution during 1901–2000

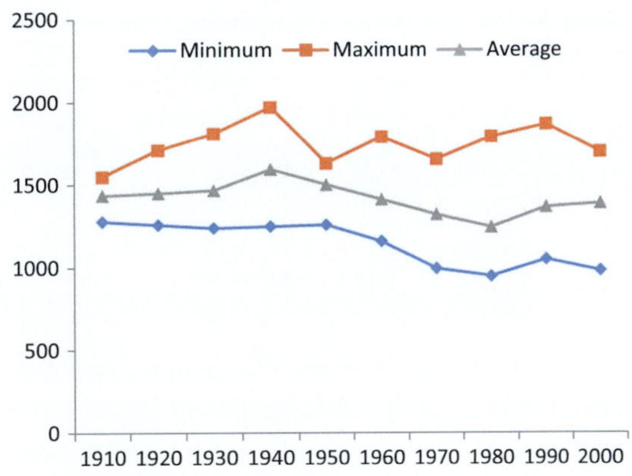

Fig. 2 Decadal distribution of annual rainfall during 1901–2000

Table 1 Descriptive statistics of annual rainfall (mm) in different periods of the last century

Period	1901–1940	1941–1969	1970–2000	1901–2000
Observations	40	29	31	100
Mean	1,487.25	1,403.33	1,344.51	1,418.67
Standard deviation (SD)	167.59	180.67	244.48	205.18
Skewness	0.84	−0.20	0.22	−0.01
Kurtosis	0.79	−0.20	−0.50	0.17
Minimum	1,240	997	950.7	950.7
P10	1,270	1,160	1,012.9	1,169.9
Q1	1,395	1,296	1,179.9	1,286.6
Median	1,470	1,417.9	1,325.6	1,420
Q3	1,555	1,530	1,535.9	1,545
P90	1,745	1,620	1,654	1,662
Maximum	1,970	1,790	1,865.8	1970
Sapiro-wilk	0.942	0.984	0.970	0.989
$p < w$	0.042	0.922	0.521	0.581

Table 2 Descriptive statistics of decadal annual rainfall (mm) in the last century

Decade	Minimum rainfall	Maximum rainfall	Average rainfall	SD
1910	1,280	1,550	1,437	97.64
1920	1,260	1,710	1,450	127.28
1930	1,240	1,810	1,468	178.75
1940	1,250	1,970	1,594	215.88
1950	1,260	1,630	1,501	114.74
1960	1,160	1,790	1,412	202.75
1970	997	1,654	1,322.07	190.37
1980	951	1,791	1,246.81	249.54
1990	1,052	1,866	1,367.09	242.08
2000	985	1,700	1,388.69	233.19

Results and Discussion

Rainfall Distribution During 1901–2000 in Orissa

In the last century during 1901–2000, Orissa has experienced 99 extreme events. These include 49 floods, 30 droughts, 10 cyclones, and 1 super cyclone. The locally weighted regression technique (LOWESS) indicated that the spatially averaged annual rainfall of Orissa experienced three significant regime shifts (Fig. 1). Hence, the climatology is experiencing an erratic distribution in recent years. The detail descriptive statistics of annual rainfall in different regimes of the last century are presented in Table 1.

The decadal maximum, minimum, and average rainfall of Orissa in the last century indicated a decreasing trend as illustrated in Fig. 2. More clearly, Table 2 shows that the decade-wise maximum, minimum, and average rainfall amounts follow a decreasing trend, and the standard deviations of rainfall follow an increasing trend.

Study of Structural Breaks in Different Regimes During 1901–2000 in Orissa

The parametric method of structural breaks was studied using the theory presented in the theory

Table 3 Test of structural breaks during 1901–1940 and 1941–1969

Id	Parameters	SE	t-cal	P value	F-cal	P
α_1	−8,725.99	4,128.30	−2.11	0.04	22.05	0.003
α_2	23,435.41	6,809.50	3.44	0.00		
β_1	5.32	2.15	2.47	0.02		
β_2	−11.27	3.48	−3.24	0.00		

Table 4 Test of structural breaks during 1941–1969 and 1970–2000

Id	Parameters	SE	t-cal	P value	F-cal	P
α_1	23,435	9,036	2.59	0.01	7.89	0.103
α_2	−4,381	8,300	−0.53	0.60		
β_1	−11	4.62	−2.44	0.02		
β_2	2.88	4.18	0.69	0.49		

section. In the last century, the annual rainfall of Orissa experienced three significant breaks. The averages of annual rainfall during these three regimes were 1,487 mm with standard deviation 167 mm during 1901–1940, 1,412 mm with standard deviation 184 mm during 1940–1969, and 1,354 mm with standard deviation 241 mm during 1970–2000, respectively. This indicates that the average rainfall is decreasing, but the standard deviation is increasing over rainfall regimes.

The study of structural breaks/regime shifts during the period 1901–1940 and 1941–1969 indicated a significant test of structural breaks both in slope and the intercepts (Table 3). During the period 1901–1940, the annual rainfall exhibited a significant increasing trend having positive slope ($\beta_1 = 5.32$), whereas during 1941–1969, it experienced a significant decreasing trend with a negative slope ($\beta_2 = -11.27$). Further, the intercepts are significant in both the regimes of the study period.

The results of structural break analysis during the period 1941–1969 and 1970–2000 indicated a nonsignificant test of structural breaks (Table 4). During the period 1941–1969, the annual rainfall exhibited a significant decreasing trend having slope ($\beta_1 = -11$), whereas during 1941–1969, it experienced a nonsignificant increasing trend with a slope ($\beta_2 = 2.88$). Furthermore, the intercepts are significant in both the regimes of the study period.

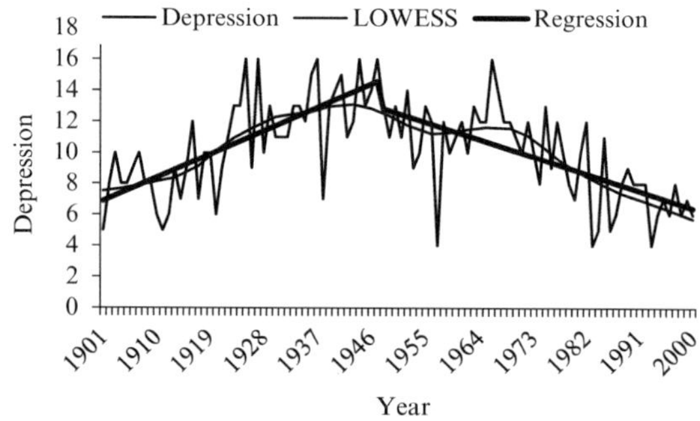

Fig. 3 Structural breaks of annual depression over the Bay of Bengal during 1901–2000

Study of Structural Breaks in Annual Cyclonic Disturbances (Depression) During 1901–2000 in Orissa

The annual cyclonic disturbances (depression) over the Bay of Bengal during 1901–2000 indicated a structural change in 1948 (Fig. 3)

Descriptive statistics indicated that an average of 10.7 depressions occurred during 1901–1948 in comparison to 9.6 during 1949–2000. It is evident that the depression frequency increased significantly during 1901–1948. In contrast, a marked decrease took place from 1949 onwards. This may have decreased the rainfall in recent decades in Orissa and the adjacent regions. Since structural change provides adequate evidence of climate change, more research is needed to understand the plausible reasons for these breaks in climatic variables.

References

CES (2007) Center of Environmental Studies (CES), Forest and Environment Department, Government of Orissa, Bhubaneswar. ENVIS Newsletter 8(1). http://www.envisorissa.org. Accessed 24 Feb 2010

Johnston J, Dinardo J (1997) Econometric methods, 4th edn. McGraw-Hill International Editions, New York

Mirza MQ (2003) Climate change and extreme weather events: can developing countries adapt? Clim Policy 3:233–248

Swiss Re (2002) Natural catastrophes and man-made disasters in 2001: man-made losses take on a new dimension. Swiss Reinsurance Company Publication, Zurich, Sigma No. 1/2002

Meteorological Drought Analysis Based on Rainfall Data of Coastal Odisha

S. Mohanty, Tipu Sultan, D.K. Panda, and Ashwani Kumar

Abstract

A 20-year analysis (1990–2009) of rainfall data of Cuttack district of coastal Odisha indicated an average annual rainfall of 1,649.8 mm with a standard deviation of 375.9 mm. The standard deviation of rainfall is higher in the monsoon months, whereas the coefficient of variation is higher in the non-monsoon months. The rainfall data was analyzed to study the monthly and yearly drought of the study area. Out of the 20 years, there were three drought years with the year 1996 being the most severe drought year wherein only an annual rainfall of 797 mm was received. The dry spell and wet spell analysis by the Markov chain model was done, and it was found that there is a high probability of availability of assured water for irrigation water during the 24th to 38th week. The probability analysis of the monthly rainfall data indicated that the two-parameter log-normal distribution was found to be best fit to the rainfall data of January, February, March, and December; Pearson type III distribution for April and November; log-Pearson type III distribution for May, June, August, and October; Gumbel type 1 extremal distribution for the month of July; and normal distribution for the month of September.

Keywords

Drought analysis • Dry spell • Markov chain model • Probability distribution

Introduction

In recent times, water scarcity is frequently occurring in different parts of the world which can be partly attributed to the increase in water demand due to population growth and expansion of agricultural and industrial sectors or partly due to climate change and contamination of water supplies (Bates et al. 2008). This is further

S. Mohanty (✉) • T. Sultan • D.K. Panda • A. Kumar
Directorate of Water Management, Bhubaneswar, Odisha 751 023, India
e-mail: smohanty.wtcer@gmail.com

A.K. Singh et al. (eds.), *Climate Change Modelling, Planning and Policy for Agriculture*,
DOI 10.1007/978-81-322-2157-9_4, © Springer India 2015

compounded by droughts which can lead to reduced water supply, deteriorated water quality, and crop failure (Riebsame et al. 1991). Drought is defined as the deficiency of available water that affects the crop growth and productivity, causes temporary scarcity of water, and influences the economic renewable resources (Pandey et al. 2010). Therefore, understanding drought and modeling its components have drawn attention of ecologists, hydrologists, meteorologists, and agricultural scientists. Several workers have done meteorological drought analysis based on rainfall data. Sharma et al. (1979, 1987) analyzed rainfall using the definition of drought month as a month in which the actual rainfall is less than 50 % of the average monthly rainfall. Kumar and Kumar (1989) and Dabral (1996) analyzed the weekly, monthly, seasonal, and yearly rainfall data of Pantnagar and Ranchi, respectively, as part of the meteorological drought analysis based on rainfall data. In the present study, attempt has been made to study the frequency of drought and analyze dry spell and wet spell characteristics of Cuttack district of coastal Odisha.

Materials and Methods

Daily rainfall data of 20 years (1990–2009) were collected from the meteorological observatory at Central Rice Research Institute (CRRI), Cuttack. The rainfall data was analyzed for investigating annual and monthly variations of rainfall. Analysis of rainfall data was done for occurrence of drought, wet spell, and dry spell analysis by the Markov chain model and best-fit probability distribution functions.

Drought Analysis

Drought implies a deficiency of precipitation of sufficient magnitude over a prolonged duration. But unlike floods, no systematic methods have been developed for the complete understanding and prediction of drought (Salas 1986). Depending upon climate, drought varies from place to place. In the current study, drought analysis was done for monthly and yearly rainfall. A total number of drought months and years were determined using the following definitions:

Drought month if actual rainfall is less than 50 % of average monthly rainfall (Sharma et al. 1979)

Drought year if actual rainfall is deficient by 25 % of the average yearly rainfall-drought year (IMD guideline)

Dry and Wet Spell Analysis by the Markov Chain Model

The method discussed in this section is based on the Markov chain, described by Pandharinath (1991), who used weekly data to establish drought frequencies on a weekly basis. A standard meteorological week having less than 20 mm rainfall was considered as dry week and having 20 mm or more rainfall was considered as wet week. Basing upon the experience, the concept used by Pandharinath (1991) was followed. Using these formulae, probabilities (P) of dry weeks (D) and wet weeks (W), conditional probabilities of dry week preceded by dry week (DD) and wet week preceded by wet week (WW), dry week preceded by wet week (WD), wet week preceded by dry week (DW), and probability of consecutive 2 or 3 dry (2D, 3D) or wet weeks (2W, 3W) starting with the week being dry or wet have been calculated for the study area.

Probability Distribution of Monthly Rainfall

The frequency analysis of the monthly rainfall data was performed considering the probability distribution functions of normal, 2-parameter log-normal, 3-parameter log-normal, Pearson type III, log-Pearson type III, Gumbel type 1 extremal, and generalized extreme value, using SMADA 6.0 software. All seven probability distribution functions were compared by Chi-squared (χ^2) test of goodness of fit given by the equation

$$\chi^2 = \sum \frac{(O - E)^2}{E} \qquad (1)$$

Where, O is the observed value and E is the estimated value by probability distribution function. Using the best-fit probability distribution functions for different months, the monthly rainfall at different probabilities was found out.

Results and Discussion

A 20-year analysis (1990–2009) of rainfall data in the study area indicated an average annual rainfall of 1,649.8 ± 375.9 mm. The variation of mean monthly rainfall over the 20-year period along with the standard deviation bars is shown in Fig. 1. It is apparent from this figure that the highest mean monthly rainfall (402.8 mm) with a standard deviation of 193.6 mm is observed in the month of August. Though the rainfall events are distributed throughout the year, the rainy season usually starts from mid-June and lasts up to mid-October. November through May is usually characterized as a dry period. The most reliable months for rainfall are July, August, and September. Thus, the bulk of the rainfall is concentrated in a relatively short time span, which increases the potential for both surface runoff and recharge to the aquifer but limits them to short periods of a year. Relatively large standard deviations in the months of May, June, July, August, September, and October indicate that the magnitude of monthly rainfall varies appreciably from year to year.

Drought Characteristics

Drought Year

Figure 2 shows the variation of annual rainfall over the basin along with the 20-year mean annual rainfall and 75 % of mean annual rainfall lines. It is obvious from this figure that years 1990, 1993, 1995, 1997, 1999, 2001, 2003, 2005, 2006, 2007, and 2008 have received more than the mean annual rainfall, and the rest of the years have received less than the mean annual rainfall. It is observed from the figure that the years 1996, 2000, and 2002 can be characterized as drought years, with the year 1996 being the most severe drought year.

Table 1 shows the total rainfall and rainfall in monsoon months (June to September) for the drought years. The rainfall of drought years as percentage of average annual rainfall and percent monsoon rainfall in drought years are also shown in the table. The amount of rainfall received in the drought years 1996, 2000, and 2002 are 797.3 mm, 1,192.7 mm, and 1,076.1 mm, respectively. The rainfall of drought years as percentage of average annual rainfall varies from

Fig. 1 Variation of mean monthly rainfall in the study area

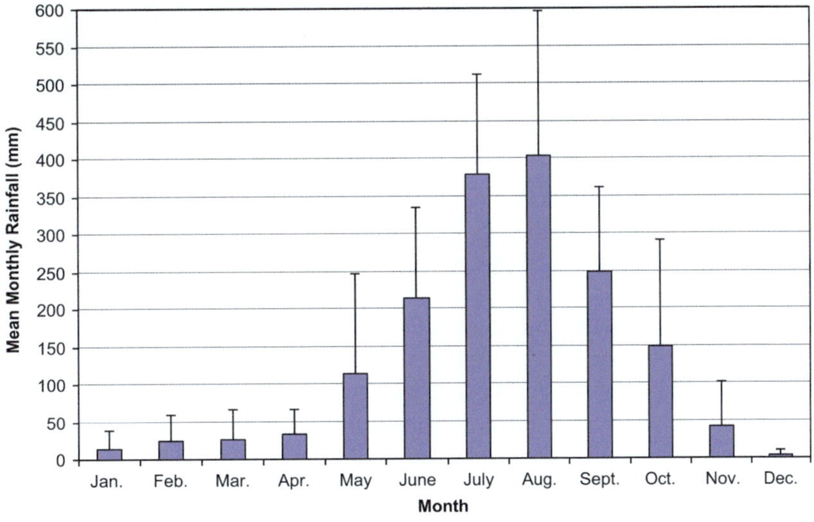

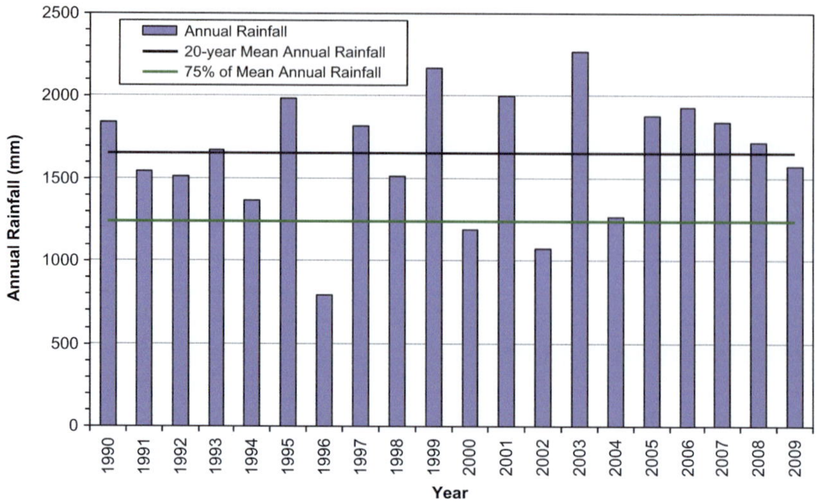

Fig. 2 Variation of annual rainfall during the 1990–2009 period

Table 1 Drought years in the study area

Drought year	Yearly rainfall (mm)	Rainfall during monsoon period (mm)	Yearly rainfall of drought years expressed as % age of av. annual rainfall	Percent of monsoon rainfall of drought years	Time interval between two successive drought years
1996	797.3	635.2	48.33	79.67	–
2000	1,192.7	960.6	72.29	80.54	4
2002	1,076.1	900.7	65.23	83.70	2

48.33 % in the year 1996 to 72.29 % in the year 2000. The percent monsoon rainfall in drought years varies from 79.67 % in the year 1996 to 83.70 % in the year 2002. There is no systematic difference between the occurrences of two consecutive drought years. The time gap varies from 2 to more than 7 years. So, nothing can be said regarding the frequency of drought years.

Drought Month

Table 2 shows the average rainfall, half of average rainfall, and a number of drought months for the 12 months during the period 1990–2009. The month having less than 50 % of average monthly rainfall was termed as drought month. There were 16 drought months in December followed by 12 drought months in January. Out of the 20 years, there were no drought months in July and 2 drought months each in June and August. Figure 3a–c shows the occurrence of drought months and normal months in the drought years of 1996, 2000, and 2002.

Table 2 Number of drought months between 1990 and 2009

Month	Average	Half of average	No. of drought month
Jan	13.79	6.83	12
Feb	25.05	12.53	11
Mar	26.30	13.15	11
April	33.36	16.68	8
May	113.35	56.68	9
June	208.54	104.27	2
July	378.53	189.27	0
Aug	402.84	201.42	2
Sep	237.71	118.8	3
Oct	146.86	73.43	7
Nov	41.51	20.75	11
Dec	3.01	1.50	16

Dry Spell and Wet Spell Occurrence

Table 3 shows the summary of dry and wet spell analysis of the study area by the Markov chain model. It is clear from the table that the probability of occurrence of a dry week is high (60–100 %) in the first 23 weeks and from the

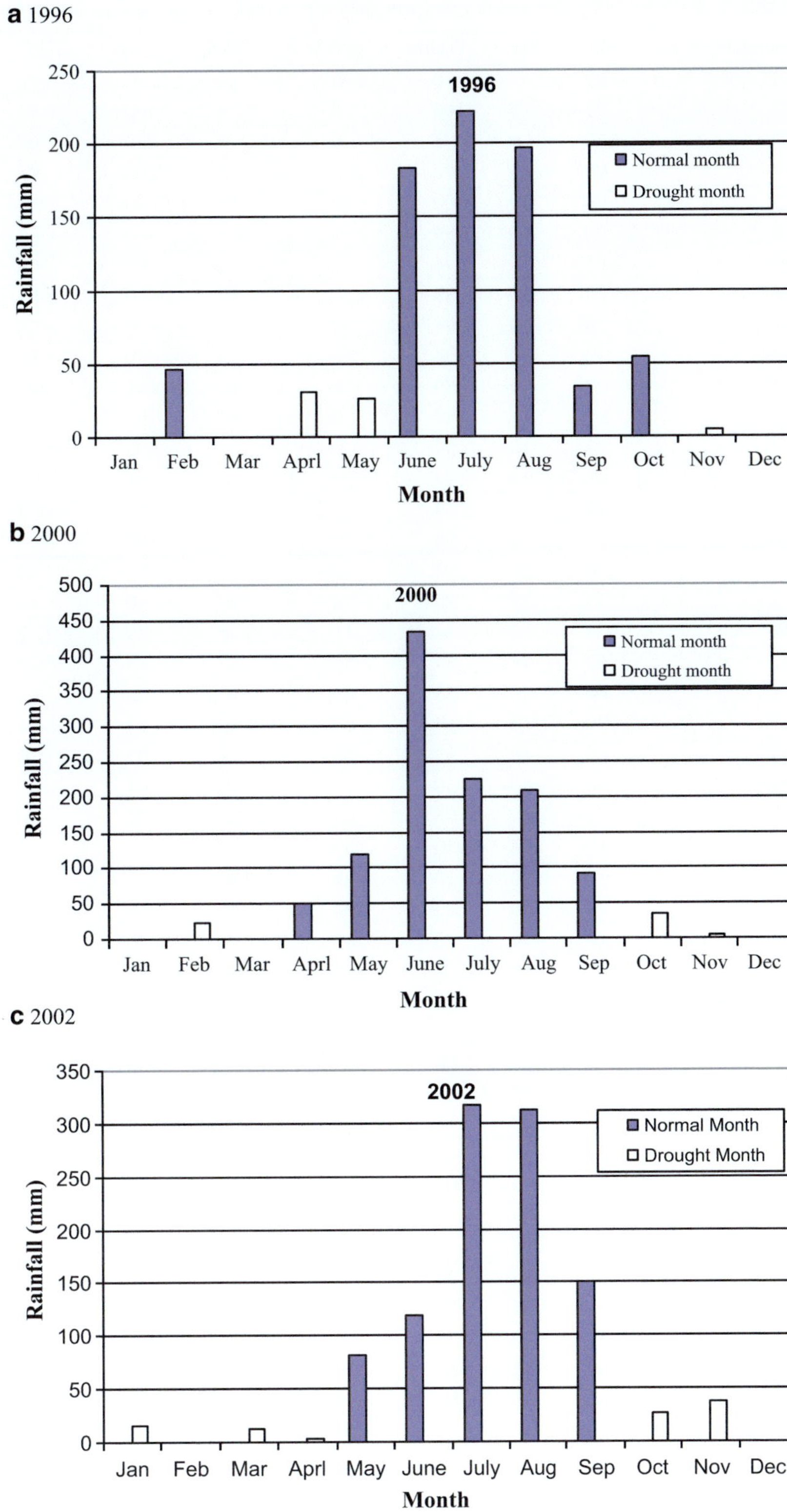

Fig. 3 (**a**–**c**) Monthly rainfall during drought years

Table 3 Probability of different dry and wet spell

Standard week	PD	PW	P(DD)	P(WW)	PWD	PDW	2D	2W	3D	3W
1	95	5	0	0	0	0	95	0	90.25	0
2	95	5	100	0	0	100	90.25	0	90.25	0
3	100	0	95	0	5	100	100	0	100	0
4	95	5	100	0	0	100	95	2.5	84.55	0
5	90	10	100	50	0	50	80.1	0	75.29	0
6	90	10	89	0	11	100	84.6	3.3	74.44	1.65
7	85	15	94	33	6	67	74.8	7.5	70.31	3.75
8	90	10	88	50	12	50	84.6	5	79.52	2.5
9	90	10	94	50	6	50	84.6	5	75.29	0
10	90	10	94	50	6	50	80.1	0	71.29	0
11	90	10	89	0	11	100	80.1	0	75.29	0
12	95	5	89	0	11	100	89.3	0	83.94	0
13	90	10	94	0	6	100	84.6	3.3	71.06	0
14	85	15	94	33	6	67	71.4	0	67.11	0
15	95	5	84	0	16	100	89.3	0	83.94	0
16	95	5	94	0	6	100	89.3	0	71.44	0
17	80	20	94	0	6	100	64	4	40.96	0.44
18	75	25	80	20	20	80	48	2.75	30.72	1.51
19	55	45	64	11	36	89	35.2	24.75	19.36	10.89
20	55	45	64	55	36	45	30.25	19.8	19.36	13.27
21	60	40	55	44	45	56	38.4	26.8	35.33	16.88
22	70	30	64	67	36	33	64.4	18.9	32.2	6.24
23	60	40	92	63	8	37	30	13.2	22.5	9.11
24	40	60	50	33	50	67	30	41.4	15	34.36
25	20	80	75	69	25	31	10	66.4	0	59.10
26	10	90	50	83	50	17	0	80.1	0	75.29
27	10	90	0	89	100	11	2.5	84.6	0	67.68
28	20	80	25	94	75	6	0	64	0	64
29	0	100	0	80	100	20	0	100	0	94
30	5	95	0	100	100	0	0	89.3	0	75.01
31	20	80	0	94	100	6	0	67.2	0	63.84
32	5	95	0	84	100	16	0	90.25	0	90.25
33	0	100	0	95	100	5	0	100	0	94.00
34	15	85	0	100	100	0	0	79.9	0	70.31
35	10	90	0	94	100	6	0	79.2	0	69.69
36	20	80	0	88	100	12	13.4	70.4	13.4	66.88
37	10	90	67	88	33	12	10	85.5	1.4	85.50
38	5	95	100	95	0	5	0.7	95	0.20	29.45
39	35	65	14	100	86	0	10.15	20.15	3.15	11.48
40	35	65	29	31	71	69	10.85	37.05	6.30	9.26
41	65	35	31	57	69	43	37.7	8.75	26.01	2.18
42	60	40	58	25	42	75	41.4	10	38.50	5
43	80	20	69	25	31	75	74.4	10	53.57	5
44	70	30	93	50	7	50	50.4	15	44.35	0
45	90	10	72	50	28	50	79.2	0	62.57	0
46	80	20	88	0	12	100	63.2	0	60.04	0
47	95	5	79	0	21	100	90.25	0	90.25	0

(continued)

Table 3 (continued)

Standard week	PD	PW	P(DD)	P(WW)	PWD	PDW	2D	2W	3D	3W
48	100	0	95	0	5	100	100	0	100	0
49	100	0	100	0	0	100	100	0	95.00	0
50	90	10	100	0	0	100	85.5	0	85.50	0
51	100	0	95	0	5	100	100	0	0	0
52	100	0	100	0	0	100	0	0	0	0

41st week to the 52nd week (65–100 %). The conditional probability of dry week preceded by dry week is also high up to the 25th week, but the probability of occurrence of two consecutive dry weeks is high only up to the 22nd week and that of three consecutive dry weeks is high only up to the 17th week of the year. These probabilities of occurrence also increase remarkably from the 43rd to 52nd week.

Similarly the probabilities of occurrence of wet weeks is very low up to the 23rd week and then starts increasing gradually up to the 38th week (60–95 %) and then starts decreasing to the level of zero percent. The conditional probability of wet week preceded by wet week is high during the period of 25th to 39th week (69–100 %) and is highest in the 30th, 34th, and 39th week of the year (100 %). The probability of occurrence of two consecutive wet weeks is significant from the 25th week up to the 38th week, but decreases suddenly from the 39th week onwards. It is highest in the 29th and 33rd week (100 %). The probability of occurrence of three consecutive wet weeks is very poor up to the 23rd week and is significant during the period of the 24th to 37th week. The maximum probability of occurrence is 94 % in the 33rd week. After the 38th week, it decreases sharply to reach 0 % by the 44th week. The analysis shows high probability of availability of assured water for irrigation water during the 24th to 38th week.

Best-Fit Probability Distribution of Monthly Rainfall

The probability analysis of the monthly rainfall data indicated that the two-parameter log-normal distribution was found to be best fit to the rainfall data of January, February, March, and December; Pearson type III distribution for April and November; log-Pearson type III distribution for May, June, August, and October; Gumbel type 1 extremal distribution for the month of July; and normal distribution for the month of September. The best-fit probability distribution can be used for predicting rainfall events at higher return periods.

Conclusion

During the 20-year period of 1990–2009, there were three drought years in the Cuttack district of coastal Odisha with 1996 being the most severe drought year. Month-wise maximum frequency of droughts was experienced in the month of December and minimum in the month of July. The dry spell analysis showed that there is a high probability of assured rainfall in the 24th to 38th week. The best-fit probability distribution function for different months was determined which are useful for predicting rainfall at higher return periods.

References

Bates BC, Kundzewicz ZW, Wu S, Palutikof S (eds) (2008) Climate change and water. Technical paper of the Intergovernmental Panel on Climate Change, IPCC Secretariat, Geneva, 210 pp

Dabral PP (1996) Meteorological drought analysis based on rainfall data. Indian J Soil Conserv 24(1):37–40

Kumar D, Kumar S (1989) Drought analysis based on rainfall data. Indian J Soil Conserv 17(1):55–60

Pandey RP, Pandey A, Galkate RV, Byun HR, Mal BC (2010) Integrating hydro-meteorological and physiographic factors for assessment of vulnerability to drought. Water Resour Manag 24:4199–4127

Pandharinath N (1991) Markov chain model probability of dry, wet during monsoon period over Andhra Pradesh. Mousum 42(4):393–400

Riebsame WE, Changnon SA, Karl TR (1991) Drought and natural resource management in the United States: impacts and implications of the 1987–89 drought. Westview Press, Boulder, 174 pp

Salas JD (1986) State of the art statistical techniques for describing drought characteristics. WARRDCC, International Seminar on drought analysis, Perugia, Italy

Sharma HC, Chauhan HS, Ram S (1979) Probability analysis of rainfall for crop planning. J Agric Eng XVI(3):87–94

Sharma HC, Shrivas RN, Tomar RKS (1987) Agricultural planning on the basis of rainfall. J Indian Water Resour Soc 7(2):17–27

Climate Change Impacts on *Rainfed* Soybean Yield of Central India: Management Strategies Through Simulation Modelling

M. Mohanty, Nishant K. Sinha, Sangeeta Lenka, K.M. Hati, J. Somasundaram, R. Saha, R.K. Singh, R.S. Chaudhary, and A. Subba Rao

Abstract

Soybean [*Glycine max* (L.) Merrill] has emerged as one of the major rainy season oilseed cash crops in central India. Despite its phenomenal growth in this agro-climatic zone, the average productivity of soybean has remained more or less at 1 t ha^{-1} due to several abiotic, biotic and socio-economic factors. The climate change (increase in temperature, CO_2 concentration and rainfall) will affect this *rainfed* crop in the future. So, proper management practices which include crop management (use of nutrients, planting time and plant population) will play a major role in future productivity in these regions. Simulation models with demonstrated accuracy and reliability provide an alternative method of investigating both short- and long-term agricultural practices with less time requirements and low cost. They have been evaluated and used as a research tool to study risks associated with various management strategies and to assist in decision-making. Hence, the present study aims at using the APSIM model in the decision-making process to evaluate the impact of climate change on soybean yield.

For the simulation study, the optimum date of sowing was chosen based on the literature available for this region. A well-calibrated and validated APSIM model was used for a long-term simulation study on the impact of rainfall pattern on soybean yield. The long-term prediction revealed that there was an interannual variation in soybean yield due to the variation in rainfall pattern. The distribution of rainfall rather than the amount during the soybean growing season is important for soybean yield. There was a significant decrease in soybean yield (as high as 96 %) when the rainfall receded during the initiation of flowering to maximum pod stage. The yield reduction was 56 % when a drought spell of around 2 weeks occurs during mid-vegetative stage. There was a significant

M. Mohanty (✉) • N.K. Sinha • S. Lenka • K.M. Hati • J. Somasundaram • R. Saha • R.K. Singh • R.S. Chaudhary • A. Subba Rao
Indian Institute of Soil Science, Nabibagh, Berasia Road, Bhopal 462038, MP, India
e-mail: mmohanty_bpl@yahoo.co.in

A.K. Singh et al. (eds.), *Climate Change Modelling, Planning and Policy for Agriculture*, DOI 10.1007/978-81-322-2157-9_5, © Springer India 2015

decrease in yield (37 %) from the maximum when the drought spell occurs at some parts of the growing season. The validated APSIM model was also used to simulate the impact of climate change on soybean production in central India. The projected temperature scenarios for the Indian subcontinent as reported by IPCC have been used in the present study. There was a decrease (ranging between 20 and 35 %) in soybean yield when the effect of the rise in surface air temperature during soybean growing season was considered. The simulation results obtained on the mitigatory option for reducing the negative impacts of temperature increases indicate that delaying the sowing dates would be favourable for increased soybean yields for this region. This will help in recommending a better alternative management options to improve the productivity of soybean in the region.

Keywords
Soybean • Rainfed • Grain yield • Temperature • APSIM model

Introduction

The increasing CO_2 concentration in the atmosphere and the anticipated temperature rise due to global warming are also likely to affect agricultural production in the world through changes in plant growth and transpiration. Soybean [*Glycine max* (L.) *Merrill*] is one of the most important oilseed crop cultivated in India. The area under soybean cultivation has steadily increased over the years in the states like MP, Chhattisgarh and Maharashtra. Future climatic change is likely to have substantial impact on soybean production depending upon the magnitude of variation in temperature. Increased temperature significantly reduces the grain yield due to faster growth and decreased time to accumulate grain weight (Baker et al. 1989). There have been a few studies in India and elsewhere aimed at understanding the nature and magnitude of gains/losses in yields of soybean crop at different sites under elevated atmospheric temperature conditions.

In this study, an attempt has been made (a) to study the long-term rainfall pattern on simulated grain yield of soybean and (b) to explore the possibilities of addressing different adaptation options for soybean cv JS 335 to alleviate the impacts of increased temperature in central Indian condition using APSIM simulation model. The long-term observed daily weather data on rainfall, maximum and minimum temperatures and solar radiation at the Bhopal station have been used in this study.

Methodology

The APSIM Model

The APSIM is a crop growth model that combines biophysical and management factors to simulate growth of different crops and cropping systems in a landscape (Keating et al. 2003). The model simulates crop growth in a given environment subject to limitations due to temperature, solar radiation, water and N supply but does not consider pest and diseases. The APSIM model has been applied to simulate maize (*Zea mays* L.)-fertilizer experiments (Shamudzarira and Robertson 2002) and a velvet bean (*Mucuna pruriens* L.)-maize cropping sequence (Robertson et al. 2005), decomposition of sugarcane (*Saccharum officinarum* L.) surface residue using the APSIM surface organic matter module (surface residue) (Thorburn et al. 2001) and elsewhere in studies ranging from soil-water relations to crop-weed interactions (Grenz et al. 2006; Stewart et al. 2006). In this study, APSIM (version 6.0) was configured with the soybean module (Mohanty et al. 2012).

Input Data

The model requires input data on soil, crop and weather for its calibration and validation in different environments. Weather (solar radiation, maximum and minimum temperatures and rainfall), soil (albedo, first-stage evaporation, drainage, USDA Soil Conservation Service Curve Number for runoff and layer-wise information on saturation, field capacity, wilting point, texture and hydraulic conductivity) and crop management data (dates of sowing, plant and row spacing, irrigation, fertilizer, etc.) were collected for the study.

Calibration and Validation of the Model

All calibration data required to derive genetic coefficients were obtained from a field experiment conducted at Bhopal. In this experiment, the soybean crop was sown at row spacing of 30 cm, and the seeding depth was maintained at 5 cm. A net of 20 kg urea was applied as the basal dose at the time of sowing. Plant population was kept at 50 plants/m^2. The genetic coefficients determined in the model using the identical conditions as in the field experiment for 'JS 335' variety of soybean are reported by Mohanty et al. (2012). These coefficients were used in the subsequent validation and application. The well-calibrated and validated APSIM model for soybean has been used for this study (Mohanty et al. 2012).

The Climate Change Scenario

The Intergovernmental Panel on Climate Change (IPCC 2001) reported that the average global surface temperature will increase between 1.4 and 3 °C above 1990 levels by 2100 for low emission scenarios and between 2.5 and 5.8 °C for higher emission scenarios of greenhouse gases and aerosols in the atmosphere. Over the land regions of the Indian subcontinent, the projected (area-averaged) annual mean surface temperature rise by the end of the twenty-first century has been estimated to range between 3.5 and 5.5 °C depending upon the future trajectory of anthropogenic radiative forcing (Lal 2001). The projected temperature increase has a large seasonal and spatial dependency over India. During the monsoon season, the temperature rise over south India is projected to be less than 1.5 °C by 2050s, while the increase in surface temperature is more pronounced over north, central and east India (>2 °C). Probable changes in precipitation, cloudiness and solar radiation under the climate changes scenarios were not taken into consideration in this analysis in view of the significant uncertainties associated with non-linear, abrupt and threshold rainfall events projected by GCMs over the Indian subcontinent.

Results

Long-Term Simulated Yield of Soybean

The APSIM model was used to predict the long-term simulated grain yield of soybean. The results revealed that due to the rainfed nature of the soybean crop, the planting of soybean in India depends on the onset of the rainy (monsoon) season. The seasonal arrival of the monsoon of the target region is from 10th to 20th of June. In the simulation, six dates of sowing (10th, 20th, 30th of June and 10th, 20th and 30th of July) were considered. Depending on the weather conditions over 16 years in Bhopal, considerable variation in soybean yield was observed. When averaged over the simulation period, the mean grain yield ranged from 1.0 to 1.6 t ha^{-1} for the different dates of sowing, with the 20th of June date of sowing yielding maximum. There was sharp decline in average yield after the 10th of July sowing date. There was a wide variability in the minimum and maximum yields recorded over the simulation period. The yield gap of soybean ranged from 0.02 to 0.62 t ha^{-1} for the region.

There was interannual variation in soybean yield, and these variations were due to the rainfall received during different growth stages of the crops. The rainfall pattern/distribution rather than the amount played an important role in

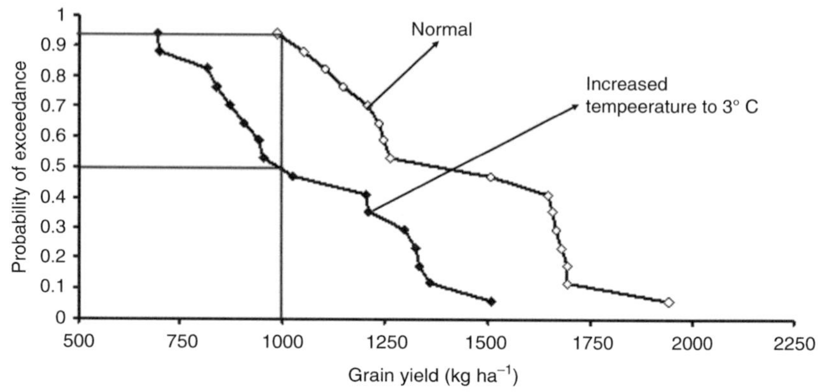

Fig. 1 Probability distribution of soybean yield (cv JS 335) under long-term simulation

soybean growth. The long-term analysis of the weather revealed that there was a significant decrease in soybean yield (as high as 96 %) when the rainfall receded during the initiation of flowering to maximum pod stage. The yield reduction was 56 % when a drought spell of around 2 weeks occurs during the mid-vegetative stage. There was a significant decrease in yield (37 %) from the maximum when the drought spell occurs at some parts of the growing season.

Impact of Increase in Surface Air Temperature on Soybean Yield

The increase in surface air temperature has a tremendous effect on soybean as observed in the long-term simulation by the APSIM model. The result revealed that there was significant decrease in yield of soybean from the normal (base line followed) (Fig. 1). The base line yield for this study was the long-term simulated yield of 16 years under recommended management practices followed for soybean. The probability distribution of soybean yield, presented in Fig. 1, described the probability of obtaining a specific amount of grain yield under the increased temperature scenario. Fig. 1 presented that there is 50 % probability of obtaining <1 t ha^{-1} of soybean grain yield under climate change scenario as predicted by the model, whereas the probability of getting the same amount of grain yield under normal condition is observed to be <10 %. So, the increased surface temperature by 3° during soybean growing season decreased the yield significantly which suggested that in the future, due to climate change, the yield of soybean is going to decrease to the tune of about 30 %. To reduce the impact of climate change, it is necessary to adopt new management practices so that the soybean crop can adapt to the new environment. This is possible either by changing the dates of sowing or by changing the plant density.

Adaptation Strategies for Soybean Crop to Increased Temperature Scenario

To reduce the impact of climate change on soybean crop, it is necessary that the new management practices should be followed in the future. It is possible to reduce further the reduction in grain yield of the same cultivar by adopting new management practices or by total replacement of the cultivar either by another cultivar which can sustain growth under increased temperature condition or by breeding to develop new cultivars. Here in this study, we developed adaptation strategies to increased temperature of soybean crop by changing the dates of sowing and plant population.

The increased temperature to the tune of 3 °C resulted in the reduction in grain yield by 33 % when the sowing date of soybean was the 20th of June. But by advancing the sowing date beyond the 20th of June, it was observed that the reduction in yield was in the range of −23 to −13 % from the normal. By adopting the sowing date to

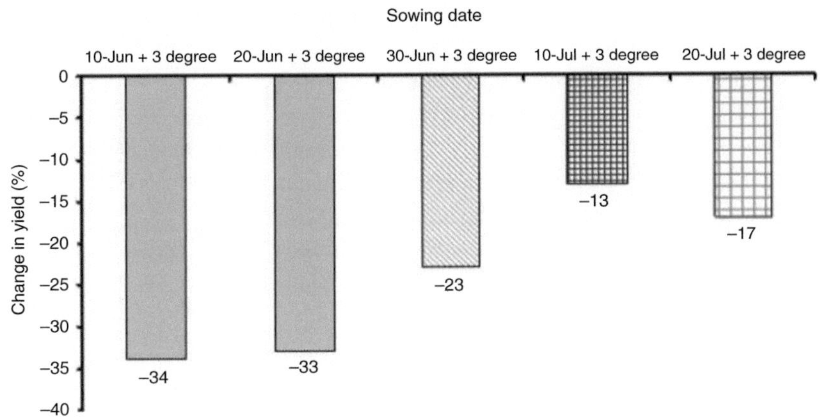

Fig. 2 Yield change (%) under increased temperature scenario of the soybean cv JS 335 under different sowing dates grown during the rainy season at Bhopal (Plant density = 50 plants/m^2)

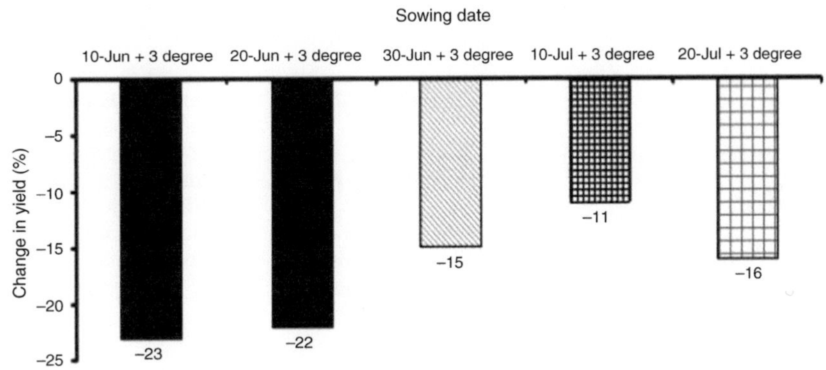

Fig. 3 Yield change (%) under increased temperature scenario of the soybean cv JS 335 under different sowing dates grown during the rainy season at Bhopal (Plant density = 60 plants/m^2)

the 10th of July, it was possible to reduce the impact of temperature change on soybean yield. A least reduction of about −13 % was observed in the 10th of July sowing date. Sowing after the 20th of July did not have any mitigating effect due to temperature change on soybean yield (Fig. 2.)

The change in plant density has a significant effect on grain yield of any crop. So, attempt has been made by the authors to change the plant density of soybean crop to simulate the impact of increased temperature on yield. Here, the chosen plant density of 60 plants/m^2 was considered after a series of simulations and used for adaptation study.

It was observed that by increasing the plant population from 50 to 60 plants/m^2 increased the grain yield of soybean under the adverse impact of climate change (Fig. 3). There was a further increase in grain yield of soybean by coupling advanced sowing dates and increased plant population. The decrease in grain yield due to increased temperature was reduced to −15, −11 and −16 % in sowing dates of the 30th of June, 10th of July and 20th of July, respectively.

So, from the study, it was revealed that agronomic management practices such as changing the sowing dates and plant population could reduce the impact of climate change on crop yield to a considerable extent.

Conclusion

Long-term simulation study indicated the effect of rainfall pattern and distribution on soybean yield in central Indian condition. The water stress during the reproductive stage poses a serious threat to *rainfed* soybean yield as predicted by the APSIM model. In general, the simulation results indicated that the increasing temperature levels could pose a serious threat to the growth of soybean crop and hence the yield. The projected increase in temperature to 3 °C could reduce the yield of soybean crop by about 30 %.

Our findings suggest that delaying the sowing dates of soybean crop and increasing the plant density to 60 plants/m^2 could be viable options to mitigate the detrimental effect of thermal stress due to climate change.

References

Baker JT, Allen LH Jr, Boote KJ, Jones P, Jones JW (1989) Response of soybean to air temperature and carbon dioxide concentration. Crop Sci 29:98–105

Grenz JH, Manschadi AM, Voil P, Meinke H, Sauerborn J (2006) Simulating crop-parasitic weed interactions using APSIM: model evaluation and application. Eur J Agron 24:257–267

IPCC (2001) Climate change: the scientific basis, summary for policymakers. Cambridge University Press, Cambridge, 20 pp

Keating BA, Carberry PS, Hammer GL, Probert ME, Robertson MJ, Holzworth D, Huth NI, Hargreaves JNG, Meinke H, Hochman Z, Mclean G, Verburg K, Snow V, Dimes JP, Silburn M, Wang E, Brown S, Bristow KL, Asseng S, Chapman S, Mccown RL, Freebairn DM, Smith CJ (2003) An overview of APSIM, a model designed for farming systems simulation. Eur J Agron 18:267–288

Lal R (2001) Soil erosion and carbon dynamics on grazing lands. In: Follett RF, Kimble JM, Lal R (eds) The potential of U.S. grazing lands to sequester carbon and mitigate the greenhouse effect. CRC/Lewis Publishers, Boca Raton, pp 231–247

Mohanty M, Probert ME, Sammi Reddy K, Dalal RC, Singh M, Subba Rao A, Menzies NW (2012) Simulating soybean-wheat cropping system of Madhya Pradesh, central India: APSIM model testing an devaluation. Agric Ecosyst Environ 152:68–78

Robertson MJ, Sakala W, Benson T, Shamudzarira Z (2005) Simulating response of maize to previous velvet bean (*Mucuna pruriens*) crop and nitrogen fertilizer in Malawi. Field Crops Res 91:91–105

Shamudzarira Z, Robertson MJ (2002) Simulating response of maize to nitrogen fertilizer in semi-arid Zimbabwe. Exp Agric 38:79–96

Stewart LK, Charlesworth PB, Bristow KL, Thorburn PJ (2006) Estimating deep drainage and nitrate leaching from the root zone under sugarcane using APSIM-SWIM. Agric Water Manag 81:35–334

Thorburn PJ, Probert ME, Robertson FA (2001) Modelling decomposition of sugarcane surface residues with APSIM-Residue. Field Crops Res 70:223–232

Impact of Climatic Variability on Wheat Yield Predication using DSSAT v 4.5 (CERES-Wheat) Model for the different Agroclimatic zones in India

P.K. Singh, K.K. Singh, A.K. Baxla, and L.S. Rathore

Abstract

The simulation models, with its complete ability, the soil-plant-atmospheric system, offer an ideal tool to analyze the response of wheat to the changing climatic conditions. Crop simulation model not only saves the accurate time but also the huge cost of experimentation. The CERES-Wheat vs. 4.5 crop growth simulation models was calibrated and evaluated for the different agroclimatic conditions of the wheat-growing area of India. In the present study, efforts are made to estimate the wheat yield for the years 2009–2010 using CERES-Wheat crop growth simulation model embedded in the DSSAT v4.5 software. Simulations were made under irrigated condition for the 15 wheat-growing locations representing different agroclimatic zones of the country. Daily weather data on maximum and minimum temperatures, rainfall, and radiation were used for 15 locations for the two winter seasons, current season (2009–2010) and previous season (2008–2009). Solar radiation required by the crop model was calculated from bright sunshine hours. Two dates of sowing representing normal (north India, 15 November; Uttar Pradesh, 10 December, and East India, 1 December) and late sown (north India, 25 November; Uttar Pradesh, 25 December, and East India, 15 December) conditions for each zone were taken up for the study. The genetic coefficient required for running the CERES-Wheat model was derived for the commonly grown cultivars, i.e., PBW-343, RAJ-3765, Malviya-234, HUW-234, and C-306, at different locations.

Model simulation results indicate that wheat crop yield in Uttar Pradesh during the crop season (2009–2010) is slightly affected due to the rise in temperature encountered at the grain-filling stage in the month of March. The simulated average yield for Uttar Pradesh was 47.8 and 46.8 q/ha for the years 2008–2009 and 2009–2010, respectively, under

P.K. Singh (✉) • K.K. Singh • A.K. Baxla • L.S. Rathore
Agromet Service Cell, India Meteorological Department,
Mausam Bhavan, Lodi Road, New Delhi 110003, India
e-mail: pksingh66@gmail.com

A.K. Singh et al. (eds.), *Climate Change Modelling, Planning and Policy for Agriculture*,
DOI 10.1007/978-81-322-2157-9_6, © Springer India 2015

normal sown condition. Similarly in the case of East India, the simulated average yield is 34.7–42.3 q/ha and 52.2 and 54.5 q/ha for Northwest India. Under normal sown condition, wheat yield prediction during 2009–2010 is ~1 % less compared to 2008–2009 in Uttar Pradesh while there was an increase in grain yield by ~4 % in Northwest India and ~18 % in East India, respectively.

In case of a late sown condition, the crop model shows reduction in grain yield in Northwest India and Uttar Pradesh except for East India. During March 2010, maximum and minimum temperatures remained above normal almost throughout the month over most parts of the country. Maximum temperatures were above normal by 4–8 °C over northwest, east, central, and adjoining peninsular India.

Simulated average yield and deviation percentage for Northwest India and Uttar Pradesh were 53.6 and 51.2 q/ha (~4 %) and 50.2 and 38.3 q/ha (~23 %) for the years 2008–2009 and 2009–2010, respectively. Wheat yield production realized during 2009–2010 is higher compared to 2008–2009 in east India that shows 32.1 and 35.4 q/ha (~8 %). The objective of this study is to analyze the temperature trend so as to help the government agencies to design and adopt a suitable corrective measure of agricultural system.

Keywords

CERES-Wheat model • Genetic coefficients • Yield prediction • Climatic variability and extreme weather

Introduction

Wheat (*Triticum* sp.) is the most important grain crop both in regard to its antiquity and its use as a source of human food. Wheat serves as a staple food for about one billion people in as many as 43 countries in the world. So it is the most widely cultivated food crop in the world.

In developing countries like India, climate change could represent an additional stress on ecological and socioeconomic systems that are already facing tremendous pressure due to urbanization, industrialization, and economic development. Thus, the Indian economy, which is closely tied to its natural resource base, is considerably vulnerable to the impacts of climate change (DOD 2002). Lal et al. (1995) predicted an increase in the annual mean maximum and minimum surface air temperature of 0.7 and 1.0 °C over land in 2040 with respect to 1980.

Saseendran et al. (1998) have reported a decrease in rice yields by 3–15 % under a scenario of 1.5 °C rise in temperature and 42 mm/day increases in precipitation. Likewise, Lal et al. (1995) have reported that the yield of soybean in India would vary between −22 and 18 % under different climate scenarios considering +2 and +4 change in temperature and ±20 and ±40 % change in precipitation.

In Rajasthan state, it was observed that the maximum temperature rises abruptly in the end of February to the first fortnight of March and causes force maturity in wheat crop. The productivity of wheat in the state has come down to 2,405 kg/ha (2000–2001) from 2,741 kg/ha in 1997–1998, which may be due to the effect of climatic changes. Intra- and interdistrict variations in rainfall were more pronounced. Likewise due to the reduction in rainfall in Jaipur district, double-cropped areas have reduced to single-cropped area. The study of rainfall trend in Jaisalmer district is quite amazing. Thus, the climate change is adversely affecting agricultural productivity directly as well as indirectly.

Wheat (*Triticum aestivum* L.), one of the important staple foods in India, is grown under diverse sets of agroclimatic conditions. It is grown in the region within latitudes 15° to 32° N and longitudes 72° to 92° E both under irrigated and rainfed conditions. Punjab, Haryana, Uttar Pradesh, Uttaranchal, Madhya Pradesh, Rajasthan, Delhi, Chhattisgarh, and Bihar are the leading states of India contributing more than 90 % of the wheat for the country. India is now the 2nd largest wheat producer in the world after China. Wheat is sown during November to December and harvested during March to April. The wheat marketing season in India is assumed to begin from April every year.

Climatic variability and weather abnormality increase like hailstorm, frost, high wind, extreme temperatures, and foggy condition during wheat-growing (Rabi) season lead to natural disasters affecting wheat productivity. Over the past few years, the per hectare yield of wheat in India has fallen due to the temperature rising steadily in January, February, and March, a time most crucial for the wheat crop.

Current year 2010 began with dense fog and cold day conditions prevailing over most parts of Indo-Gangetic plains in the month of January. However, February was near normal in respect to both maximum and minimum temperatures which was favorable and beneficial for the wheat productivity. March is generally known as the transition month from winter to summer conditions with a gradual increase in both the maximum and minimum temperatures of the country particularly over the northern and central parts. However, on some occasions, this transition is abrupt in day temperatures crossing 40 °C leading to the development of heat wave conditions in some parts of the country.

During the month of March, 2010, maximum and minimum temperature remained above normal, almost throughout the month over most parts of the country in general and wheat grown region in particular. The maximum temperatures increased considerably from the beginning of the third week of March when they crossed 40 °C mark at some places over Gujarat and south Rajasthan. Subsequently during the third week, the region of high maximum temperatures (40 °C and more) has extended into the remaining parts of Gujarat and Rajasthan and many parts of west Madhya Pradesh, Chhattisgarh, Jharkhand, and south Uttar Pradesh. Maximum temperatures were above normal by 4–8 °C over northwest, east, central, and adjoining peninsular India (Figs. 1, 2, and 3).

Wheat crop response to temperature varies with the stage of development, mainly germination, tillering, booting (late ear development), anthesis (pollination and fertilization), and grain growth (a week after anthesis to maturity). In wheat, day temperature (>31 °C) after anthesis can decrease the rate of grain filling (Al-Khatib and Paulsen 1984; Randall and Moss 1990; Stone et al. 1995; Wardlaw and Moncur 1995; Aggarwal 1992, 1993), while highest temperature imposed before anthesis can also decrease yield (Wardlaw et al. 1989; Tashiro and Wardlaw 1990; Hunt et al. 1991). In wheat season, high-temperature episodes occurring near to anthesis can reduce the number of grains per ear and the subsequent rate of increase in the harvest index, resulting in smaller grain yields (Wheeler et al. 1996). Crop models have been used to evaluate climatic yield potential for different regions (Aggarwal and Kalra 1994). Wheat is a crop, which needs low temperature for the kernel to form during booting stage. During the current season, crop has not encountered increased temperature at booting and anthesis stages, but the temperature was significantly higher than normal during grain-filling period thus adversely affecting the grain-filling process.

They are also capable for evaluating long-term management strategies (Hoogenboom 1991), environmental characterization and agroecological zoning (Aggarwal 1992, 1993), defining research priority and technology transfer, estimating production potential (Aggarwal 1988), strategic and tactical decision-making (Singh et al. 2010; Rathore et al. 1994), and predicting effects of climate change and variability (Lal et al. 1997, 1998).

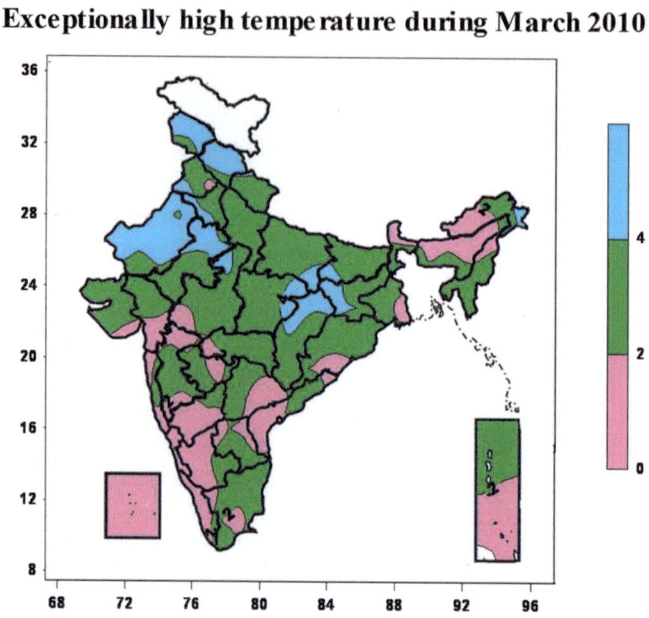

Fig. 1 Maximum temperature (°C) anomaly in India for the month of March 2010

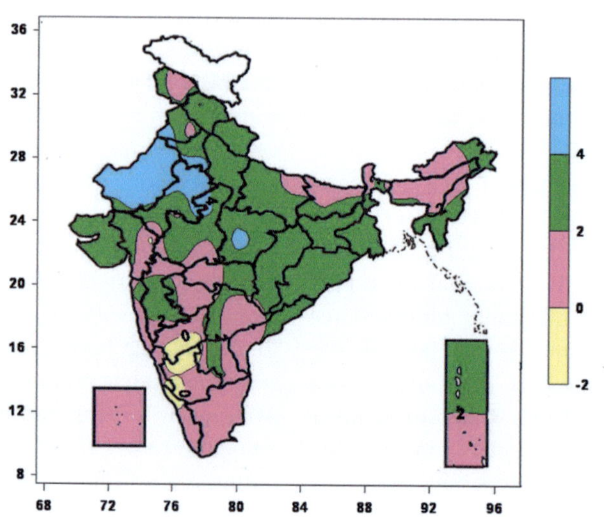

Fig. 2 Minimum temperature (°C) anomaly in India for the month of March 2010

In the present study, Agromet Service Cell, IMD, has attempted to make a preharvest estimate of the wheat yield for the years 2009–2010 using DSSAT crop growth simulation model CERES-Wheat v 4.0. Crop yield simulation runs were made under irrigated condition for 15 wheat-growing locations in India. As wheat cultivation is mostly irrigated and recommended NPK is applied, the productivity of a given cultivar is primarily governed by the weather, particularly temperature and radiation.

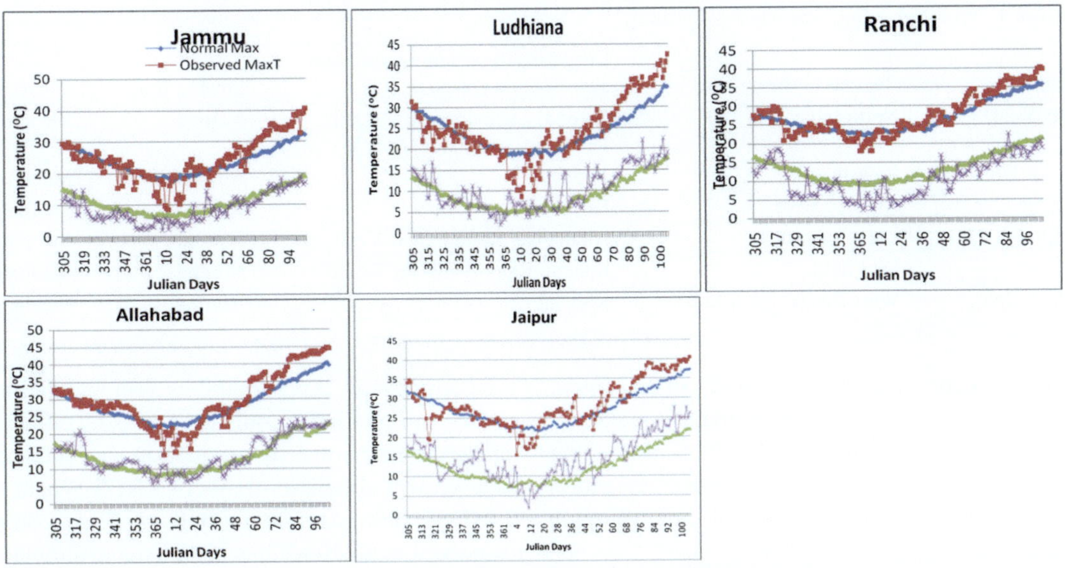

Fig. 3 Relation between observed and normal temperature (°C) of representative stations

Data and Methodology

CERES-Wheat Model

Crop models which share a common input and output data format have been developed and embedded in a software packages called the Decision Support System for Agrotechnology Transfer (DSSAT). DSSAT is a shell that allows the users to organize and manipulate crop, soil, and weather data, to run crop models in various ways, and to analyze their outputs (Jones and Kiniry 1986; Tsuji et al. 1994). The models running under DSSAT include the CROPGRO (CROP GROWTH) model for chickpea, cowpea, dry bean, faba bean, lentil, peanut, soybean, cotton, and velvet bean; CANEGRO for sugarcane crop; SUBSTOR for potato, cassava, tanier, and taro; and the CERES (Crop Estimation through Environment and Resource Synthesis) model for rice, wheat, maize, sorghum, pearl millet, and barley.

The CERES-Wheat model simulates the following process on a daily basis: (1) phonological development of the crop as effected by the genetic characters of the crop variety studied and weather; (2) growth of leaves, stems, and roots; (3) biomass accumulation and partitioning among the leaves, stem, spikelet, grains, and roots; (4) soil water balance and water use by the crop; and (5) soil nitrogen transformations and uptake by the crop (Alocilja and Ritchie 1988). The phonological stages simulated by the model are (1) sowing, (2) germination, (3) emergence, (4) panicle initiation, (5) end of vegetative phase, (6) beginning of grain filling, (7) end of grain filling, and (8) physiological maturity. The model provides simulations at daily time steps from sowing until physiological maturity. Simulation of the duration of each phonological stage makes use of the concept of thermal time or degree days and photoperiod as defined by the genetic characteristics of the crop. The soil water balance has been adapted from the model of Ritchie and Otter (1985). The CERES-Wheat model includes detailed soil and plant nitrogen balance components which simulated nitrogen uptake, nitrogen fixation, and nitrogen mobilization (Hoogenboom et al. 1991).

Input Data

The input data required to run the CERES-Wheat v4.0 model include daily weather data, soil albedo, soil water drainage constant, field capacity, wilting point, and initial soil moisture in different layers as well as maximum root depth,

Table 1 Wheat-growing locations for the study

Cultivar	Regions	Locations (representing different agroclimatic zones)	Sowing time	Duration
PBW-343	North India	Jammu, Ludhiana, Hisar, Delhi, Palampur, and Pantnagar	Normal	Medium
RAJ-3765	Rajasthan	Jaipur	Normal	Medium
HUW-234	Uttar Pradesh	Faizabad	Normal	Medium
Malviya-234	Uttar Pradesh	Allahabad, Kanpur, and Varanasi	Normal	Medium
C-306	East India	Ranchi, Raipur, and Jagdalpur	Normal	Medium
PBW-343	West Bengal (East India)	Kalyani	Normal	Medium

crop genetic coefficients, and management practices such as sowing date, plant population, plant row spacing, planting distribution, seeding depth, planting method, irrigation date, irrigation soil depth, available soil water, and nitrogen application (Tsuji et al. 1994).

Weather Data Base

Daily observed weather data on maximum and minimum temperatures and rainfall and radiation for 15 locations representing different regions of the country have been used for the two winter seasons, current season (2009–2010) and previous season (2008–2009) for different locations (Table 1). Daily solar radiation was derived from sunshine hours' data using the Angstrom's formula (1924). Global radiation has been computed from sunshine hours in case radiation data were not available using the following Angstrom formula:

$$R_s = \left\{a + b\frac{n}{N}\right\}R_A$$

Where:

R_S = global solar radiation at the surface (MJ/m^2/day)

R_A = extraterrestrial radiation (MJ/m^2/day)

n = actual duration of sunshine (hours)

N = maximum possible duration of sunshine (hours)

a and b = constant (0.25, 0.5)

For the current season 2009–2010, the observed data at all the locations is taken up to 20 March. After 21 March, 1-week (7 days) medium range weather forecast was incorporated, and thereafter 28 March onwards daily normal data was spliced in for the remaining period to complete the crop season in the current year. Agricultural system simulation models are predictive tools for assessing climate change impacts on crop production. A validated crop growth simulation model CERES-Wheat v 3.5 (DST 2006; Attri 2000; Mall and Singh 2000; Aggarwal et al. 2000) has been used to simulate attributes of popular wheat genotypes at wheat-growing locations in India during 2009–2010. The details of commonly grown cultivars at different locations used in the study and the normal weather year selected for different stations and cultivars are presented in Table 1.

Soil Data

Soil input date for CERES model includes soil texture, soil local classification, soil family SCS system, soil depth (m), color (moist), albedo (fraction), evapotranspiration limit (cm), drainages rate (fraction day^{-1}), runoff curve number, mineralization factor (0–1 scale), photosynthesis factor (0–1), pH in buffer, determination method, amount (%), and determination method of nitrogen, phosphorus, and potassium. The terms "lower limit" and "drained upper limit" correspond to the permanent wilting point and field capacity, respectively (Ritchie 1991). Total extractable soil water is function of soil physical depth of wheat crop.

Crop Data

Crop cultivars considered in the study are major dominant medium duration (varying from 115 to 140 days) varieties grown by the farmers of the region as depicted in Table 1. Water and nitrogen management in the model is as per agronomical recommendations widely accepted in different agroclimatic zones. Other crop management

Table 2 Genetic coefficient for different genotype in the agroclimatic zones of India

	Genetic coefficient				
Growth and development aspects of wheat crop	PBW-343	RAJ-3765	Malviya-234	HUW-234	C-306
Development aspects					
Optimum vernalizing temp (P1V)	20	20	20	20	20
Critical photoperiod(P1D)	80	80	80	90	85
Grain-filling duration coefficient (P5)	650	750	700	770	595
Growth aspects					
Kernel number coefficient (G1)	23	20	20	23	20
Kernel size coefficient (G2)	45	47	45	34	40
Single tiller weight (G3)	1.5	1.7	1.7	1.3	1.4
Tip leaf appearance interval (PHINT)	95	95	91	95	95

conditions are chosen as per the current field practices at the selected sites. Two dates of sowing for each zone were taken up for the study. Normal sowing dates are 15 November, 10 December, and 01 December and late sown condition, i.e., 25 November, 25 December, and 15 December, respectively, for North West India, Uttar Pradesh, and East India. The management practices considered for this study are plant population 90-plants/m^2, row spacing 20 cm, and planting depth 5 cm.

Genetic Coefficients

Crop genetic coefficients input data, which explains how the life cycle of a wheat cultivar responds to its environment, has been developed for different *cultivars* PBW-343, Raj-3765, Malviya-234, HUW-234, and C-306 in Table 2.

Results and Discussion

CERES-Wheat model, a part of DSSAT4.5 shell, was calibrated using the data on weather parameters, crop biometric parameters, and management practices. A methodology was developed for a large-area yield forecast using a crop simulation model and a discrete technology trend and was applied to the coherent wheat yield variability zones of different regions in India. Two dates of sowing representing normal (north India, 15 November; Uttar Pradesh, 10 December, and East India, 1 December) and late sown (north India, 25 November; Uttar Pradesh, 25 December; and East India, 15 December) conditions for each zone were taken up for the study. The CERES-Wheat model runs were made for the wheat seasons, 2008–2009 and 2009–2010, for different locations, i.e., North India (Jammu, Ludhiana, Hisar, Delhi, Palampur, Pantnagar, and Jaipur), Uttar Pradesh (Faizabad, Allahabad, Kanpur, and Varanasi), and East India (Kalyani, Ranchi, Raipur, and Jagdalpur) under normal and late sown conditions. The simulated grain yield, grain number, and maturity dates for the different locations are given in Figs. 4, 5, 6, 7, 8, and 9, respectively. Yield percentage deviation from year 2008–2009 is sown in Fig. 6.

The grain yield predicated for both seasons under normal sown condition lies in the range of 47 q/ha at Jammu to 62 q/ha at Hisar in Northwest India. Many stations show the increase in yield over previous season except Hisar and Palampur where the trend is reversed, shown in Fig. 4. The grain number of the current season at all of the locations is higher due to the favorable temperature in the month of February 2010 compared to the month of February 2009. The month of February is the period of booting and the anthesis stages, shown in Fig. 4. There is a sign of increase in days to maturity predicated in the current season but marginally by 1–3 days depending on the respective weather at individual stations in Northwest India, shown in Fig. 4.

Wheat crop in Uttar Pradesh during the crop season 2009–2010 was slightly affected due to the rise in temperature encountered at the grain-filling stage during March. The simulated average yield for Uttar Pradesh was 47.8 and 46.8 q/ha for

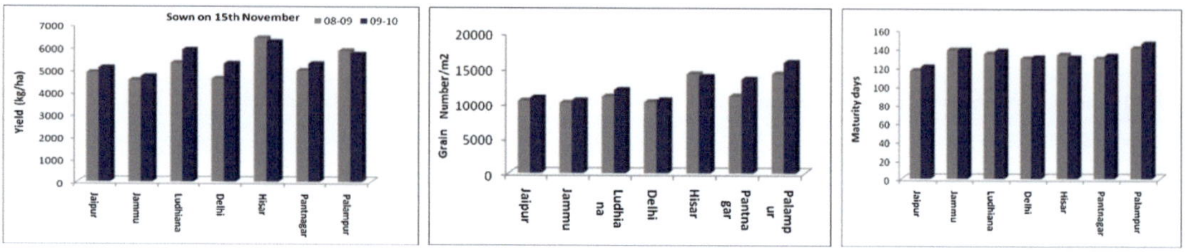

Fig. 4 Simulated grain yield, grain number, and maturity period for current and previous years (normal sown): Northwest India

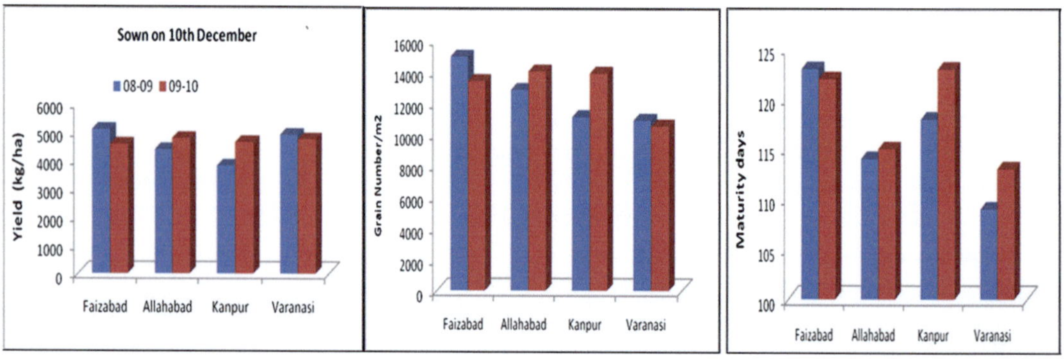

Fig. 5 Simulated grain yield, grain number, and maturity period for current and previous years (normal sown): Uttar Pradesh

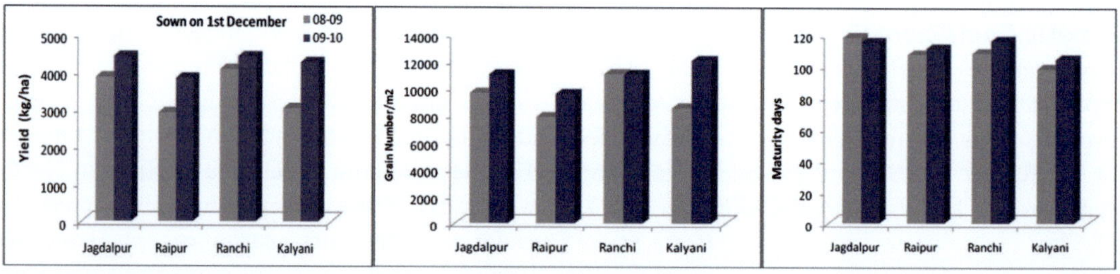

Fig. 6 Simulated grain yield, grain number, and maturity period for current and previous years (normal sown): East India

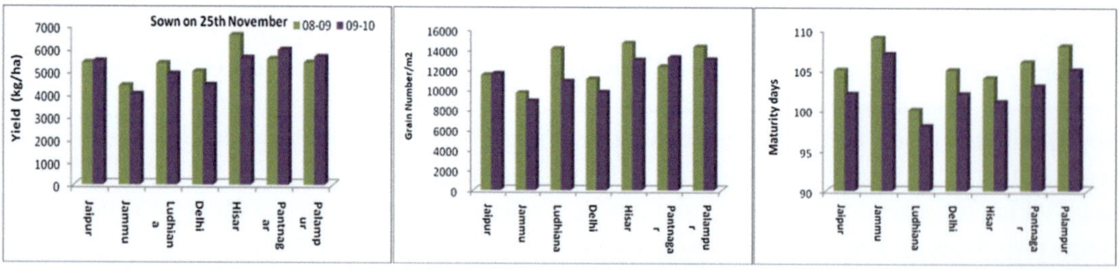

Fig. 7 Simulated grain yield, grain number, and maturity period for current and previous years (late sown): Northwest India

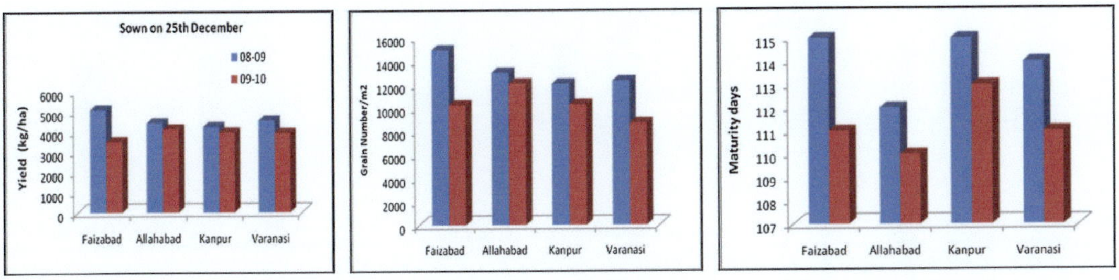

Fig. 8 Simulated grain yield, grain number, and maturity period for current and previous years (late sown): Uttar Pradesh

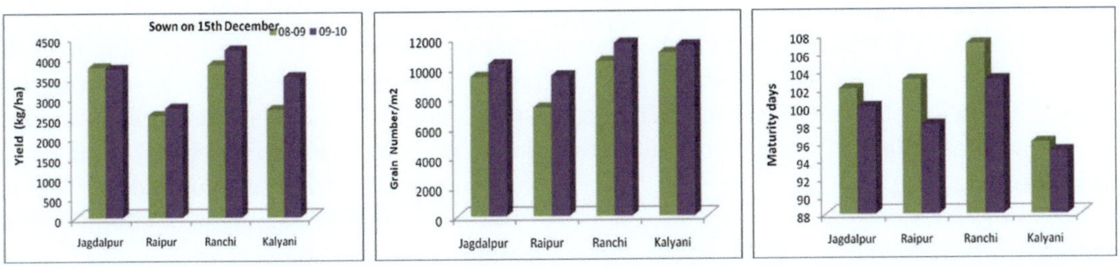

Fig. 9 Simulated grain yield, grain number, and maturity period for current and previous years (late sown): East India

the years 2008–2009 and 2009–2010, respectively (Figs. 4, 5, and 6); similarly for East India, the simulated average yield was 34.7 and 42.3 and 52.2 and 54.5 q/ha for Northwest India.

Wheat yield production realized during 2009–2010 was ~1 % less compared to 2008–2009 in Uttar Pradesh. As a combined effect, Figs. 4, 5, and 6 depict increase in grain yield by ~4 % in Northwest India and ~18 % in East India, respectively.

In Uttar Pradesh, the model predicts higher yield in the current season while decrease in yield at Faizabad and Varanasi as compared with the previous season. Similar trend is predicted for grain number per m^2 in Uttar Pradesh. Days to maturity are higher at Kanpur and Varanasi by 3–4 days over the previous season while different in Faizabad and Allahabad in Uttar Pradesh shown in Fig. 5.

The grain yield predicated in Fig. 6 is higher at all four locations in east India during the current season compared with the previous season, while increase in yield is more pronounced at Kalyani followed by Raipur and Jagdalpur. Similarly, the predication of grain no. per m^2 is noticed. Maturity period is longest at all locations except at Jagdalpur. Post-anthesis period varies from 28 days at Raipur to 32 days at the remaining three locations shown in Fig. 6.

The grain yield predicated for both seasons under late sown, i.e., 25 November in Hisar in 2008–2009 and Pantnagar in 2009–2010, due to the temperature also increases. The yield increased in Pantnagar (2009–2010) due to increase the grain number, but in case of Hisar and Ludhiana yield decreased 2009–2010 compared to 2008–2009 due to temperatures increase in grain filling stage.

Late sown at Faizabad in 2008–2009 as compared to 2009–2010 grain-filling stages is favorable for the weather. Grain number increased in 2008–2009 compared to all location stations except Faizabad. But Allahabad's grain number increased as compared to Faizabad. Maturity stages also increased in 2008–2009 compared to all location stations of Uttar Pradesh in 2009–2010. Kanpur has more maturity days as compared to all location stations of Uttar Pradesh, shown in Fig. 8.

Late sown condition at East India (15 December) as compared current year 2009–2010 is more yields for all station except Jagdalpur as compared to 2008–2009. Grain number of East India for all station increased in the current year as compared to 2008–2009. Maturity days

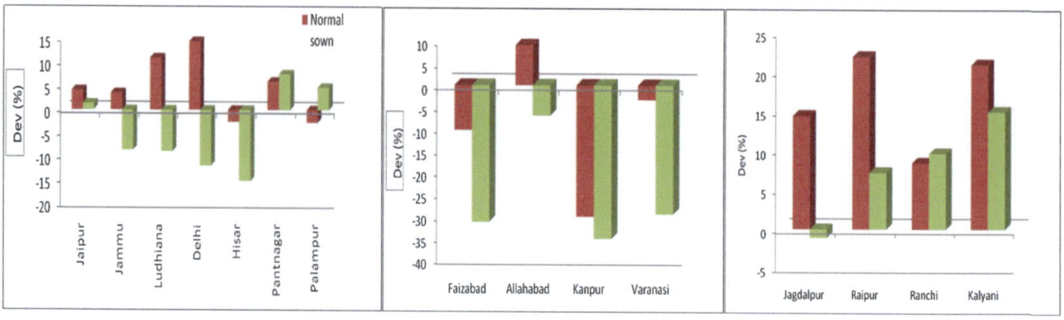

Fig. 10 Percentage (%) deviation in yield predication compared to wheat season of 2008–2009

increased in 2008–2009 as compared to the years 2009–2010 in East India. It is observed that a maturity day in Ranchi is more than other station, in current year (2009–2010) as compared to 2008–2009 Fig. 10.

In Northwest India during crop season in 2009–2010, there was lower yield as compared to the years 2008–2009 except in East India. During March of this year, the maximum and minimum temperatures remained above normal almost throughout the month over the most parts of the country. Maximum temperatures were above normal by 4–8 °C over northwest, east, central, and adjoining peninsular India.

Daily temperature (maximum, minimum, mean) departures over western and central Himalayan regions were even more than 10 °C on some days. The simulated average yield for Northwest India was 53.6 and 51.2 q/ha for the years 2008–2009 and 2009–2010, respectively; similarly for Uttar Pradesh, the simulated average yield was 50.2 and 38.3 q/ha. Wheat yield production during 2009–2010 is higher (35.4 q/ha) compared to production in 2008–2009 in East India (32.1 q/ha). Per cent deviation in wheat yield prediction compared to season 2008–2009 shows decrease of ~4.0 % in North West India, ~23.0 % in Uttar Pradesh while there is ~10.0 increase in East India.

Conclusion

Model simulation indicates increase in the grain yield per hectare under the early sown condition by ~10 % in Northwest India while ~16 % in Uttar Pradesh and ~21 % in East India over the previous years of 2008–2009. In case of late sown condition, the crop model predicts a reduction in grain yield per hectare by ~4 % in Northwest India while ~36 % in Uttar Pradesh. There is ~9 % more yield in East India, while in case of normal sown condition, the model shows increase in grain yield per ha by ~4 % in Northwest India and ~18 % in East India and reduction by ~10 % in Uttar Pradesh. A number of grains/m^2 are increased in all three wheat-growing zones ranging from 9 to 46 % in early sown conditions due to the cold condition prevailing over most part of the wheat-growing region in the month of January and February. The model simulation indicates that crops mature earlier by 2–3 days in different zones in the late sown condition.

Acknowledgments We thank the Nodal office of Agromet Field Units (Jammu, Ludhiana, Hisar, Delhi, Palampur, Pantnagar, Jaipur, Faizabad, Allahabad, Kanpur, Varanasi, Kalyani, Ranchi, Raipur, and Jagdalpur) for providing the weather data.

References

Aggarwal PK (1988) Wheat production in south-east Asia-potential and constraints. Outlook Agric 17:49–53

Aggarwal PK (1993) Agro-ecological zoning crop growth simulation models: characterization of wheat environments of India. In: De Vries FP, Paul T, Klaas M (eds) Systems approaches for agricultural development. Kluwer Academic Publishers, London, pp 97–109

Aggarwal PK (1992) Agro-ecological zoning using crop simulation models: characterization of wheat environments of India. In: Penning de Varies FWT, Teng PS (eds) Systems approach to agricultural development. Kluwer Academic Publishers, Dordrecht, pp 97–109

Aggarwal PK, Kalra N (1994) Analyzing the limitation set by climatic factors, genotype and water and nitrogen availability on productivity of wheat II. Climatic potential yield and management strategies. Field Crops Res 38:93–103

Aggarwal PK, Talukdar KK, Mall RK (2000) Potential yields of rice-wheat system in the Indo Gangetic plains of India. In: Rice wheat consortium paper series 10. RWCIGP, CIMMYT, New Delhi, p 16

Al-Khatib G, Paulsen GM (1984) Mode of high temperature injury to wheat during grain development. Physiol Plant 61:363–368

Alocilja EC, Ritchie JT (1988) Upland rice simulation and its use in multicriteria optimization, IBSNAT research project series 01, IBSNAT

Angstrom A (1924) Solar and terrestrial radiation. Q J Roy Meteorol Soc 50:121–126

Attri SD (2000) Weather based crop management using dynamic simulation model. PhD thesis, GJ University of Science and Technology, Hisar (Haryana), India

DOD (2002) Annual report 2001/02. Department of Ocean Development, New Delhi

DST (2006) Report on wheat yield predication using CERES-wheat model for the season 2005–06. In: Singh KK, Baxla AK, Rathore LS, Kaushik S, Singh PK

Hoogenboom G, Jones JW, Boote KJ (1991) A decision support system for predication of crop yield, evaporation and irrigation management. In: Transactions of ASAE, irrigation and drainage proceedings, Honolulu

Hunt LA, van der Poorten G, Pararajasingham S (1991) Postanthesis temperature effects on duration and rate of grain-filling in some winter and spring wheats. Can J Plant Sci 71:609–617

Jones CA, Kiniry JR (1986) CERES maize: a simulation model of maize growth and development. Texas A &M University Press, College Station

Lal M, Cubasch U, Voss R, Waszkewitz J (1995) Effect of transient increase in greenhouse gases and sulphate aerosols on monsoon climate. Curr Sci 69(9):752–763

Lal M, Singh KK, Rathore LS, Srinivasan G, Saseendran SA (1998) Vulnerability of rice and wheat yields in NW India to future changes in climate. Agr Forest Meteorol 89:101–114

Lal M, Singh KK, Rathore LS, Srinivasan G, Saseendran SA (1997) Vulnerability of rice and wheat yields in NW India to future change in climate. Agric For Meteorol 89:1–13

Mall RK, Singh KK (2000) Climate variability and wheat yield progress Punjab using the CERES-wheat and WTGROWS models. Yayu Mandal 30(3–4):35–41

Randall PJ, Moss HJ (1990) Some effects of temperature regime during grain filling on wheat quality. Aust J Agric Res 41:603–617

Rathore LS, Srinivasan G, Singh KK (1994) Soil–moisture and evapotranspiration simulations for irrigated wheat using SPAW model. Mausam 45 (1):63–68

Ritchie JT (1991) Wheat phasic development. In: Hanks J, Ritchie JT (eds) Modeling plants and soil system, Agronomy monograph 31. ASA, CSSA, Madison, pp 31–54

Ritchie JT, Otter SN (1985) Description and performance of CERES-Wheat, a user oriented wheat yield model, USDA, ARS, ARS-38, pp 159–175

Saseendran SA, Singh KK, Rathore LS, Prasad RGSLHV, Mendiratta N, Narayan Rao L, Singh SV (1998) Evaluation of the CERES-Rice v. 3.0 model for the climate conditions of the state of Kerala. Indian Meterol Appl 5:385–392

Singh PK, Rathore LS, Singh KK, Baxla AK, Mall RK (2010) Modeling the effect of planting date on Maize (*Zea mays* L.) crop in Sabour region of Bihar using DSSATv3.5 crop simulation model. J Mausam 61 (1):75–80

Stone PJ, Savin R, Wardlaw IF, Nicolas ME (1995) The influence of recovery temperature on the effect of a brief heat stroke on wheat I. Grain growth. Aust J Plant Physiol 22:945–954

Tashiro T, Wardlaw IF (1990) The response to high temperature shock and humidity changes prior to and during the early stages of grain development in wheat. Aust J Plant Physiol 17:551–561

Tsuji GY, Uehara G, Balas SS (1994) DSSATv3. University of Hawaii, Honolulu, 284 p

Wardlaw IF, Dawson IA, Munibi P, Fewster R (1989) The tolerance of wheat to high temperature during reproductive growth. II. Survey procedures and general response patterns. Aust J Plant Physiol 40:1–13

Wardlaw I, Moncur L (1995) The response of wheat to high temperature following anthesis. I. The rate and duration of kernel filling. Aust J Plant Physiol 22:391–397

Wheeler TR, Hong TD, Ellis RH, Batts GR, Morison JIL, Hadley P (1996) The duration and rate of grains growth and harvest index of wheat (*Triticum aestivum*) in response to temperature and CO_2. J Exp Bot 47:623–630

Multispectral Remote Sensing to Distinguish the Little Seed Canary Grass (*Phalaris* Minor) from Wheat Crop Under Field Conditions for Environmental Sustainability and Precision Weed Management

Ramanjit Kaur, R.K. Mahey, and P.K. Kingra

Abstract

To take full advantage of site-specific variable rate technology (VRT) systems, highly accurate digital mapping of weed infestations within fields via scouting, GPS, GIS and remote sensing technologies will be necessary. When combined, these tools can increase weed control efficiency and reduce herbicide use and residues, thereby avoiding excess applications that lead to increased costs, potential herbicide resistance in the field and runoff into the environment. Keeping this in view, a field experiment was conducted at the Research Farm, Department of Agronomy, Punjab Agricultural University, Ludhiana, Punjab, for 2 years to study the multispectral remote sensing to distinguish the little seed canary grass (*Phalaris minor)* from wheat crop under field conditions for environmental sustainability and precision weed management. The experimental site during both the seasons were sandy loam in texture, with normal soil reaction and electrical conductivity, low in organic carbon and available nitrogen and medium in available phosphorus and potassium. The experiment consisted of five treatments, viz, T_1,, control (weedy check); T_2, half of the recommended dose of herbicide for partial control of *Phalaris minor*; T_3, recommended dose of herbicide to obtain economic threshold level to control *Phalaris minor*; T_4, manual weeding (partial), done after a month of sowing of crop; and T_5, weed free (manual). The treatments T_3 (recommend dose of herbicide) and T_5 (weed free, manual) being at par with one another recorded highest plant height, dry matter accumulation and number of tillers per plant by wheat at all observational dates during both the years in experiments, whereas minimum plant height, dry matter accumulation and number of tillers per plant were recorded in control treatment (T_1). Reduction in dry matter production,

R. Kaur (✉)
Indian Agricultural Research Institute, New Delhi 110012, India
e-mail: ramaan180103@yahoo.com

R.K. Mahey • P.K. Kingra
Department of Agronomy, Punjab Agricultural University, Ludhiana, Punjab, India

A.K. Singh et al. (eds.), *Climate Change Modelling, Planning and Policy for Agriculture*,
DOI 10.1007/978-81-322-2157-9_7, © Springer India 2015

number of tillers as well as effective number of tillers and ultimately yield of wheat are mainly attributed to the reduction in the number of effective tillers, lesser number of grains per spike and lesser 1000-grain weight. The weed-free treatments (T_3 and T_5) had lower red reflectance percentage as compared to other weed control treatments. The control treatment recorded the highest red reflectance. On the other hand, the two weed-free treatments T_3 and T_5 had higher IR reflectance percentage as compared to other three weed control treatments, and the lowest IR reflectance was recorded under control treatment. Highest RR and NDVI values were obtained in treatments T_3 and T_5 where there was no competition between wheat and weeds, and control treatment had the lowest RR value amongst all the treatments during both years. Differences in RR between these three treatments are mainly due to dark green colour of wheat, more leaf area index (LAI) and more biomass of wheat as compared to *Phalaris minor*. The RR value increases in the early stages of crop growth which is maximum at maximum crop canopy cover and after that decreases as the leaves senesce. The highest RR values were obtained at 95 days after sowing almost in all the treatments. It is feasible to distinguish pure wheat from weeds just 34 days after sowing, but amongst different weed control treatments, i.e. pure *Phalaris minor* plot and less/partial *Phalaris minor* weeds 52 days after sowing amongst themselves, and they remain distinguished up to 107 days after sowing based on their NDVI values. After 52 DAS, the differences in the NDVI of different weed control treatments were very clear. So, from such type of information, we can discriminate/define the areas which are heavily or partially infested with weeds so that timely weed control measures can be taken which can help the farmers in preventing yield losses due to weeds.

Keywords

Wheat • Weeds • Sustainability • Canary grass • Remote sensing

Introduction

Wheat is the most important and widely cultivated cereal crop in the world. In Punjab, it is cultivated over an area of 3.52 million hectares with a total production of 15.2 million tones and an average yield of 4.3 tones per hectare. Major detriments to higher productivity of wheat are stiff competition from weeds, multiple nutrient deficiencies, insect and pests and incidence of diseases. Severe weed competition has resulted from the change in morphological characteristics of wheat varieties and shift in sowing paradigm over a period of four decades. Moreover, high soil moisture and low temperature at the time of sowing lead to severe infestation by grassy and non-grassy weeds. Uncontrolled weeds reduce wheat yield to the tune of 20–57 % (Singh et al. 1997). Amongst different weed species, the infestation by little seed canary grass (*Phalaris minor*) in wheat is rampant, and it alone divests the crop of 42.2 kg N, 6.5 kg P and 71.6 kg K per hectare (Walia and Gill 1985) and lowers the crop yield by 30–50 % (Pandey and Singh 1997). Similarly wild oats (*Avena ludoviciana*) reduces small grain harvest efficiency, yield and crop quality, costing North American growers an estimated one billion dollars annually (Sharma 1979).

In Punjab, about 80–85 % of area under wheat is treated with herbicides, as chemical method of weed control has become very popular amongst farmers being very effective, economical and quick. Though herbicides have been very effective against these weeds, most of them are narrow in selectivity and give partial control. The continuous and indiscriminate use of same herbicide year after year has lead to pollution problems such as ground water and soil as well as the development of resistant biotypes of *P. minor* (Malik and Singh 1995; Walia et al. 1997) which necessitates that only need-based use of herbicides should be done. In order to achieve this, if major weeds like *Phalaris minor* and *Avena ludoviciana* are identified through spectral measurements, timely and rational weed control measures can be followed, the estimates of loss will be more reliable and accurate, and timely forecast of production will enable the government to plan its agricultural economy with greater degree of reliability. Also decision models can be developed and used for spatially variable herbicide application in precision agriculture.

In cultivated systems, early studies suggested that due to the differences in chlorophyll content, colour, leaf area, intercellular spaces and surface characteristics, it was feasible to separate weeds from crops at a level where the weed has an economic impact on crop yield (Menges et al. 1985). Estimates of 40–94 % reduction in herbicide use may be realised using site-specific technology. The ability to selectively apply herbicides through the use of weed mapping and variable rate sprayers offers the potential to not only increase producer profitability but also to reduce harmful effects on the environment (Brown and Steckler 1995; Mortensen et al. 1995). Most theoretical analyses indicate that cost savings from the use of precision technologies are likely to be small (Christensen and Walter 1995; Olesen 1995). Models allowing variable herbicides rate applications, threshold or multiyear optimisation based on population simulation have shown greater potential economic benefits in applying herbicides based on the presence/absence of weeds (Maxwell and Colliver 1995).

Materials and Methods

Field experiment was conducted at Punjab Agricultural University, Ludhiana (30°56′N latitude, 75°52′E longitude and 247 m altitude), during the *rabi* seasons of 2006–2007 and 2007–2008. Ludhiana is characterised by subtropical semi-arid type of climate with hot summer and very cold winters. The mean maximum and minimum temperatures therefore show considerable fluctuation during summer and winter seasons. The meteorological data recorded during crop growing seasons indicate that almost normal weather conditions prevailed during the two crop seasons except for rainfall, which was highly variable during the 2-year study. Maximum temperature ranged between 19.5 and 36.8 °C during the first year and 16.9–34.1 °C during the second year against the normal range of 18.3–34.2 °C from November to April. Minimum temperature ranged between 4.2 and 19.4 °C during the first year and 4.5–17.7 °C during the second year against a normal range of 4.2–16.9 °C. The first year received 185 mm rainfall during November to April, whereas the second season received 89 mm of rainfall against a normal value of 111 mm.

The soil of the experimental fields was normal in reaction (pH 7.52), and electrical conductivity (0.16 dS/m) was low in organic carbon (0.34 %) and available nitrogen (229.1 kg/ha) and medium in available phosphorus (19.05 kg/ha) and potassium (205.35 kg/ha). Wheat seed (PBW 343) treated by using 4 ml chlorpyrifos per kg seed was sown on November 15 in 2006 and November 14 in 2007 with the *kera* method using a seed rate of 100 kg/ha in a randomised block design. The gross plot size kept for experiment was 5 m × 4 m net plot size was 4.8 m × 3.8 m. The crop was supplied with 125 kg N, 60 kg P_2O_5 and 30 kg K_2O per hectare through urea, di-ammonium phosphate and muriate of potash, respectively. Half the dose of nitrogen and whole of phosphorus and potassium were applied at the time of sowing by the broadcast method, while the remaining half dose of nitrogen was top dressed after the first irrigation. The first irrigation to wheat was applied 21 days after sowing, and subsequent irrigations were given as per the requirement of the crop. The experiment was conducted during the *Rabi* seasons of 2006–2007 and 2007–2008 with five weed control treatments, viz., control, where no herbicide was applied for controlling weeds (T_1), partial control means half of the recommended dose of herbicide was applied (T_2), recommended dose of herbicide was applied (T_3), partial but manual control (T_4) and complete (weed free, manual) control of *Phalaris minor* (T_5). In addition to the periodic observations on biological parameters, the observation on yield and yield contributing characters of the crop were taken as follows:

Yield and Yield Contributing Character

The following yield attributes of wheat crop were recorded at the time of harvesting:

Ear length: For recording the spike length, ten spikes were selected at random from each plot, and then the length excluding awns was measured and averaged. The average length is presented in cms.

Number of grains per spike: The number of grains per spike was counted from ten randomly selected earheads from each plot, and the mean values were calculated for presentation.

Thousand grain weight: Thousand grain weight (g) was recorded by taking two random samples of grains from the production of each plot during threshing, and thousand grains were counted and weighed.

Biological yield: The bundle weight of each net plot was recorded before threshing and presented in qha^{-1} as the biological yield.

Grain yield: After threshing, the net plot grain yield (kg/plot) was recorded, and the data are presented as qha^{-1}.

Remote Sensing Observations

Spectral reflectance in two wavebands, i.e. red (625–689 nm) and infrared (760–897 nm), was recorded at fortnightly interval with the help of a hand-held ground truth spectroradiometer, and remote sensing parameters were calculated as under:

Radiometric observations: Spectral reflectance in two wavebands, i.e. red (625–689 nm) and infrared (760–897 nm), was recorded at fortnightly intervals with the help of a hand-held ground truth spectroradiometer, and remote sensing parameters were calculated as under:

Red reflectance (per cent) and IR reflectance (per cent): The observations were taken with hand-held ground truth radiometer to determine the spectral signature characteristics of *Phalaris minor* and *Avena ludoviciana* and also to know their critical population for identification in crops. Periodic observations on radiometric reflectance in two wavebands, i.e. (600–700 nm) and (720–900 nm), were recorded at fortnightly interval throughout the crop growth cycle, and red reflectance (per cent) and infrared reflectance (per cent) were calculated.

Radiance ratio (RR) and normalised difference vegetation index (NDVI): Radiance ratio (RR) and normalised difference vegetation index (NDVI) were derived from red and IR band reflectance by the following formulae:

$$RR = \frac{\text{Infrared Reflectance (IR)}}{\text{Red Reflectance (R)}} \tag{1}$$

$$NDVI = \frac{(IR - R)}{(IR + R)} \tag{2}$$

Results and Discussion

Yield Contributing Characters of Wheat

Number of grains per year: Significant differences in number of grains per year were observed between different weed control treatments during both the years of study (Table 1). The maximum number of grains per year were recorded in treatment T_3 (recommended dose of herbicide) which was statistically at par with

Table 1 Yield attributes of wheat under different weed control treatments

	Grains per year (No.)		1000 grain wt (g)		Grain yield (qha^{-1})		Straw yield (qha^{-1})	
Treatment	2006–2007	2007–2008	2006–2007	2007–2008	2006–2007	2007–2008	2006–2007	2007–2008
Control (weedy check) (T_1)	37.2	38.8	35.9	36.2	40.9	40.9	37.9	44.1
Half recomm. dose (T_2)	43.1	43.7	40.4	40.8	50.2	48.5	45.0	44.7
Full recomm. dose (T_3)	48.6	44.9	44.2	45.1	56.3	55.6	42.2	40.6
Partial control (manual, T_4)	41.5	40.9	39.8	38.7	48.3	45.4	45.1	43.7
Weed free (manual, T_5)	47.9	46.8	44.3	45.5	56.5	56.1	43.7	43.4
C.D. ($\alpha = 0.05$)	**4.50**	**2.00**	**3.2**	**2.45**	**4.27**	**3.10**	**2.65**	**1.93**

T_5 (weed free, manual) treatment but significantly superior to the rest of weed control treatments in 2006–2007. In 2007–2008, a maximum number of grains per year were recorded under T_5 which was statistically at par with T_3 treatment but significantly superior to the remaining three weed control treatments. The lowest number of grains per year were recorded in treatment T_1 (control) during both the years of study which was due to the stiff competition posed by *Phalaris minor* to crop plants at all stages of growth.

1000-grain weight: The 1000-grain weight of wheat presented in Table 1 showed significant differences in 1000-grain weight between different weed control treatments. Treatments T_3 and T_5 being at par with one another produced significantly higher 1000-grain weight than the other weed control treatments in 2006–2007 and 2007–2008. The treatments T_2 and T_4 being at par with one another were significantly superior to the control (T_1). The lowest 1000-grain weight was recorded in the control treatment (T_1), where no herbicide was applied in 2006–2007 and 2007–2008.

Grain yield: Grain yield of wheat showed significant differences between different weed control treatments studied during 2 years of investigation. Highest grain yield was recorded in weed-free (manual) treatment (T_5) which was statistically at par with the treatment where full herbicide dose was applied (T_3) during both the years (Table 1). Both these treatments performed significantly better than the remaining weed control treatments. The minimum grain yield was recorded under control which was significantly inferior to all other treatments during both the years.

The treatment T_5 recorded 27.6 and 27.1 % higher grain yield over control (T_1) in 2006–2007 and 2007–2008, respectively, whereas treatment T_3 produced 27.4 and 26.4 % higher grain yield than control (T_1) during the 20 years of study. The higher grain yield produced in treatments T_3 and T_5 was due to the absence of competition between weeds and the crop plants resulting in higher dry matter accumulation and higher number of effective tillers which ultimately contributed towards the higher grain yield.

Straw yield: The highest straw yield was recorded under T_4 partial control (manual) treatment, which was superior to T_1 (no herbicide) and T_3 (recommended dose of herbicide) in 2006–2007, whereas the highest straw yield of wheat was recorded in T_2 in 2007–2008 which was at par with all weed control treatments except T_1, with which it was statistically at par (Table 1).

Fig. 1 (**a**) Red reflectance (per cent) of wheat and little seed canary grass under field conditions in 2006–2007. (**b**) Red reflectance (per cent) of wheat and little seed canary grass under field conditions in 2007–2008

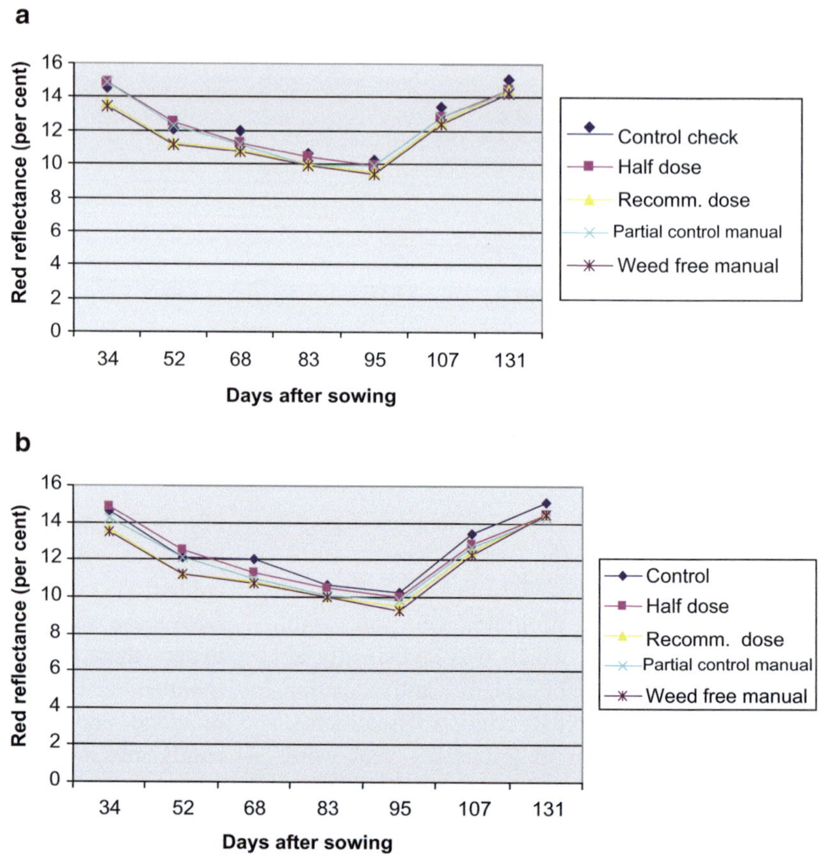

Remote Sensing Parameters

A review regarding the potential of remote sensing techniques for crop production in the field suggested that one way to distinguish between weeds and crops was by examining the temporal patterns of vegetation indices throughout the growing season (Hatfield and Pinter 1993). This was also supported by (Brown et al. 1994) who reported the potential for distinguishing weeds from agricultural crops on the basis of their relative spectral reflectance characteristics through time.

Red Reflectance and IR Reflectance

Red reflectance (per cent) data are presented in Fig. 1a, b for the years 2006–2007 and 2007–2008. In all the treatments irrespective of wheat and *Phalaris minor*, the red reflectance (per cent) values decreased from 34 days after sowing to 95 days after sowing in 2006–2007 and 2007–2008, and thereafter a sharp increase was observed in all the five weed control treatments. This trend might be due to the increasing chlorophyll content as red reflectance is reduced by chlorophyll absorption. The figures clearly showed that both weed-free treatments (T_3 and T_5) had lower red reflectance percentage as compared to the other three weed control treatments. The control treatment recorded the higher red reflectance, and partial control treatments (T_2 and T_4) had intermediate percentage of red reflectance during both years.

On the other hand, the IR reflectance (per cent) increased up to 95 days after sowing, and then it decreases till maturity in all the five weed control treatments (Fig. 2a, b). The sharp fall in IR reflectance is due to the reduction in the chlorophyll of leaves. The trend in case of IR reflectance is

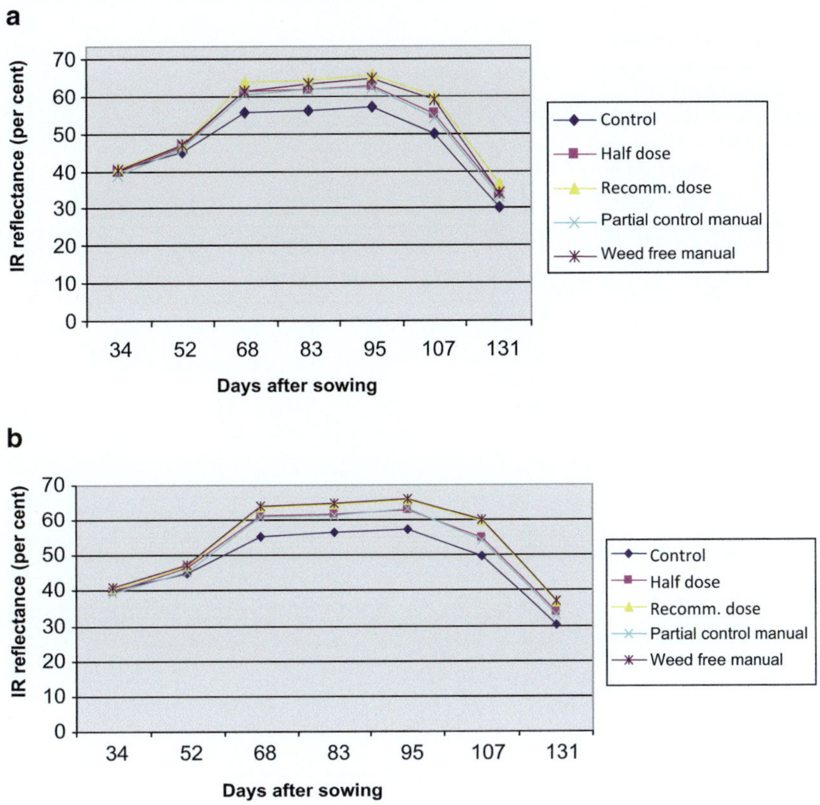

Fig. 2 (**a**) Infrared reflectance (per cent) of wheat and little seed canary grass under field conditions in 2006–2007. (**b**) Infrared reflectance (per cent) of wheat and little seed canary grass under field conditions in 2007–2008

totally reverse to that of red reflectance. The two weed-free treatments (T_3 and T_5) had higher IR reflectance percentage as compared to the other three weed control treatments. The control treatment recorded the lowest IR reflectance, and partial control treatments (T_2 and T_4) had intermediate percentage of IR reflectance during both years.

It was reported (Jurado-Exposito et al. 2003) that the spectral window between 750 and 950 nm, which corresponds to near-infrared wavelength, was able to discriminate amongst wheat stubble, and sunflower and the most problematic "hard to control" weeds in no-till sunflower, i.e. little mallow (*Malva parviflora* L.) and *Ecballium elaterium* L. (Bajwa and Tian 2001), used an airborne digital colour-infrared sensor to acquire remotely sensed images for mapping weed intensity.

Near-infrared band (850 nm) and wavelength region of 750–1,000 nm showed promising results for discriminating between weed-free cotton canopy and cotton canopy with morning glory (Iqbal et al. 2006).

Radiance Ratio (RR) and Normalised Difference Vegetation Index (NDVI)

The radiance ratio (RR) presented in Fig. 3a, b showed that it increases with the advancement of crop growth and reaches maximum at maximum leaf area growth stage and then it decreases till maturity. Highest RR value was reached at 95 days after sowing in all the treatments and then it decreases. The highest RR value was obtained in the treatment with complete control of *Phalaris minor* (T_3 and T_5) compared to partially controlled *Phalaris minor* (T_2 and T_4) and where no herbicide was used (control, T_1) as treatment. The control treatment had the least RR value amongst all the treatments. Differences in RR between these three treatments are mainly due to dark green colour of wheat, more LAI and

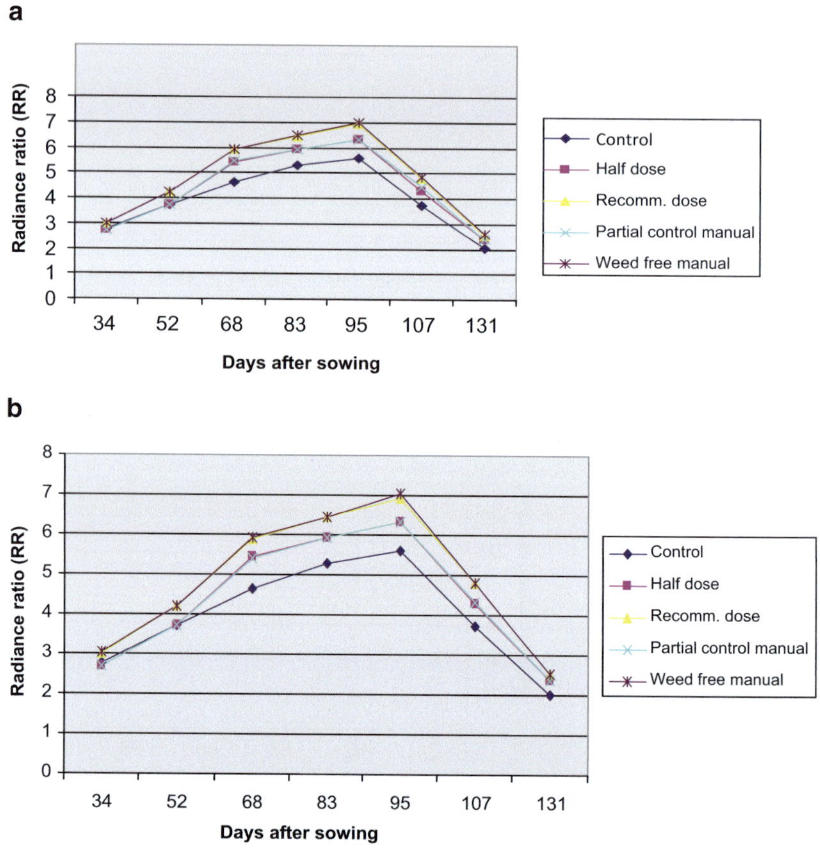

Fig. 3 (**a**) Radiance ratio (RR) of wheat and little seed canary grass under field conditions in 2006–2007. (**b**) Radiance ratio (RR) of wheat and little seed canary grass under field conditions in 2007–2008

biomass, as compared to *Phalaris minor*. So, from such type of information, we can discriminate the areas which are heavily or partially infested with weeds. We can discriminate between pure wheat and weeds just after 34 days after sowing, but amongst pure weed and less/partial weeds after about 52 days after sowing, and they remain distinguished up to 107 days after sowing. The same trend was also observed in case of NDVI during both the years of investigation. The data of NDVI are presented in Fig. 4 for the years 2006–2007 and 2007–2008, respectively. The decrease in NDVI after 95 days after sowing is mainly due to senescence of leaves in all the five treatments. The highest NDVI was recorded in weed-free treatments (T_3 and T_5) and the lowest in control T_1 treatment, and intermediate NDVI was attained by the rest of the treatments (T_2 and T_4). It is feasible to distinguish pure wheat from weeds just after 34 days after sowing, but amongst different weed control treatments, i.e. pure *P. minor* plot and less/partial *P. minor* weeds, after about 52 days after sowing amongst themselves, and they remain distinguished up to 107 days after sowing. After 52 days after sowing, the differences in the NDVI of different weed control treatments were very clear.

The results of Gibson et al. (2004) showed that remote sensing has potential for weed detection in soybean, particularly when weed management system do not require differentiation amongst weed species. In another study, threshold populations of ten or more *S. obtusifolia* or *I. lacunose* plants/m^2 were generally classified with at least 85 % accuracy (Medlin et al. 2000).

The visible and NIR bands on the satellite multispectral sensors allow monitoring of the greenness of vegetation. Stressed vegetation is less reflective in the IR channel than non-stressed vegetation and also absorbs less energy in the visible band. The NDVI varies

Fig. 4 (**a**) NDVI of wheat and little seed canary grass under field conditions in 2006–2007. (**b**) NDVI of wheat and little seed canary grass under field conditions in 2007–2008

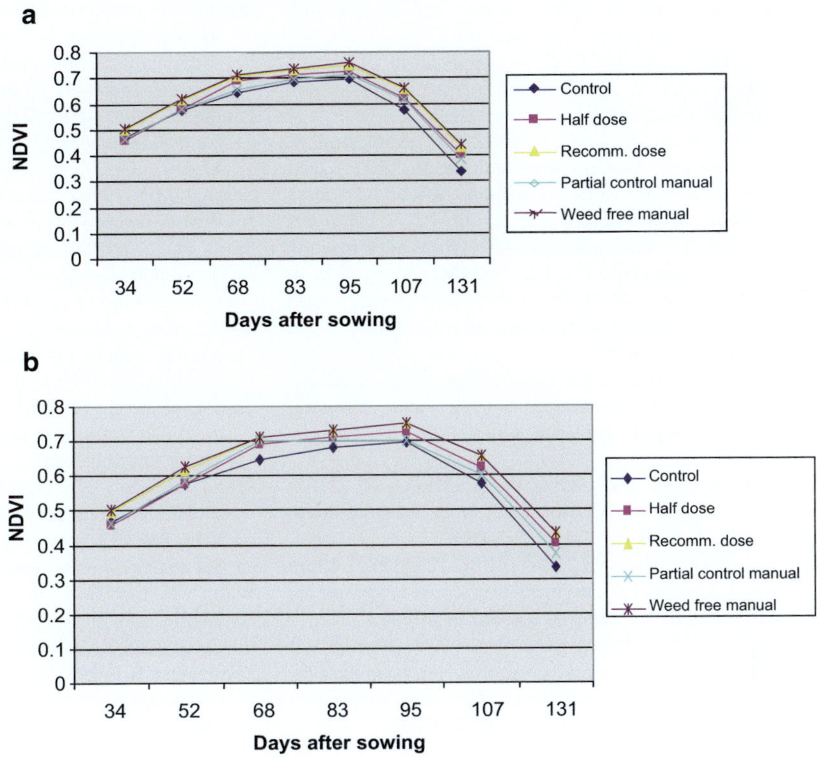

with the magnitude of green foliage (green leaf area index, green biomass) brought about by phenological changes or environmental stresses. The temporal pattern of NDVI is useful in diagnosing vegetation conditions (Das 2000). Review on several studies regarding the use of spectral properties for discriminating crop plants from weeds was done, and it was observed that although there was evidence that spectral properties can be used to discriminate between a certain set of crops and weeds, frequently unique wavebands are selected for classifying specific species (Zwiggelaar 1998). This suggests that site-specific calibration techniques might be required to distinguish crops from weeds using spectral reflectance.

Herbicide application amount could be reduced by up to 48 % by using site-specific weed control in real time (Tian et al. 1999). A ground-based weed mapping system was developed to measure weed intensity and distribution in a cotton field, and this system performed well during field evaluation. Weed intensity in the field was also estimated based on remotely sensed imagery (Sui et al. 2008).

References

Bajwa SG, Tian LF (2001) Aerial CIR remote sensing for weed density mapping in a soybean field. Trans Am Soc Agric Eng 44(6):1965–1974

Brown RB, Steckler JPGA (1995) Prescription maps for spatially variable herbicide application in no-till corn. Trans Am Soc Agric Eng 38:1659–66

Brown RB, Steckler JPGA, Anderson GW (1994) Remote sensing for identification of weed in no till corn. Trans Am Soc Agric Eng 37:297–302

Christensen S, Walter AM (1995) Site specific weed management. In: Olesen SE (ed) Proceedings of the seminar on site specific farming. Danish Institute of Plant and Soil Science, Tjele, pp 151–60

Das HP (2000) Monitoring the incidence of large scale droughts in India. In: Wilhite DA (ed) Drought a global assessment, vol I. Routledge, London/New York, pp 181–195

Gibson KD, Richard D, Medlin CR, Johnston L (2004) Detection of weed species in soybean using multispectral digital images. Weed Technol 18(3):742–749

Hatfield JL, Pinter PJ (1993) Remote sensing for crop protection. Crop Prot 12(6):403–13

Iqbal J, Phillip R, Ali I (2006) Application of remote sensing data to assess weed infestation in cotton. Agric J 1(4):186–191

Jurado-Exposito M, Lopez-Granados F, Atenciano S, Garcia-Torres L, Gonzalez-Andujar JL (2003) Discrimination of weed seedlings, wheat (*Triticum aestivum* L.) stubble and sunflower (*Helianthus annuus*) by near infrared reflectance spectroscopy (NIRS). Crop Prot 22(10):1177–1180

Malik RK, Singh S (1995) Little seed canary grass (*P. minor*) resistance to isoproturon in India. Weed Technol 9:419–425

Maxwell BD, Colliver CT (1995) Expanding economic thresholds by including spatial and temporal weed dynamics. Brighton Crop Prot Conf Weeds 3:1069–76

Medlin CR, Shaw DR, Gerard PD, LaMastus FE (2000) Using remote sensing to detect weed infestations in *Glycine max*. Weed Sci 48:393–398

Menges RM, Nixon PR, Richardson AJ (1985) Light reflectance and remote sensing of weeds in agronomic and horticultural crops. Weed Sci 33:569–81

Mortensen DA, Johnson GA, Wyse DY, Martin AR (1995) Managing spatially variable weed populations. Chapter 27 in site-specific management for agricultural systems. ASA-CSSA-SSSA, Madison

Olesen SE (1995) Conclusion of the seminar on site specific farming. In: Proceedings of the seminar on site specific farming. Danish Institute of Plant and Soil Science, Madison, pp 98–200

Pandey J, Singh RP (1997) Weed control is key on weed control in wheat (*Triticum aestivum*). Indian Farm 47:4–7

Sharma MP (1979) Wild oat a billion dollar problem. Weeds Today 10(3):5–6

Singh PK, Singh DK, Singh RP (1997) Weed crop competition in wheat as affected by different weed species. Indian J Weed Sci 29:69–71

Sui R, Alex J, Hanks J, wooten J (2008) Ground-based sensing system for weed mapping in cotton. Comput Electron Agric 60(1):31–38

Tian L, Reid J, Hummel J (1999) Development of a precision sprayer for site specific weed management. Trans Am Soc Agric Eng 42(4):893–900

Walia US, Gill HS (1985) Influence of nitrogen and substituted urea, herbicide on the uptake of N, P and K by *Phalaris minor* Retz. and wheat. Indian J Weed Sci 17:12–17

Walia US, Dhaliwal BK, Brar LS (1997) Resistance to isoproturon in *Phalaris minor* in Punjab. Plant Prot Quart 12:138–40

Zwiggelaar R (1998) A review of spectral properties of plants and their potential use for crop/weed discrimination in row crops. Crop Prot 17(3):189–206

Modeling of Weather Parameters Using Stochastic Methods

Deodas T. Meshram, V.T. Jadhav, S.D. Gorantiwar, and Ram Chandra

Abstract

The study was carried out to developed stochastic model for weekly temperature, humidity and precipitation in Solapur (Latitude 17°40′N, Longitude 75°54′E and altitude 483.50 m amsl) station of western part of Maharashtra, India. In the present study, 42 years data (1969–2010) of daily temperature, relative humidity and precipitation of Solapur station have been used for time series analysis. Weekly mean temperature, relative humidity and monthly precipitation values were used to fit the ARIMA class of models for different orders. ARIMA models of first and second orders were selected based on autocorrelation function (ACF) and partial autocorrelation function (PACF) of the time series. The parameters of the selected models were obtained with the help of maximum likelihood method. The diagnostic checking of the selected models was then performed with the help of three tests (i.e. standard error, ACF and PACF of residuals and AIC) to know the adequacy of the selected models. The ARIMA models that passed the adequacy test were selected for forecasting. One year ahead forecast (i.e. for 2010) of temperature, relative humidity and precipitation values were obtained with the help of these selected models and compared with the values of temperature, relative humidity and precipitation obtained from the climatological data of 2010 by root mean square error (RMSE). According to the Seasonal ARIMA model, ACF, PACF and evaluation of all eventual parameters, the results from analysis show that the model fitted is weekly temperature: ARIMA $\mathbf{(111)\ (011)_{52}}$, weekly relative humidity: ARIMA $\mathbf{(111)\ (111)_{52}}$ and monthly precipitation: ARIMA $\mathbf{(211)\ (201)_{12}}$ and hence are the best stochastic model for generating and forecasting of weekly temperature, relative humidity and monthly precipitation values for Solapur station, Maharashtra, India.

D.T. Meshram (✉) • V.T. Jadhav • S.D. Gorantiwar • R. Chandra
ICAR-NRC on Pomegranate, Kegaon, Solapur 413 255, Maharashtra, India

Department of Irrigation and Drainage Engineering, MPKV, Rahuri 413722, Maharashtra, India
e-mail: gomesh1970@rediffmail.com

A.K. Singh et al. (eds.), *Climate Change Modelling, Planning and Policy for Agriculture*, DOI 10.1007/978-81-322-2157-9_8, © Springer India 2015

The studies reveal that if sufficient spread and depth of data are used in model building, frequent updating of model may not be necessary. The study also showed the utility of forecast of climatic parameter values in estimating the irrigation quantity and monitoring the insect pest and disease 1 year ahead for pomegranate orchards. It is concluded that seasonal ARIMA model is a viable tool which can successfully be used for generation and forecasting of climatic parameters having inbuilt seasonal patterns.

Keywords
Climate change • Rainfall • Temperature • Humidity and stochastic model

Introduction

Stochastic models have been proposed as an important technique to generate and forecast of future climate change. Climate change in the world has been one of the most important issues in water resources, insect pest and diseases management studies. In climatological study temperature, relative humidity and precipitation are significant factors for forecasting in making decisions, risk management and optimum uses of water resources. These three parameters have undeniable effects on hydrological cycle, production of horticulture/agriculture crops, water usage, people activities and the environments. The purpose of this study is simulation and modeling of temperature, humidity and precipitation using stochastic methods. Time series analysis has two goals, modeling random mechanism and forecasting for future series quantities according to the past.

Several stochastic models have been developed in the past (Box and Jenkins 1994) for modeling hydrological time series mainly rainfall, runoff and evaporation. These include autoregressive (AR) models of different orders (Davis and Rappoport 1974; Salas et al. 1980; Kamte and Dahale 1984; Gorantiwar et al. 1995; Narulkar 1995; Mutua 1998; Singh 1998; Reddy and Kumar 1998; Subbaiah and Sahu 2002; Patil 2003), moving average (MA) models for different orders (Gupta and Kumar 1994; Verma 2004), autoregressive moving average(ARMA) models of different orders (Katz and Skaggs 1981; Chhajed 2004; Katimon and Demun 2004) for annual stream flow. For monthly or intra-seasonal flows, seasonal or periodic autoregressive integrated moving average (ARIMA) model (Bender and Simonovle 1994; Montanari et al. 2000; Trawinski and Mackay 2008), Thomas-Fiering models (Srinivasan 1995) and fractionally difference ARIMA models (Montanari et al. 1997) were used. Often the historical series is short and inadequate for irrigation planning. Hence, stochastic models are useful for generation of long-term temperature; humidity and precipitation data needed for irrigation planning, insect pest and diseases management studies.

The models used for generation and forecasting of the annual evaporation series were Thomson-Firing, AR, MA, ARMA and ARIMA models of different orders. The models SARIMA, PARMA and FARIMA were used for seasonal and periodic evaporation series. The studies indicated that stochastic models can be successfully used for the generation of the synthetic sequence of rainfall, runoff and evaporation. ARIMA class of models was also used for forecasting of runoff/evaporation few time periods ahead. For appropriate planning of the water resources available to farmers, they must match with the water requirement. The reasonable forecast of water requirement at least 1 year ahead water would enable them to manage water resources efficiently, as the ARIMA models showed possibility to forecast the other hydrological events.

In this study, the applicability of the ARIMA models to forecast temperature, humidity and precipitation for Solapur station were investigated; and finally the appropriate ARIMA model was identified for the forecast of temperature, humidity and precipitation for Solapur station.

Materials and Methods

This study was concerned with the forecast of weekly values of temperature; relative humidity and monthly precipitation were summed up to obtain the weekly and monthly values by using ARIMA class of models.

Development of ARIMA Model

Seasonal autoregressive integrated moving average (SARIMA) is useful for modeling seasonal time series in which the mean and other statistics for a given season are not stationary across the year. The basic ARIMA model in its seasonal form is described as (Hipel et al. 1976; Box and Jenkins 1994) a straightforward extension of the non-seasonal ARIMA models.

The different approaches involved in fitting of ARIMA models to historical hydrological series as suggested by Hipel et al. 1976 and Box and Jenkins 1994 are standardization and normalization of time series, identification of the models, determination of the parameters, diagnostic checking and selection of the best model.

Standardization and Normalization of Time Series Variables

The first step in time series modeling is to standardize and transform the time series. In general, standardization is performed by normalizing the series as follows.

$$y_{i,j} = \frac{x_{i,j} - x_i}{\sigma_i} \qquad (1)$$

where $y_{i,j}$ is the stationary stochastic component in the mean and variables for week i or the year j, $x_{i,j}$ the weekly temperature, humidity and monthly precipitation in the week i of the year j,
x_i the weekly and monthly mean and
σ_i the weekly and monthly standard deviation

Identification of the Model

An important step in the modeling is the identification of a tentative model type to be fitted to the data set. In the present study, the procedure stated by Hipel and McLeod (1994) was adopted for identifying the possible ARIMA models. A time series with the seasonal variation may be considered stationary, if the theoretical autocorrelation function (ρ_k) and theoretical partial autocorrelation function (ρ_{kk}) are zero after a lag $k = 2s + 2$ (where, s is the seasonal period; in this study, $s = 52$). The requirement of identification procedure is as: i.e. Plot of the original series, Plot of the standardized series, ACF analysis and PACF analysis. The estimates of theoretical autocorrelation function (e_m) i.e. r_m is given by Eq. (2).

$$r_m = \frac{\sum_{i=1}^{n-m} (x_i - \bar{x})(x_{i+m} - \bar{x})}{\sum_{i=1}^{n} (x_i - \bar{x})^2} \qquad (2)$$

where n is the number of observations $\bar{x}$ the average of the observations and r_m the autocorrelation function at lag m

The estimate of theoretical partial autocorrelation function (e_{kk}) i.e. Φ_{mm} is given by the Eq. (3). The partial autocorrelation function varies between −1 and +1, with values near ± 1 indicating stronger correlation. The partial autocorrelation function removes the effect of shorter lag autocorrelation from the correlation estimates at longer lags.

$$\Phi_{mm} = \frac{r_m - \sum_{i=1}^{m-1} (\Phi_{m-1})(r_{m-1})}{1 - \sum_{i-1}^{m-1} (\Phi_{m-1})_r} \qquad (3)$$

where

Φ_{mm} is the Partial autocorrelation function at lag k

It is considered that ρ_k and ρ_{kk} equal to zero if (Maier and Dandy 1995)

$$\boldsymbol{\rho}_{\mathbf{k}} = 0\,\mathbf{i.e.}\,|r_k| \leq \frac{2}{T^{0.5}} \quad (4)$$

$$\boldsymbol{\rho}_{\mathbf{kk}} = \mathbf{0}\,\mathbf{i.e.}\,|r_{kk}| \leq \frac{2}{T^{0.5}} \quad (5)$$

where r_k is the sample autocorrelation at lag k, r_{kk} the Sample partial autocorrelation at lag k and T the number of observation

If the sample autocorrelation function (ACF) of analyzed series does not meet the above condition, the time series needs to the transformed into a stationary one using different differencing schemes. For example, for ($d = 0$, $D = 1$, $s = 52$) according to the expression given by equation

$$y_t = (1 - B)^d (1 - B^s)^D x_t = (1 - B^s) E_{0,t} \quad (6)$$

where y_t is the original time series, d the order of non-seasonal differencing operator, D the order of seasonal differential operator, B the Backshift operator, s – the seasonal length, t the discrete time, $E_{0,t}$ the evaporation series, k thelag period and x_t the stationary series formed by differencing series.

The time series y_t is stationary if the ACF and PACF cut off at lags less than $k = (2s + 2)$ seasonal periods. Thus, it is necessary to test the stationary status of the transformed time series obtained by differencing the original times series according to different orders of differencing (seasonal and non-seasonal). The differenced series that pass the stationary criteria needs to be considered for further analysis. The following guidelines were used for selecting the orders of AR and MA terms (Gorantiwar 1984).

- If the autocorrelation function cuts off, fit ARIMA (0, d, q) (0, 1, Q)$_{52}$ model to the data, where q is the lag after which the autocorrelation function first cuts off and Q is the lag after which seasonal ACF cutoff.
- If the autocorrelation function cuts off, fit ARIMA (p, d, 0) (P, 1 ,0)$_{52}$ model to the data, where p is the lag after which the partial autocorrelation function first cuts off and P is the lag after which seasonal PACF cuts off.
- If neither the autocorrelation nor partial autocorrelation functions cuts off, fit the ARIMA (p, d, q) (P, 1, Q)$_{52}$ model for a grid of values of p, P, q and Q.

Thus on the basis of information obtained from the ACF and PACF, several forms of the ARIMA model need to the identified tentatively.

Estimation of Parameters of the Model

After the identification of model, the parameters of the selected models were estimated by the statistical analysis of the data series. The most popular approaches of parameters estimation is the method of maximum likelihood.

Diagnostic Checking of the Model

Once a model has been selected and parameters calculated, the adequacy of the model has to be checked (diagnostic checking). The following three tests were used in this context.

Examination of Standard Error

A high standard error in comparison with the parameter values point out a higher uncertainty in parameter estimation which questions the stability of the model. The model is adequate if it meets the following condition.

$$t = \frac{cv}{\mathrm{se}} > 2 \quad (7)$$

Where, cv – parameter value, se – standard error

ACF and PACF of Residuals

If the model is adequate at describing behavior of a time series (evaporation), the residuals of the model should not be correlated i.e. all ACF and PACF should lie within the limits calculated by Eqs. (4) and (5) after lag $k = 2s + 2$, where s = number of periods.

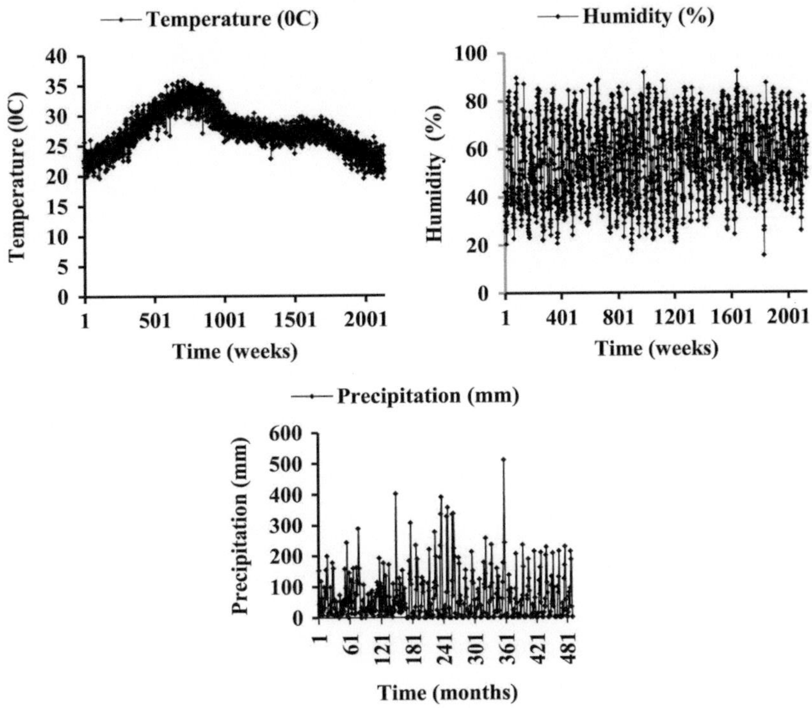

Fig. 1 Weekly temperature, humidity and monthly precipitation series in Solapur station from 1969 to 2010

Akaike Information Criteria (AIC)

The AIC (Akaike 1974) are computed as

$$\mathrm{AIC} = 2k + \left(\left(\ln\frac{2\pi v_r}{T}\right) + 1\right)T \qquad (8)$$

where k is the number of model parameters, υ_γ the residual variance and T the total number of observations.

Selection of the Most Appropriate Model

The root mean square error (RMSE) were estimated for each model

$$\mathrm{RMSE} = \sqrt{\frac{\sum_{i=1}^{n} \left(E_{\mathrm{act}} - E_{\mathrm{for}}\right)^2}{N}} \qquad (9)$$

Where, E_{act} – Actual value of temperature (°C), relative humidity (%) and precipitation (mm); E_{for} – Forecast value of temperature (°C), relative humidity (%) and precipitation (mm); N – Total number of observation

Results and Discussion

Temperature, Humidity and Precipitation Analysis

Weekly temperature, humidity and precipitation (Fig. 1) show a seasonal cycle. The ACF and PACF of the original temperature, humidity and precipitation data are not stationary. The descriptive statistics for temperature, humidity and precipitation data are shown in Table 1.

Fitting of ARIMA Model

The weekly temperature, humidity and precipitation data were used for generating and

Table 1 Basic statistics for Solapur station of weekly temperature (°C), humidity (%) and precipitation (mm) data

Parameters	No. of observations	Mean	St. Dev.	Variance	Min	Max
Temperature	2,132	27.4	11.3	3.4	19.4	35.9
Humidity	2,132	55.3	15.39	236.95	15.4	92.3
Precipitation	492	59.9	75.44	5,691.66	0.0	509.3

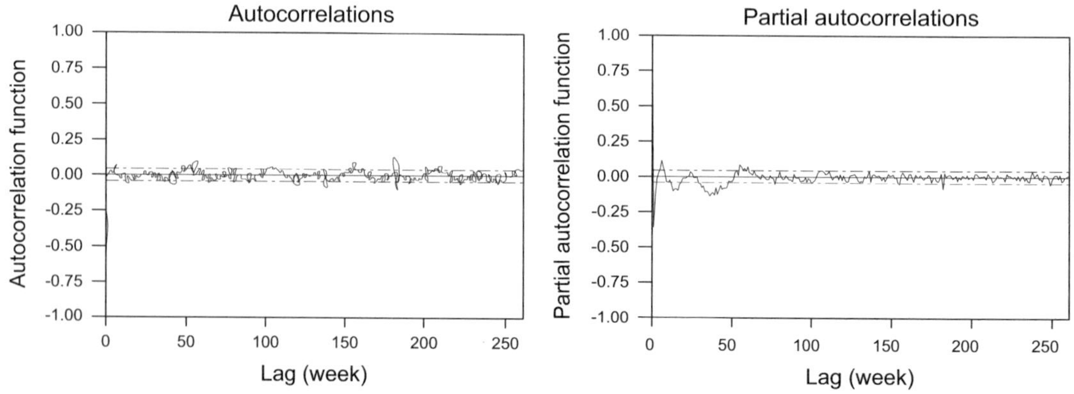

Fig. 2 Autocorrelation and partial autocorrelation pattern of the differenced time series of temperature ($d = 1, D = 0$)

forecasting for development and validation of stochastic model. The results obtained from the study have been presented and discussed under the following heads.

Standardization and Normalization of Time Series Variables

The ARIMA model has the provision to differentiate the time series. Hence, standardization and normalization was not performed.

Identification of the Model

The ACF and PACF of weekly temperature, humidity and precipitation time series were estimated for different lags. These are shown with upper and lower limits. It is seen from Fig. 2 that ACF lie outside the limit after lag $k = 2s + 2$ i.e. 106. Thus, ARIMA model cannot be applied to the original time series of temperature, humidity and precipitation. Therefore, the time series was transformed differencing schemes by using $d = 0$; $D = 1$, $d = 1$; $D = 1$, $d = 1$; $D = 0$, $d = 0$; $D = 0$; $d = 2$, $D = 0$; $d = 2$, $D = 1$; $d = 1$, $D = 2$ and $d = 2$, $D = 2$. The ACF and PACF along with the upper and lower limits were estimated by Eqs. (4) and (5). It is observed from the Figs. 3, 4, and 5 that ACF of $d = 1, D = 0$, $d = 0, D = 1$ and $d = 0, D = 0$ lie within the limits of range specified by Eqs. (4) and (5) after lag 104. However, for $d = 0$, $D = 1$; $d = 0$, $D = 0$; $d = 1$, $D = 1$; $d = 2$, $D = 0$; $d = 2$, $D = 1$; $d = 1$, $D = 2$ and $d = 2$, $D = 2$. ACF does not lie within the limits after the lag 104. Therefore, differencing schemes (i.e. $d = 1$; $D = 0$, $d = 1$, $D = 0$ and $d = 0$; $D = 0$) were used for developing ARIMA model for weekly temperature, humidity and monthly precipitation time series.

On the basis of information obtained from ACF and PACF, the orders of autoregressive (AR) and moving average (MA) terms were identified as one. Based on this, 36 forms for temperature and humidity and 256 forms for precipitation of ARIMA models were identified and parameters computed.

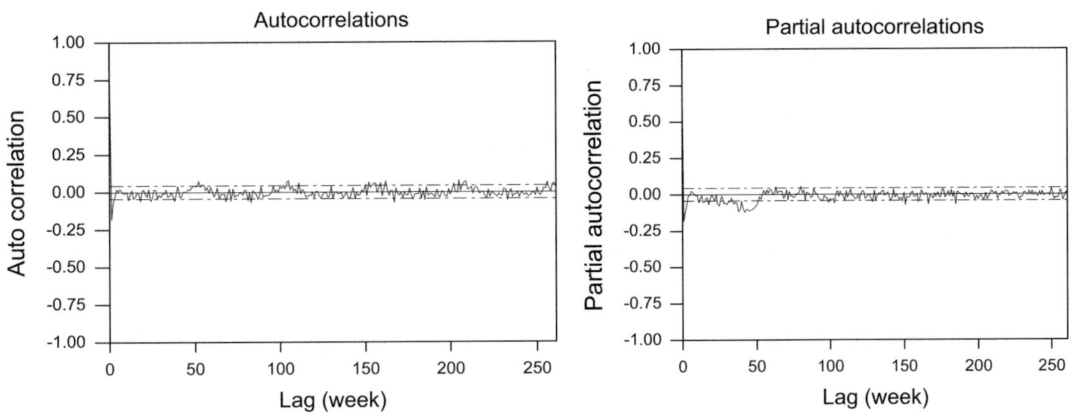

Fig. 3 Autocorrelation and partial autocorrelation pattern of the differenced time series of humidity ($d = 1, D = 0$)

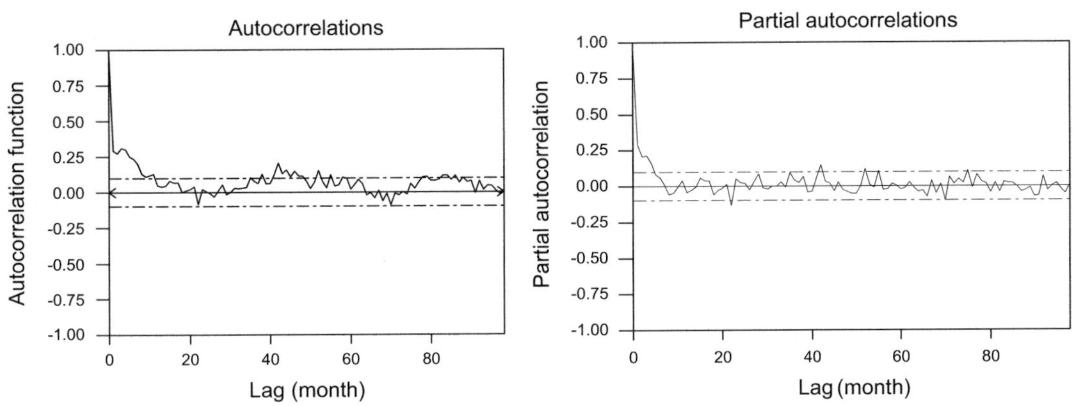

Fig. 4 Autocorrelation and partial autocorrelation pattern of the differenced time series of precipitation ($d = 0$, $D = 0$)

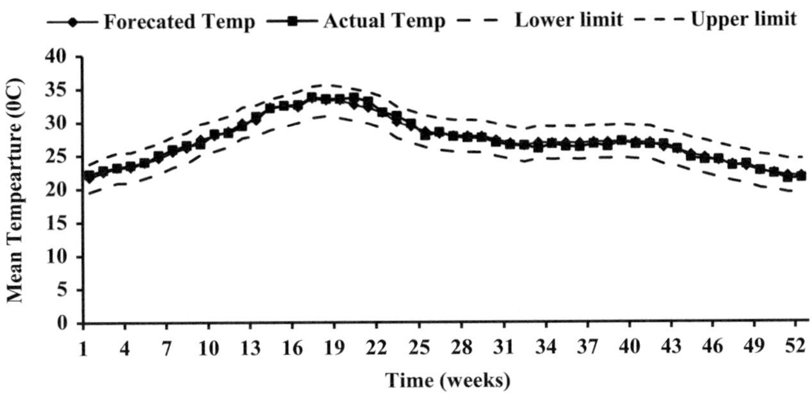

Fig. 5 Comparison of forecasted and actual values of temperature by using ARIMA (1, 1, 1) (0, 1, 1)$_{52}$ model

Table 2 Parameter estimates, standard error, corresponding *t* values and AIC values for different ARIMA models for temperature, humidity and precipitation

Temperature											
Models	ϕ_1	θ_1	Φ_1	Θ_1	c	Models	ϕ_1	θ_1	Φ_1	Θ_1	c
ARIMA (1,1,1) (0,1,1)$_{52}$						**ARIMA (1,0,0) (1,0,1)$_{52}$**					
Estimate	0.0035	0.269	0.9468		0.9399	Estimate	27.281	0.4235		0.998	0.925
SE	0.0002	0.023	0.0078		0.0109	SE	0.452	0.0202		0.001	0.008
t-Value	2.11	12.47	120.21		86.19	*t*-Value	60.3	20.92		558.8	111.8
AIC	7,203.9					AIC	7,342.4				
ARIMA (1,0,1) (0,1,1)$_{52}$						**ARIMA (0,0,1) (1,1,1)$_{52}$**					
Estimate	0.0095	0.709	0.3988		0.944	Estimate	0.0092		0.3173	0.034	0.935
SE	0.0061	0.035	0.0466		0.011	SE	0.0042		0.0208	0.024	0.011
t-Value	1.56	19.84	8.56		85.07	*t*-Value	2.16		15.24	1.42	78.61
AIC	7,218.2					AIC	7,321.3				
ARIMA (1,0,1) (1,0,1)$_{52}$						**ARIMA (0,0,1) (1,0,1)$_{52}$**					
Estimate	27.265	0.800	0.4851	0.9986	0.93941	Estimate	27.3		0.3302	0.998	0.908
SE	0.534	0.028	0.0363	0.0020	0.00793	SE	0.443		0.0206	0.001	0.009
t-Value	51.03	30.35	13.36	486.04	118.41	*t*-Value	61.64		16.07	630.4	100.6
AIC	7,292.2					AIC	7,458.9				
Humidity											
ARIMA (1,1,1) (1,1,1)$_{52}$						**ARIMA (0,1,1) (1,0,1)$_{52}$**					
Estimate	0.0003	0.509	0.9726	0.054	0.9602	Estimate	0.011		0.4413	0.988	0.938
SE	0.0009	0.021	0.0058	0.0239	0.0118	SE	0.246		0.0198	0.007	0.014
t-Value	0.34	24.29	167.65	2.26	81.68	*t*-Value	0.05		22.33	130.2	67.12
AIC	14,726					AIC	14,966				
ARIMA (1,0,0) (1,1,0)$_{52}$						**ARIMA (0,0,1) (1,1,1)$_{52}$**					
Estimate	0.165	0.540		0.4715		Estimate	0.211		0.4474	0.075	0.967
SE	0.292	0.018		0.0196		SE	0.024		0.0197	0.023	0.011
t-Value	0.56	28.72		24.07		*t*-Value	8.81		22.76	3.18	82.87
AIC	15,342					AIC	14,888				
ARIMA (1,0,1) (0,1,1)$_{52}$						**ARIMA (0,0,1) (0,1,1)$_{52}$**					
Estimate	0.2128	0.588	0.0624		0.944	Estimate	0.2109		0.4491		0.932
SE	0.038	0.035	0.0402		0.0109	SE	0.0273		0.0196		0.011
t-Value	5.6	18.08	1.56		86.66	*t*-Value	7.73		22.9		84.51
AIC	14,745					AIC	14,894				

Table 3 Root mean square error values of first five models

Temperature		Humidity		Precipitation	
Models	RMSE	Models	RMSE	Models	RMSE
ARIMA(1,0,1)(0,1,1)$_{52}$	7.34	ARIMA(1,1,1)(1,1,1)$_{52}$	1.27	ARIMA(0,0,1,)(0,1,2)$_{12}$	45.38
ARIMA(1,1,1)(0,1,1)$_{52}$	0.99	ARIMA(1,0,1)(1,0,0)$_{52}$	42.27	ARIMA(0,0,2)(0,1,2)$_{12}$	57.94
ARIMA(0,0,1)(1,1,1)$_{52}$	7.20	ARIMA(0,1,1)(1,0,1)$_{52}$	31.25	ARIMA(1,1,1)(2,1,2)$_{12}$	87.04
ARIMA(0,0,1)(1,0,1)$_{52}$	7.39	ARIMA(1,0,1)(0,1,1)$_{52}$	43.18	ARIMA(0,1,2)(1,0,1)$_{12}$	33.83
ARIMA(1,0,1)(0,0,1)$_{52}$	7.26	ARIMA(0,0,1)(1,1,1)$_{52}$	38.02	ARIMA(2,1,2)(2,1,1)$_{12}$	16.20

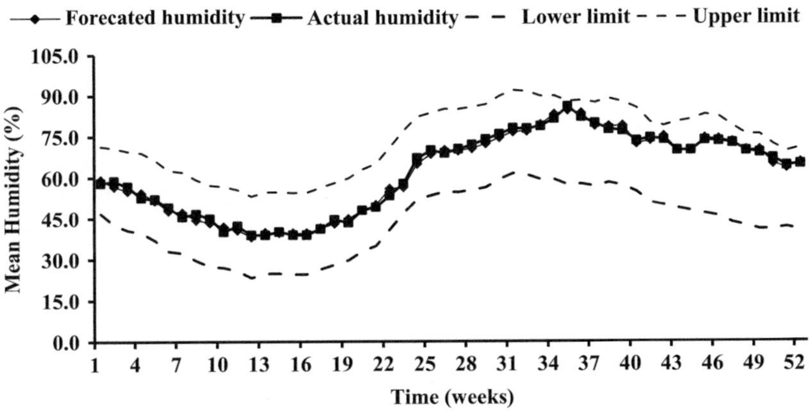

Fig. 6 Comparison of forecasted and actual values of humidity by using ARIMA (1, 1, 1) (1, 1, 1)$_{52}$ model

Determination of Parameters of Model and Diagnostic Checking

The following parameters of the selected models were calculated by maximum likelihood method.

(1) ϕ_1 (2) θ_1 (3) Φ_1 (4) Θ_1 (5) c

Out of the 36 and 256 possibilities, ten ARIMA models satisfied the test for all parameters. Standard error and t values for these ten models are given in Table 2.

ACF and PACF of Residual Series

For a model to be considered adequate at describing behavior of temperature, humidity and precipitation time series, the residuals of model should be correlated i.e. all ACF and PACF should lie within the limits calculated by Eqs. (4) and (5) after lag $k = 2s + 2$, where $s =$ number of periods. In this case, for the value of k is 106, computations showed that ACF and PACF residual series of 15 each and 20 models lie within the prescribed limits.

Selection of the Best Model

First 5 models with less AIC that satisfied the standard error and ACF and PACF of residuals criteria were finally used (Table 2) for generation of weekly temperature, humidity and precipitation values. For this purpose, the temperature, humidity and precipitation values were forecast for 1 year with the help of identified ARIMA models. These values were compared with the actual values for 1 year by calculating the root mean square error (RMSE) between them (Table 3). It is observed from Figs. 5, 6, and 7 that the seasonal pattern of temperature, humidity and precipitation series is maintained in generated values by all the ARIMA models.

Based on RMSE, the ARIMA (1,1,1) (1,0,1)$_{52}$, ARIMA (1,1,1) (1,1,1)$_{52}$ and ARIMA

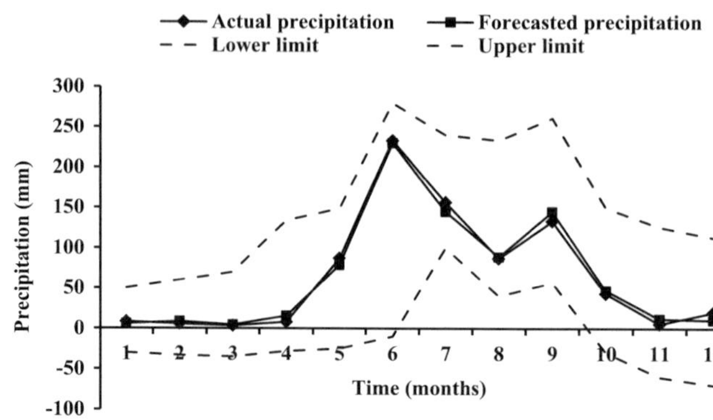

Fig. 7 Comparison of forecasted and actual values of precipitation by using ARIMA (2, 1, 2) (2, 0, 1)$_{12}$ model

(2,1,1) (2,0,1)$_{12}$ of models are selected for forecasting. The values of the parameters of the ARIMA model which is finalized for forecasting of parameters are: $\phi_1 = 0.0035$, $\theta_1 = 0.269$, $\Phi_1 = 0.0$, $\Theta_1 = 0.9468$, $c = 0.9399$; $\phi_1 = 0.0003$, $\theta_1 = 0.509$, $\Phi_1 = 0.97$, $\Theta_1 = 0.054$, $c = 0.9602$ and $\phi_1 = 0.0035$, $\phi_2 = 0.233$, $\theta_1 = 0.0899$, $\theta_2 = 0.0817$ $\Phi_1 = 0.8685$, $\Theta_2 = 0.00$, $c = 0.9399$;

Comparison of Forecast and Actual Values of Temperature, Humidity and Precipitation

The ARIMA models that were finalized to forecast the values of temperature, humidity and precipitation for Solapur station are presented in Figs. 5, 6, and 7. These values were developed using the climatological data up to 2009. The temperature, humidity and precipitation values were forecast with the help of the best model, and forecast weekly temperature, humidity and monthly precipitation values were calculated with the help of weekly temperature, humidity and monthly precipitation series. Forecast values were compared (Figs. 5, 6, and 7) with actual values of temperature, humidity and precipitation of 2010.

Conclusion

The study indicates that the seasonal ARIMA model is a viable tool for forecasting temperature, humidity and precipitation at Solapur location. The system studies reveal that, if sufficient length of data is used in model building, frequent updating of model may not be necessary. This forecast temperature, humidity and precipitation can be advantageously used in deriving the optimal irrigation system and monitoring insect-pest and other diseases. The **ARIMA (1,1,1) (1,0,1)$_{52}$**, **ARIMA (1,1,1) (1,1,1)$_{52}$** and **ARIMA (2,1,1) (2,0,1)$_{12}$** gave the lower values of RMSE and hence is the best stochastic model for generation and forecasting of weekly temperature, humidity and monthly precipitation values for Solapur, Maharashtra, India. It is concluded that seasonal ARIMA models can be successfully used for forecasting of temperature, humidity and precipitation having inbuilt seasonal pattern.

References

Akaike H (1974) A new look at the statistical model identification. IEEE Trans Autom Control AC 19(6): 716–723

Bender M, Simonovle S (1994) Time–series modeling for long-range stream flow forecasting. J Water Resour Plan Manag 120(6):857–870

Box GEP, Jenkins GM (1994) Time series analysis, forecasting and control, Revised edn. Holden-Day, San Francisco

Chhajed N (2004) Stochastic modeling for forecasting Mahi river inflows. M.E. thesis, MPUAT, Udaipur

Davis JM, Rappoport PN (1974) The use of time series analysis techniques in forecasting meteorological drought. Mon Weather Rev 102:176–180

Gorantiwar SD (1984) Investigating applicability of some operational hydrology models to WB streams. M.Tech. thesis, IIT, Kharagpur

Gorantiwar SD, Majumdar M, Pampattiwar PS (1995) Application of autoregressive models of different orders to annual stream flows of Barkar river with their logarithmic transformation. J Appl Hydrol 8: 33–39

Gupta RK, Kumar R (1994) Stochastic analysis of weekly evaporation values. Indian J Agric Eng 4(3-4): 140–142

Hipel KW, McLeod AI (1994) Time series modeling of water resources and environmental systems. Elsevier, Amsterdam

Hipel KW, McLeod AI, Lennox WC (1976) Advances in Box Jenkins modeling: 1. Model construction. Water Resour Res 13:567–575

Kamte PP, Dahale SD (1984) A stochastic model on drought. Mausam 35:387–390

Katimon A, Demun AS (2004) Water use trend at university technology Malaysia: application of ARIMA model. J Technol 41(B):47–56

Katz RW, Skaggs RH (1981) On the use of autoregressive moving average processes to model meteorological time series. Mon Weather Rev 109:479–484

Maier HR, Dandy GC (1995) Comparison of Box-Jenkins procedure with artificial neural network methods for univariate time series modeling, Research report no. R 127. Department of Civil and Environmental Engineering, University of Adelaide, Adelaide

Montanari A, Rosso R, Taqqu MS (1997) Fractional differenced ARIMA models applied to hydrological time series: identification, estimation and simulation. Water Resour Res 33(5):1035–1044

Montanari A, Rosso R, Taqqu MS (2000) A seasonal fractional ARIMA model applied to the Nile river monthly flows at Aswan. Water Resour Res 36(5): 1249–1259

Mutua FM (1998) Transfer function hydrologic modeling: a case study. J Appl Hydrol 11(2):11–15

Narulkar MS (1995) Optimum real time operation of multi-reservoir systems for irrigation scheduling. Ph.D. thesis, IIT, Bombay

Patil RM (2003) Stochastic modeling of water deficit for Rahuri region. M.E. thesis, MPUAT, Udaipur

Reddy KM, Kumar D (1998) Time series analysis of monthly rainfall for Bimo watershed of Ramganga river. J Agric Eng 36(4):19–29

Salas JD, Dellur JW, Yevjevich V, Lane WL (1980) Applied modeling of hydrological time series. Water Resources Publication, Littleton

Singh CV (1998) Long term estimation of monsoon rainfall using stochastic models. Int J Climatol 18: 1611–1624

Srinivasan K (1995) Stochastic modeling of monsoon river flows. J Appl Hydrol 8:51–57

Subbaiah R, Sahu DD (2002) Stochastic model for weekly rainfall of Junagadh. J Agrometeor 4:65–73

Trawinski PR, Mackay DS (2008) Meteorologically conditioned time series predictions of West Nile virus vector mosquitoes. Vector Borne Zoonotic Dis 8(4):505–522

Verma A (2004) Stochastic modeling on monthly rainfall of Kota, Rajasthan. M. E. thesis, GBPUAT, Pantanagar, India

Geospatial Methodology Towards Planning Adaptation/Mitigation Measures of Climate Change Impact on the Apple Orchards in India

S. Panigrahy, C.P. Singh, N.B. Bhatt, and J.S. Parihar

Abstract

Apple is the predominant horticulture crop of Himachal Pradesh and Jammu and Kashmir states in India. Efforts are in progress to further strengthen this crop by bringing more areas under cultivation and improving the condition of the existing orchards. However, future changes in the climatic parameters projected under the global climate change scenario will have significant impact on the apple orchard viability. This is mainly due to its sensitivity to availability of chilling units. Temperate fruits like apple have a specific chilling unit requirement for fruit set and quality of fruit. In the Indian context, the chilling requirement is related to the elevation range of the orchards. This study analyses the current distribution pattern of the apple orchards in relation to elevation ranges and simulates the change under the climate change scenario. Remote sensing data of IRS-P6 LISS-III and AWiFS sensor was used to map the orchards. Digital elevation model (DEM) was used to generate the elevation, slope and aspect in spatial domain. Ancillary data district/state boundary, weather data, soil and drainage were integrated using geospatial technique. Terrain analysis showed that the orchards in Jammu and Kashmir were distributed in the elevation range of 1,600–2,100 m. The equal proportion of orchards was observed in the elevation range of 1,600 − 1,800 m as well as 1,800–2,000 m. In case of Himachal Pradesh (Shimla, Kullu, Mandi), the orchards are distributed from 1,600 to 3,000 m.

To predict the suitable elevation of apple growth under the climate change scenario, a modelling method known as GARP (Genetic Algorithm for Rule-Set Production) was used. It is a genetic algorithm that creates ecological niche models for species distribution from presence-only occurrence data. The model gives final solution as environmental conditions under which the species should be able to maintain populations. The simulation showed significant upward shift of the

S. Panigrahy (✉) • C.P. Singh • N.B. Bhatt • J.S. Parihar
Space Applications Centre, Indian Space Research Organisation, Satellite Road, Ambawadi Vistar P.O., Ahmedabad 380015, India
e-mail: sushma@sac.isro.gov.in

A.K. Singh et al. (eds.), *Climate Change Modelling, Planning and Policy for Agriculture*,
DOI 10.1007/978-81-322-2157-9_9, © Springer India 2015

existing belt to higher elevation. The paper describes the results in detail. The use of geospatial technique and simulation model enhances the scope for preparing a road map to plan mitigation measures to combat climate change impact on the vast tracts of apple and stone fruit belt in India.

Keywords
Geoinformatics • Apple orchard • Remote sensing • GIS • Climate change

Introduction

Climate change and its impact on food production is one of the leading research themes today. India announced a National Action Plan on Climate Change (NAPCC) in 2008, which identifies measures that aim coexistence of the development and climate change mitigation plan. Agriculture is one of the eight National Missions identified with emphasis on adaptation and mitigation planning. Of particular concern is preparedness to mitigate the negative impact if any. Temperate fruit/orchards in India are one of most vulnerable targets of climate change, as temperature is one major factor that affects the fruit yield in these (Awasthi et al. 2001). These plants enter a dormant period followed by a growth period. During this dormant period, plants cannot resume growth until the chilling requirement is met. Trees also have heat requirement before bloom can occur. Apple is the most important temperate fruit in India. The area under apple cultivation in India increased by 24 % from 1.95 lakh ha. in 1991–1992 to 2.42 lakh ha. in 2001–2002 (NHB 2011). It is predominantly grown in Jammu and Kashmir, Himachal Pradesh and Uttarakhand, accounting for about 90 % of the total production. Its cultivation has also been extended to Arunachal Pradesh, Sikkim, Nagaland and Meghalaya in northeastern region and Nilgiri hills in Tamil Nadu.

The apple-growing areas in India do not fall in the temperate zone of the world but the prevailing temperate climate of the region is primarily due to high altitude and proximity to the snow-covered Himalayan ranges. Most of the apple varieties require 1,000–1,500 h of chilling below 7 °C during winter to break the rest period. Cumulative chilling and heat unit models and the dynamic model (Byrne and Bacon 1992) have been used to assess the optimum requirement of different cultivars of apple. These conditions are generally available at an elevation of 1,500–2,700 m above mean sea level in the Himalaya ranges. Studies have shown that the cumulative chill units of coldest months have already declined by 9.1–19.0 units per year in the last two decades in Himachal Pradesh affecting the apple productivity (Singh et al. 2009; Jindal and Mankotia 2004). It is predicted that the rise in temperature will benefit the apple orchards grown in high altitude (>2,300 m amsl). However, looking at the large geographical extent of the apple region in India, growing under varied terrain condition and the climatic condition, it is essential to use geospatial technique to secure region specific understanding.

It is now well established that satellite remote sensing (RS) data with its multispectral imaging, temporal monitoring and large area coverage has the capability of generating information on crop area, its condition and phenological events (Navalgund et al. 1991). Mapping of apple orchard using Indian remote sensing satellite (IRS) data has already been done for Himachal Pradesh (SAC 2006). Geographic information system (GIS) techniques help in the integration of a variety of ancillary data, both spatial and nonspatial, and is the most efficient way to capture, store, retrieve, analyse and display such data. Thus, RS and GIS together are best suited to create the spatial database and maps. This can help simulate spatial domains at various scales (regional to local) using growth simulation models to aid decision support system for long-term planning for orchard development including

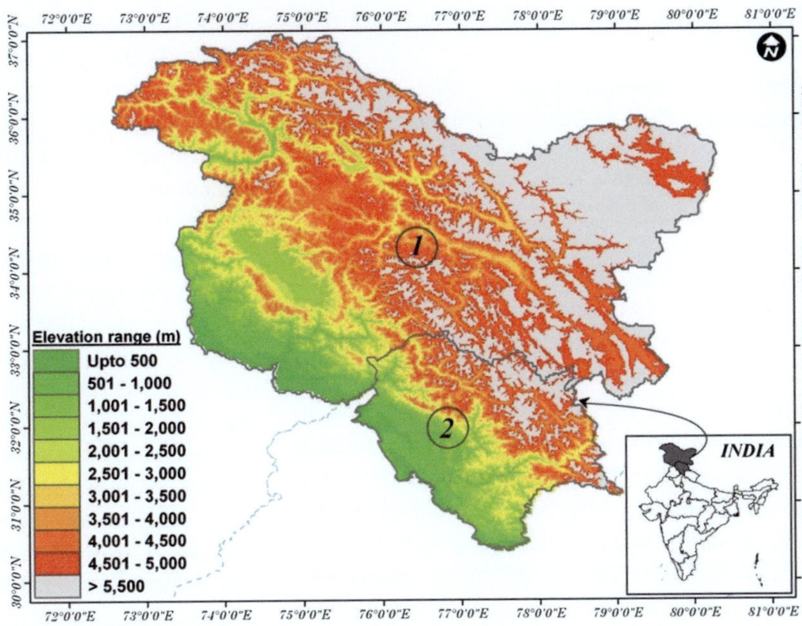

Fig. 1 Location of the two study states: *1* Jammu and Kashmir, *2* Himachal Pradesh in relation to elevation gradient in the Himalayan region (*Source*: Panigrahy et al. 2014)

measures to mitigate climate change impact. This paper highlights the methodology and the results for the apple belt covering the two major Himalayan states, viz., Jammu and Kashmir (J&K) and Himachal Pradesh (HP) in India.

Study Area

J&K and HP accounting for 76 % of the area and 94 % of the production of the apple crop in the country are the study areas (Fig. 1). The apple-growing region in these states is concentrated in few districts (NHB 2011), where the climatic condition is best suited for growth (Table 1). By and large, the average temperature in this region is around 21–24 °C during active growth period; the areas have frost-free spring and adequate sunshine during summer without wide fluctuations in temperature.

Data Used

Different types of spatial and nonspatial data were used in this study. The description of data used is given below:

Remote Sensing Data

IRS LISS-III images are the basic remote sensing data used in this study for mapping apple orchards. Multi-temporal AWiFS data were used to study the orchard phenology in relation to other vegetation classes like forest, agricultural crops, etc. and to select optimum dates for identification of apple orchards. AWiFS sensor provides 55 m spatial resolution data in Green, Red, NIR and SWIR bands. Large swath and high temporal data facilitates the selection of optimum bio-window. IRS LISS-III data provides data in four bands (Green, Red, NIR and MIR). The spatial resolution is 24 m and is thus amenable for 1:50,000 scale mapping. IRS LISS-IV data of 5.3 m spatial resolution is used on sample basis to collect ground truth and validate the classification accuracy. The satellite path-row map/information and orbit calendar are used to select the path-row and date of pass as per the need.

Terrain parameters are derived using digital elevation model (DEM) data. For this, the NASA Shuttle Radar Topographic Mission (SRTM) that provides DEM in 5° × 5° tiles with a resolution of 90 m at the equator is used. All are produced from a seamless dataset to allow easy

Table 1 The apple-growing belts in the study states (NHB 2011)

State	Growing belts
Jammu and Kashmir	Srinagar, Budgam, Pulwama, Anantnag, Baramulla and Kupwara
Himachal Pradesh	Shimla, Kullu, Mandi, Sirmour, Chamba and Kinnaur

mosaicking. The vertical error of the DEM is reported to be less than 15 m. These are available in both ArcInfo ASCII and GeoTiff format to facilitate their ease of use in a variety of image processing and GIS applications. Data can be downloaded using a browser or accessed directly from the ftp site. This is available for download from the National Map Seamless Data Distribution System (http://seamless.usgs.gov/) or the USGS ftp site. Survey of India topographic maps were also used to calibrate the DEM generated from SRTM and carry out local level bias corrections.

Ground Truth and Ancillary Data

Ground truth data or field data is an integral part of the interpretation of remote sensing data. Ground truth data is attributed to collection, verification and measurement of information about the different features on the earth surface contributing to spectral reflectance pattern during the desired period. Ground truth plan is made based on stratification of the image data, variation in vegetation signatures, etc. The information and location were noted with the help of maps, image prints and Global Positioning System (GPS). Other ancillary data were administrative boundary, crop statistics, land use/cover map, etc.

Methodology

Since multisource data were integrated in the study with different spatial resolution satellite data, it is necessary to co-register the dataset. Here, UTM projection system with WGS84 (earth model) was used as referencing scheme. A stepwise classification approach was employed to identify the apple area, viz., (1) delineation of non-vegetation classes using Normalised Difference Vegetation Index (NDVI) threshold, (2) unsupervised classification to delineate forest plantation and (3) supervised classification for the identification of apple orchards.

The confusion matrix was used to assess the classification accuracy. Since NDVI indicates the vigour of vegetation, it was used to generate three orchard classes, viz., dense, moderate and sparse based on threshold value. SRTM DEM was used to generate theme layers of elevation, slope and aspect. The orchard area map was then integrated with these data to characterise the apple orchards on the basis of slope, elevation and aspect, using logical modelling.

A modelling method known as GARP (Genetic Algorithm for Rule-Set Production) that best predicts species distribution (Stockwell and Peters 1999) was used to simulate the impact of climate change scenario on apple distribution. The model gives final solution as environmental conditions under which the species should be able to maintain populations. The apple-growing region simulation was carried out using bioclimatic indices. Nineteen bioclimatic indices in all based on the BIOCLIM model were used to approximate energy and water balances at a given location. To generate the bioclimatic indices for current conditions, weather parameters from years 1950–2000 were used from WorldClim (Hijmans et al. 2005). Similarly, to generate the future bioclimatic indices, future weather parameters projections (HadCM3 A1B scenario) for the year 2050 were used. Information about the methods used to generate the climate layers are provided at http://www.worldclim.org/methods. These 19 bioclimatic indices and DEM were used in GARP-based environmental niche model. The model was trained with the presence-only records of existing apple plantations to predict current scenario and the results were validated. The future apple belt scenarios were generated and analysed based on HadCM3 A1B projections for year 2050.

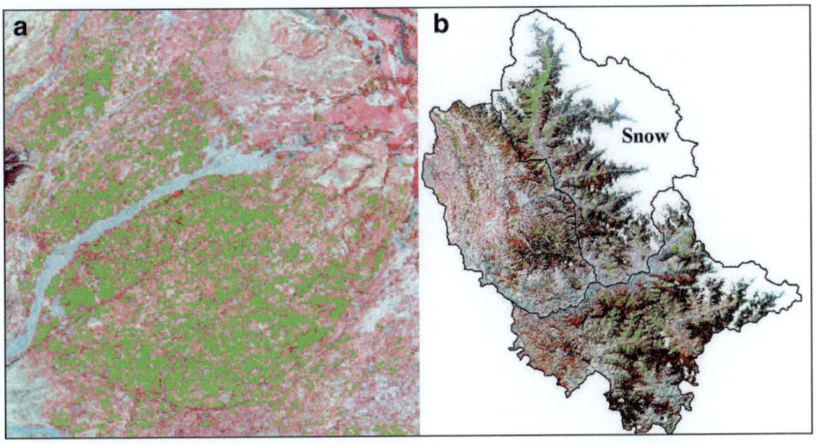

Fig. 2 Apple orchards (*green* colour) overlaid over the false colour composite satellite image (**a**) covering the districts of Anantnag, Pulwama, Budgam, Srinagar and Baramulla districts in J&K state; (**b**) covering Shimla, Kullu and Mandi districts of Himachal Pradesh

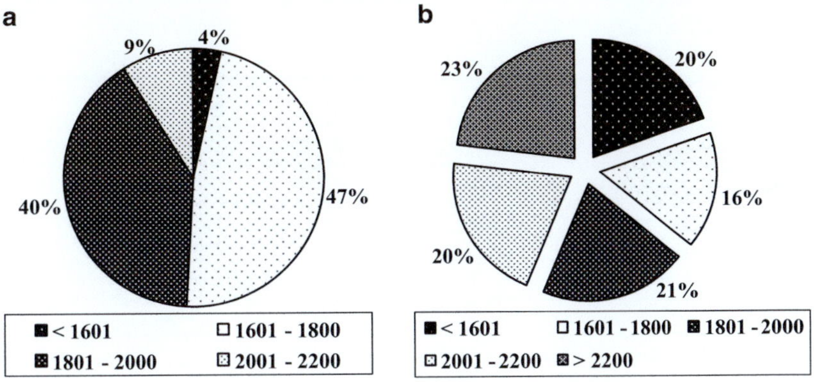

Fig. 3 Percent share of apple orchards in different elevation ranges in (**a**) Jammu and Kashmir, (**b**) Himachal Pradesh in India (*Source*: Panigrahy et al. 2014)

Results

Status of Current Orchards as Observed from Remote Sensing Data

Apple orchards of Himachal were mapped at 1:50,000 scale using IRS LISS-III data of May which was found best suited to discriminate the apple orchards from other land cover classes like forest, agriculture and scrubland. The average classification accuracy of apple orchards using LISS-III data was around 90 % in Himachal Pradesh. In case of Jammu and Kashmir, the orchards were mapped at 1:2, 50,000 scale using temporal AWiFS data. The orchards were found in large contiguous patches in J&K compared to that in Himachal Pradesh as shown in Fig. 2.

Terrain analysis showed that the orchards in J&K were distributed in the elevation range of 1,600–2,100 m. Equal proportion of orchards were observed in the elevation range of 1,600–1,800 m as well as 1,800–2,000 m. In case of Himachal Pradesh (Shimla, Kullu, Mandi), the orchards are distributed from 1,500 to 3,000 m (Fig. 3).

The distribution pattern indicates the role of elevation in meeting the chilling unit requirement of apple plants. Comparatively lower elevation range (1,600–2,000 m) of orchards in J&K is attributed to the higher latitude. In case of HP, the wide range of elevation occupied by orchards (1,400–3,000 m) also indicates the variety diversity.

Vicarious Calibration of Niche Model

The most rigorous evaluation of any model is a test of its ability to correctly predict independent data that have not been processed by the model.

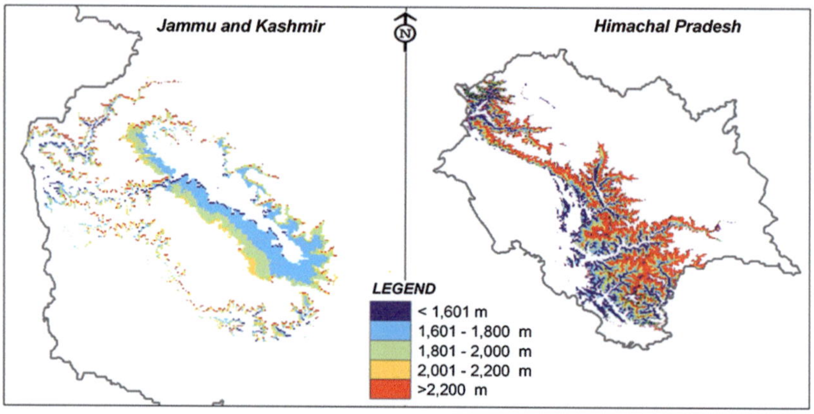

Fig. 4 Future scenario of apple niche using GARP model on bioclimatic indices derived using HadCM3 A1B projections for year 2050 (*Source*: Panigrahy et al. 2014)

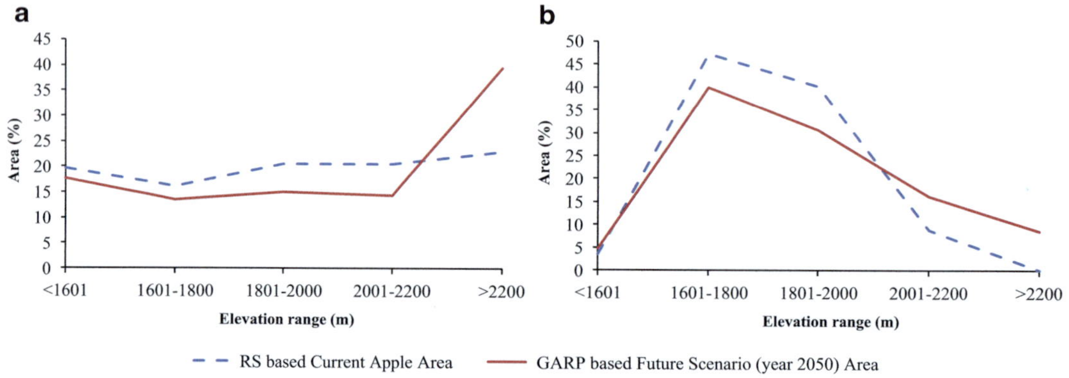

Fig. 5 Change in apple area in comparison to future scenario (HadCM3 A1B: 2050) and *w.r.t.* elevation using GARP model for (**a**) Himachal Pradesh and (**b**) Jammu and Kashmir (*Source*: Panigrahy et al. 2014)

In this work, Genetic Algorithm for Rule-Set Production (GARP) model-based result of current growing environment of the apple plants was evaluated by comparing the actual distribution of orchards derived from satellite remote sensing data. Validation of such predictions generally uses direct field measurements. In this case, we have used correlation of the predicted distribution pattern vis-a-vis remote sensing derived distribution pattern. Orchard patch sizes above the minimum mapping units (as per scale) were selected for each elevation range for this purpose. The model predicted niche matched well with the actual niche as 92 % area was correctly predicted by the GARP model. Thus, the model was used to simulate niche under the future climate change scenario and evaluate change in relation to elevation shift.

Climate Change Scenario

For the present study, the climate parameters as projected by GCM (Ramirez and Jarvis 2008) for the A1-B_2050 scenario of Hadley Centre were used to simulate the changes in the orchard suitability. The empirical downscaling of GCM data was arrived using fourth IPCC assessment and reprocessed using a spline interpolation algorithm for the anomalies and the current distribution of climates from the WorldClim database developed by Hijmans et al. (2005).

Table 2 Apple orchard distribution in relation to the elevation gradient in J&K and HP: current scenario (actual) and simulated future scenario (HadCM3 A1B: 2050)

	Jammu and Kashmir			Himachal Pradesh		
Elevation	Current area (%)	Future area (%)	±	Current area (%)	Future area (%)	±
<1,601	3.52	05.00	1.48	20.00	18.00	−2.00
1,601–1,800	47.36	40.00	−7.36	16.00	13.00	−3.00
1,801–2,000	40.16	30.00	−10.16	21.00	15.00	−6.00
2,001–2,200	8.92	16.00	7.08	20.00	15.00	−5.00
>2,200	0.04	09.00	8.96	23.00	39.00	16.00

(*Source*: Panigrahy et al. 2014)

The GARP-simulated future suitable area for apple in terms of elevation range for J&K and HP is shown in Fig. 4. There is a significant increase in the suitable area in higher elevation (>2,200 m) in case of HP (Fig. 5, Table 2). Similar observations are made for J&K but with less magnitude. This is accompanied by decrease in the area of orchards in lower elevation range (1,600–2,000 m).

However, this study was based on the physiological requirements of the normal varieties. There are new varieties developed with low-chilling requirement. These varieties perform well in areas where winter chilling is less than 800 h below 7 °C, insufficient for breaking dormancy of Delicious varieties. Introduction of low-chilling varieties has expanded the scope of apple cultivation to warmer and marginal areas.

Observations and Conclusion

Climate change irrespective of the uncertainty of various models is the reality that needs to be addressed. The latest generation of regional high-resolution model (PRECIS) carried out for the Indian subcontinent for three time slices, viz., 2020, 2050 and 2080, indicates an all-round warming. The annual mean surface air temperature rise by the end of century ranges from 3.5 to 4.3 °C. Apple orchards are identified as one of the most vulnerable crop to climate change. To realise the adaptation and mitigation strategy as identified under the National Action Plan on Climate Change (NAPCC), it is essential to look into the whole growing region holistically. Though location-specific studies give an in-depth understanding of the issues, they limit holistic planning, as the apple-growing region is quite large and spreads over diverse terrain conditions (elevation, slope and aspect). This paper has highlighted the methodology of using satellite remote sensing data and space-based terrain data to characterise the growing environment of the orchards present in two major apple-growing states. The use of niche model to define the suitable growing area in geospatial domain has been demonstrated.

The results show significant change in suitable growing environment in future scenario and ingression to higher elevation ranges. Currently, the niche model has been used as a generic one, encompassing all the varieties grown in the region, which can be improved by using sub-modules for dominant varieties. The model has been calibrated using the "presence-only" input, for which orchard locations were used as input. Further improvement needs to be done using "high productive" orchard locations as input. This advanced geospatial simulation model can be used for preparing action plan map at desired scale (regional to local) as remote sensing data from Indian remote sensing satellites is now available at various map resolutions (5–55 m), so also satellite-derived DEM.

Acknowledgements This work has been carried out under the PRACRITI programme of SAC, ISRO.

References

Awasthi RP, Verma HS, Sharma RD, Bhardwaj SP, Bhardwaj SV (2001) Causes of low productivity in apple orchards and suggested remedial measures. In: Jindal KK, Gautam DR (eds) Productivity of temperate fruits. Dr YS Parmar University of Horticulture & Forestry, Solan

Byrne DH, Bacon TA (1992) Chilling estimation: its importance and estimation. Texas Hortic 18(8):8–9

Hijmans RJ, Cameron SE, Parra JL, Jones PG, Jarvis A (2005) Very high resolution interpolated climate surfaces for global land areas. Int J Climatol 25:1965–1978

Jindal KK, Mankotia MS (2004) Impact of changing climatic conditions on chilling units, physiological attributes and productivity of apple in western Himalayas. Acta Hortic 662:111–117

National Horticulture Board (2011) Indian horticulture data base. http://nhb.gov.in/area-pro/database-2011.pdf

Navalgund RR, Parihar JS, Ajai, Rao PPN (1991) Crop inventory using remotely sensed data. Curr Sci 61:162–171

Panigrahy S, Singh CP, Bhatt NB, Parihar JS (2014) Geospatial planning towards climate change impact on the apple (Malus domestica) orchards in India. eJ Appl For Ecol 2(1):1–8

Ramirez J, Jarvis A (2008) High resolution statistically downscaled future climate surfaces. International Center for Tropical Agriculture (CIAT)/CGIAR Research Program on Climate Change, Agriculture and Food Security (CCAFS), Cali

SAC (2006) Apple orchard characterization using remote sensing and GIS for Shimla, Kullu and Mandi districts in Himachal Pradesh, Project Report, Space Applications Centre, ISRO, Ahmedabad, Report No. RSAM/ SAC/RESIPA/NEHORT/PR/04/03

Singh RR, Bhagat RM, Kalia V, Lal H (2009) Impact of climate change on shift of apple belt in Himachal Pradesh, ISPRS Archives XXXVIII-8/W3. In Workshop proceedings: impact of climate change on agriculture, pp 131–137

Stockwell DRB, Peters DP (1999) The GARP modelling system: problems and solutions to automated spatial prediction. Int J Geogr Info Syst 13:143–158

Utilization of Open-Source Web GIS to Strengthen Climate Change Informatics for Agriculture

Kamlesh Golhani, A.S. Rao, and J.C. Dagar

Abstract

Geographic information system (GIS) is used for location-based analysis and decision-making. GIS professionals in agriculture typically employ it to examine selected geographic datasets in detail, viz., land use and land cover (LULC), soil data, weather parameters with crop data such as variety distribution, and sowing date which are combined for the comprehensive study and analysis of spatial problems. Web GIS is a platform for distributing and mapping GIS data and services on the Web. It can be used to distribute geographic data to many concurrent users and allow them to do location-based analyses. Open-source GIS software and map server are rescued by the users from high-cost geospatial software and provide better functional environment through editing and customization of application programming interfaces (API).

Keywords

Geographic information system (GIS) • LULC • Web GIS • Application programming interfaces (API)

Introduction

Agriculture gets a makeover! The title of a popular cover story of Geospatial World (August 2011) "Geospatial technology, with its potential to address the complete life cycle of agriculture, is fast finding acceptance in agriculture to fulfill its responsibilities in addressing food security and as a fundamental instrument for sustainable development and poverty reduction, especially in developing nations. In the process, one of the oldest economic practices of human civilization is getting a makeover (Roy 2011)." But in this

K. Golhani (✉)
Agricultural Mechanization Division, Central Institute of Agricultural Engineering, Bhopal, India
e-mail: kamlesh.golhani@gmail.com

A.S. Rao
Central Research Institute for Dryland Agriculture, Hyderabad, India
e-mail: ashivrao@gmail.com

J.C. Dagar
Central Soil Salinity Research Institute, Karnal, Haryana, India
e-mail: dagarjc@gmail.com

A.K. Singh et al. (eds.), *Climate Change Modelling, Planning and Policy for Agriculture*,
DOI 10.1007/978-81-322-2157-9_10, © Springer India 2015

cutting-edge geospatial trend, to view any kind of information either spatial or nonspatial, the Internet is the easiest way to review information closely or of course it is a place to compare quality, authenticity, and limitation of information.

Web GIS is a source which can represent information of on Web with quality, authenticity, and advantages. It is however said, open-source software makes web GIS more advantageous and gives freedom to download, redistribute, modify, and use the API codes. It plays an important role in breaking down barriers where the cost limits the use of public spatial data and the access to GIS tools. Agriculture production, vulnerability and risk assessment, pest control, agriculture monitoring, precision farming, agricultural climatology, and land and water resource management are areas where GIS is being utilized for mapping, monitoring, and as a decision support system (DSS), and the number of activities is possible only through remote sensing and GIS; keeping this fact in mind, the utilization of open-source GIS is more recommended. Further, there is a dire need to make the Web available to all agriculture operations by using open-source web GIS. As the geographic datasets used in this work contain a huge volume of dataset collections, interdisciplinary approach, this, to some extent, is also an experiment that examines the capability of open-source web GIS solutions in building a large-scale Agro-GIS system.

Web GIS-based agricultural decision-making system is a cost-effective agricultural information dissemination system to disseminate all kinds of expert agricultural knowledge to the farming community in order to improve crop productivity. This GIS-based decision-making system for agriculture enables the farmer to provide a location-specific expert advice in a personalized and timely manner. The Web enabled a tool used in GIS environment to combine technical expertise and resource information including soil, water, agroecological regions according to the administrative boundaries and provincial agricultural data that information providers can use to make decisions regarding agricultural development. This is a web-based information system that provides access to a location-based agriculture information. It is designed to allow viewing, querying, and analyzing agricultural information geographically. As this is a software-independent system, users do not have to buy GIS software, and the system can be viewed through all Internet browsers (Jayawardena et al. 2009).

Web-Based Mapping

This paper highlights an open-source prototype system for web mapping. The different compliant map servers such as WMS, Map Server, and GeoServer use a technically competitive solution to serve a large-scale GIS database on the Web. The map clients are lightweight cross-browser spatial wares – Flash SWF, SVG, and Java applet, which enable enhanced graphics and sophisticated interactivity. This successful development shows us once more how the open-source solution benefits users, particularly those working in education and research and in developing countries. The techniques and system source code in this research work can be applied freely as a laboratory exercise or a course curriculum (Song et al. 2004).

Development of Location-Based Services

Web GIS differentiates itself from conventional DBMS and information systems in the sense that every piece of information has to be directly or indirectly associated with a location on the Earth's surface expressed as coordinates in any coordinate system. Agriculture could achieve rapid growth using computing technologies, particularly in relation to use tools developed by using GIS and GPS. The use of GIS and GPS in precision agriculture, field management, fertilizer, crop and animal transportation, and agricultural product storage is growing and flourishing, defined distances between markets and production areas along with transportation routes. Through the use of GPS, GIS, and remote sensing, the information needed for improving land and water use can be collected. Farmers can achieve additional benefits by combining better utilization of fertilizers and other soil amendments, determining the economic threshold for treating pest and weed infestations and protecting the natural resources for future use.

Web GIS Sources and Publishing Servers

Distributing geospatial information is an enforcing factor for information providers. The Internet allows all levels of society access to geospatial information and provides a media for processing geo-related information with no location restrictions. Web-based GIS evolved from different web maps and client server architectures to distributed ones (Fig. 1). As such, the Internet reshapes all functions of information system including gathering, storing, retrieving, analyzing, and visualizing data. The high cost of GIS system, the release of system-specific databases, and the enormous software developer efforts on upgrading the system are fading with the introduction of the web-based GIS. Moreover, disseminating spatial information on the Internet improves the decision-making processes (Alesheikh et al. 2002).

The development of the Web and expansion of the Internet provide two key capabilities that can greatly help agricultural scientist, researchers, and even farmers. First, the Web allows visual interaction with data by setting up a web server, which clients can produce maps. Since the maps and charts are published on the Internet, other clients can view updates, helping to speed up the evaluation process. Second, because of the near ubiquitous nature of the Internet, the geospatial data can be widely accessible. Clients can work on it from almost any location. Both of these features alter the ways agriculturist do their work in the near future.

Open-Source Web GIS

Free and open-source GIS (FOSGIS) are successfully applied for Internet GIS by many entrepreneurs and research organizations, but the field of desktop GIS is still occupied by proprietary systems. Open-source projects are widely used for web mapping; as in recent years, many interesting desktop FOSGIS projects have been initiated; FOSGIS may be an interesting alternative to proprietary software. Google Maps API, ArcGIS for Flex, ArcGIS API for Flex, QGIS 2.0, open Jump 1.2, GRASS 6, and ILWIS open 3.5 have been evaluated. Most of the products have been tested and evaluated successfully on Windows XP, Vista, and Window 7 (Steiniger and Hunter 2011). Some examples are as follows.

Google Maps API

Google Maps API has revolutionized online mapping service to display an integration of spatial data in a lot of different ways. But it is still a

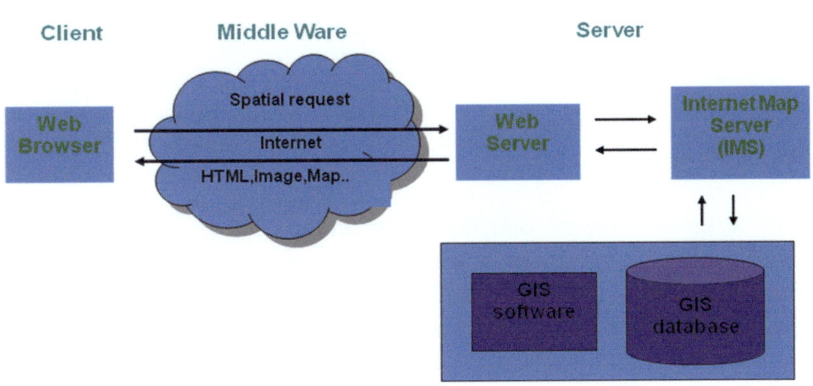

Fig. 1 Client server architecture

daunting task to display spatial data in Google map for people who do not have knowledge about programming languages. The Google Maps API for Flash allows users to add more dynamic and interactive mapping application to their website. To use Google Maps API for Flash, users must obtain a mandatory API key from Google. The process is rather simple and straightforward. Users need to provide the URL of the site that would use the application to be developed and need to have a Google account. After this process, SDK was downloaded from Google's website which contains all necessary class libraries (ActionScript library made of public class of Maps API as SWC file) necessary to develop the application utilizing the Flash environment.

ArcGIS for Flex

The ArcGIS Viewer for Flex application enables users to create GIS-enabled web mapping applications, without requiring programming. It is designed so users can configure and deploy a web mapping application easily and quickly. Viewer is designed as a stand-alone application that is downloaded onto a local web server. Its purpose is to be configured and/or customized to work with custom data content from ArcGIS Server or ArcGIS Online Web services. Web maps are published and shared on ArcGIS.com

Web GIS Server

This is a different map server and an open-source platform available for publishing spatial data and interactive mapping applications to the Web.

ArcGIS Server

This server is complete and integrated server-based geographic information system (GIS). It comes with out-of-the-box end user applications and services for spatial data management, visualization, and spatial analysis. ArcGIS Server offers open access to extensive GIS capabilities that enable organizations to publish and share geographic data, maps, analyses, models, and more. With ArcGIS Server's rich standard-based platform, centrally managed, high-performance GIS applications and services can be accessed throughout an organization using browser-based, desktop, or mobile clients. ArcGIS Server offers the following advantages: browser-based access to GIS.

Map Server

It is an open-source development environment for building spatially enabled Internet applications. It can run as a CGI program or via Map Script which supports several programming languages. Map Server was developed by Minnesota. So, it is often and more specifically referred as "UMN Map Server" to distinguish it from the commercial "map server."

Geo Server

This open-source geospatial server is mostly used for web mapping and web feature services and also having geographical user interface (GUI) for configuration.

Conclusion

In general, the purpose of using open-source web GIS is to meet the needs of free GIS for publishing the data showing impact of climate in changing agriculture. Increasing the number of users for free GIS application in natural resources management going to overcome gap between agriculture science and computer applications. To conclude, geospatial agriculture requires interdisciplinary accomplishment for enabling various aspects of agricultural and allied sciences. Learning with some web functions using code libraries, anyone can build dynamic-rich Internet applications (RIAs) on open-source web GIS. It creates interactive and expressive web applications leveraging GIS server resources and flex components.

Acknowledgments This work is partially supported by Dr. B. Venkateswarlu, Director, Central Research Institute of Dryland Agriculture (CRIDA), Hyderabad under NICRA scheme. The authors also gratefully acknowledge the helpful comments and suggestions of the reviewers, which have improved the presentation.

References

Alesheikh AA et al (2002) Web GIS: technologies and it's applications. In: Symposium on geospatial theory, processing and applications, Ottawa. Email: alesheikh@hotmail.com

Jayawardena N, Wijayarathna G, Gunathilake J (2009) GIS based decision making system for agriculture. In: 8th annual Asian conference and exhibition on geospatial information technology and applications Suntec Singapore International Convention & Exhibition Center, Singapore 18–20 Aug 2009. Email: nbmjnirmala@gmail.com

Roy D (2011) Agriculture gets a makeover!. Geospatial World, Geospatial Media and Communication Pvt Ltd., Noida

Song X, Yasuyuki K, Shibayama M (2004) The Development of Web mapping application using open source GIS solution. Paper presented in International Symposium on Geoinformatics for Spatial Infrastructure Development in Earth and Allied Sciences. Center for Southeast Asian Studies, E-mail: song@cseas.kyoto-u.ac.jp

Steiniger S, Hunter AJS (2011) Jump pilot project on free GIS. Department of Geomatics Engineering, University of Calgary, 2500, University Drive NW, Calgary, AB T2N IN4, Canada, email: ssteinig@ucalgery.ca

Design and Development of Rubber Dams for Watersheds in the Climate Change Scenario

S.K. Jena, A. Kumar, P.S. Brahmanand, A. Mishra, N. Sahoo, and D.U. Patil

Abstract

A rubber dam (flexible check dam) is an inflatable structure build across a stream used for water conservation, flood control, and regulating flow of water in the stream. When it is inflated, it serves as a check dam/weir, and when it is deflated, it functions as a flood mitigation device and sediment flushing. Generally, most of the check dams in watersheds are made of concrete, steel, stone, soil, or vegetation. The use of rubber as a construction material is a technological innovation in materials application. At the same time, the check dams are rigid one and they cannot allow more water to flow over it at times of heavy flood/runoff or store sufficient runoff to conserve the rainfall at lean season for use by farmer for different rabi crops like pulses, oilseeds, and vegetables. To give more flexibility in release and control of water flow across the streams, research efforts were made at Directorate of Water Management, Bhubaneswar, in collaboration with Indian Rubber Manufacturers Research Association, Central Institute for Research on Cotton Technology, and Kusumgar Corporate Private Limited, Mumbai, to design, fabricate, and install rubber sheets instead of cement material for check dams and to study their impact on crop performance. Five rubber dams were installed as different hydraulic structures for various uses in watersheds at different locations of Khurda district, Odisha, i.e., Mendhasal, Baghamari, Badapokharia, and Chandeswar with innovative manufacturing, fabrication, and installation technology. This is the first indigenous rubber dam in our country. The installation of rubber dams in watersheds has increased the production and productivity of rice crop, helped in taking second crop thus increasing cropping intensity and net profit of the farmers.

Keywords

Rubber dam • Check dam • Watershed management • Flood • Drought mitigation

S.K. Jena (✉) • A. Kumar • P.S. Brahmanand • A. Mishra • N. Sahoo • D.U. Patil
Directorate of Water Management (ICAR), Chandrasekharpur, Bhubaneswar 751023, India
e-mail: skjena_icar@yahoo.co.in

A.K. Singh et al. (eds.), *Climate Change Modelling, Planning and Policy for Agriculture*, DOI 10.1007/978-81-322-2157-9_11, © Springer India 2015

Introduction

In climate change scenario we experience more extreme events of cyclone, flood, drought, long dry spells, etc. The intensity, amount, and distribution of rainfall are also not as per the crop requirement. Secondly land and water are the two most vital natural resources of the world, and these resources must be conserved and maintained carefully for environmental protection and ecological balance. Prime soil resources of the world are finite, nonrenewable over the human time frame, and prone to degradation through misuse and mismanagement. Total global land degradation is estimated at 1,964.4 M ha, of which 38 % is classified as light, 46 % as moderate, 15 % as strong, and the remaining 0.5 % as extremely degraded, whereas the present arable land is only 1,463 M ha which is less than the land under degradation (Koohafkan 2000). The annual rate of loss of productive land in the whole world is 5–7 M ha, which is alarming. In India, out of 328 M ha of geographical area, 182.03 M ha is affected by various degradation problems out of which 68 M ha are critically degraded and 114.03 M ha are severely eroded whereas total arable land is only 156.15 M ha (Velayutham 2000). It was reported that in India 0.97 % of the total geographical area is under very severe erosion (>80 t ha^{-1} $year^{-1}$), 2.53 % of area under severe erosion (40–80 t ha^{-1} $year^{-1}$), 4.86 % of area under very high erosion (20–40 t ha^{-1} $year^{-1}$), 24.42 % of area under high erosion (10–20 t ha^{-1} $year^{-1}$), 42.64 % of area under moderate erosion (5–10 t ha^{-1} $year^{-1}$), and 24.58 % of area under slight erosion (0–5 t ha^{-1} $year^{-1}$) (Singh et al. 1992). Therefore, the problem of land degradation due to soil erosion is very serious, and with increasing population pressure, exploitation of natural resources, and faulty land and water management practices, it will further aggravate. Land degradation also reduces the world's freshwater reserves. It has a direct impact on river flow rates and the level of groundwater tables. The reduction of river flow rates and the lowering of groundwater levels lead to the silting up of estuaries, the encroachment of salt water into groundwater, and the pollution of water by suspended particles and salinization, which in turn reduces the biodiversity in fresh and brackish water and consequently fish catches. Lower river flows also interfere with the operation of reservoirs and irrigation channels, increasing coastal erosion and adversely affecting human and animal health.

Proper watershed management, which is a comprehensive term meaning the rational utilization of land and water resources for optimal production and minimum hazard to natural resources, could be the solutions to all these problems (Jena 2002). There are several measures such as mechanical (engineering) and biological (agricultural) which are used for soil and water conservation in watershed management. Check dams are engineering measures which are mainly used for soil and water conservation in watersheds. In India, several types of check dams are being used for regulating runoff in watersheds which in turn help in assured water supply to crops. Generally, most of the check dams in watersheds are made of steel, concrete, soil, rock (permanent), or with vegetation (temporary). The use of rubber as a construction material is a technological innovation in materials application. At the same time, the check dams are rigid one, and they cannot allow more water to pass over it at times of heavy flood/runoff or store sufficient runoff at lean season of rainfall for use in rabi season by farmer for different rabi crops like pulses, oilseeds, and vegetable. To give more flexibility in release and control of water flow across the streams, research efforts were made at the Directorate of Water Management (DWM), Bhubaneswar, in collaboration with Indian Rubber Manufacturers Research Association (IRMRA), Central Institute for Research on Cotton Technology (CIRCOT), and Kusumgar Corporate (KC), Mumbai, to fabricate and install rubber sheets instead of cement concrete/stone material for check dams and to study their impact on crop performance.

What Is Rubber Dam?

A rubber dam is an inflatable and deflatable structure used for regulating water flow and store water. Rubber dams are installed to function as weirs or barrages which are relatively low-level dams constructed across a river for the raising of river level for the diversion of flow in full, or a part, into a supply canal or conduit for irrigation, domestic, industrial use (Tam 1998). The rubber dam is also known as inflatable dam or Fabridam. Most of the civil engineering structures constructed in the history of humankind are made of steel, concrete, soil, or rock. The use of rubber as a construction material is a technological innovation in materials application (Tam and Zhang 2002).

When it is inflated, it serves as a check dam, and when it is deflated, it functions as a flood mitigation device. The head or height of the rubber dam is variable. According to the requirement its height can be increased or decreased. This variable head also regulates the depth of flow in the diversion channel for irrigation present in the upstream side of the check dam.

Components of Rubber Dam

As an innovative hydraulic structure, the rubber dam mainly consists of four parts: (1) a concrete foundation with head wall extension, sidewall, and wing wall of a normal check dam; (2) the head wall replaced by rubberized fabric dam body; (3) anchoring mechanism (anchoring of rubber sheet with bottom and side of the check dam); (4) an inlet/out let piping system for inflation and deflation by water; and (5) a pump for filling water for inflation.

Advantages of Rubber Dam

The main advantages of the rubber dam are its ability for better soil erosion control (streambed as well as stream banks or side of channel) and flood control during excess runoff water flow. It also acts as a dam/reservoir for storing water during scanty rainfall period (dry spells) so that supplemental irrigation can be provided to the crops. It also helps in groundwater recharge. This technology has a potential to benefit farmers in rainfed agroecosystems which constitute about 60 % of the net sown area.

Installation and Field Evaluation

The different steps followed in successful design and installation of rubber dam by the Directorate of Water Management (DWM), Bhubaneswar, and its impact on agricultural productions are given below.

(a) Selection of sites for check dam, impounding structure, and loose boulder structure to install rubber dam
(b) Socioeconomic survey of the location where rubber dams were to be installed
(c) Hydrological data collection and analysis
(d) Design, estimation, and execution of base structure
(e) Fabrication, installation, and testing of rubber dam
(f) Development of inflation/deflation mechanism of the rubber dam
(g) Observation of crop yield data of the farmers who got benefit from rubber dam

Five rubber dams were installed as different hydraulic structures for various uses in watersheds at different locations of Khurda district, Odisha, i.e., Mendhasal, Baghamari, Badapokharia, and Chandeswar, with innovative manufacturing, fabrication, and installation technology. These are the first indigenous rubber dams which were been fabricated and installed in our country by Indian scientists.

The dimensions of different important components of the rubber dams installed in five different sites are given in the following Table 1.

Installation of Rubber Dam

Enough care has been taken during development of the rubber composite along with nylon

Table 1 Dimension of different components of installed rubber dams

Sl. no.	Sites	Dimensions (m) Crest length (≈width of stream)	Spacing between anchoring bolts	Height of head wall
1.	Baghamari	5.00	1.20	1.00
2.	Badapokharia	2.00	1.50	1.00
3.	Mendhasal	2.00	0.50	0.50
4.	Chandeswar 1	4.15	1.50	1.50
5.	Chandeswar 2	4.15	1.50	1.35

reinforcement that when installed across streams, it does not have any adverse effect on water quality (may be due to oozing out of chemicals or any extracts from the rubber composite) and also on crop productivity. It does not have any adverse impact on environment.

Rubber composite sheet manufactured by IRMRA was fixed with concrete base structure through double rows anchoring mechanism. The angle of inclination of side anchoring to the base has been optimized by DWM to minimize wrinkles and easiness to inflate and deflate. The angle varied within 105° to 150° to the base. The spacing between bolts and also the dimension and structural strength of different bolts were tried. The dimension and strength of different anchoring bots were optimized by DWM for different dimensions of rubber dam. The structure was made leak proof (no water flow between top of the base of the concrete foundation structure and the rubber sheet) using different proportions of adhesives like silica gel, M-seal, and araldite and was tested by filling with water through inlet pipe using 1.5 hp kerosene-operated petrol start centrifugal pump. Two of the installed rubber dams at Chandeswar are presented through Plate 1.

Impact of Rubber Dam on Crop Performance

Uneven rainfall in the *kharif* season results in lower rice productivity. Rubber dam helped in providing irrigation in critical stage of paddy, i.e., in flowering stage and hence it has saved the crop. Also the water stored above the rubber dam was used for rice nursery raising which helped in transplanting during the recommended period. The stored water in rubber dam was diverted to the right side of dam through diversion irrigation channel to around 40 ha of paddy field at Baghamari. From July 15, 2010, to August 13, 2010, there was no rainfall in the rubber dam project site, i.e., Baghamari in Khurda district of Orissa. There was water scarcity and paddy fields became dry. Therefore, 19 farmers from village Baghamari got benefit by irrigating their fields with stored water by diverting the water through the diversion irrigation channel present just adjacent to the upstream side of rubber dam. Farmers also utilized water from this rubber dam in the flowering stage of rice which is very crucial period (critical growth stage) for crop. After harvesting rice, the average productivity of the above farmers in *kharif* 2010 was found to be 4.67 t ha^{-1}, whereas it was only 2.87 t ha^{-1} in *kharif* 2009, i.e., before installation of rubber dam in this area. The increase in average productivity is around 62 %. The rainfall that occurred during the 2nd week of December 2010 has been stored in rubber dam and has been used by the farmers for *rabi* pulses, oilseeds, and vegetable cultivation.

The average productivity of green gram in the *rabi* season at Baghamari enhanced from 0.63 to 0.92 t ha^{-1}, and the productivity of sunflower and cucumber in the *rabi* season is 0.84 t ha^{-1} and 4.3 t ha^{-1}, respectively. The increase in cropping intensity at Baghamari is 31 % due to cultivation of green gram, sunflower, and cucumber.

The economic analysis indicated that the intervention of rubber dam has potential to enhance the gross returns of the farmers by

Plate 1 Rubber dams installed in watersheds of Odisha by DWM, Bhubaneswar

62 % from Rs. 28,700 to 46,700 ha^{-1} if farmers grow only rice crop. At the same time, the total gross returns of the farmers may increase from Rs. 45,184 to 70,792 ha^{-1} if farmer practices rice-green gram cropping system with the additional water available through rubber dam, and the total gross returns may increase to Rs. 72,500 and 75,135 ha^{-1} if farmers practice rice-cucumber and rice-sunflower cropping system. Similarly, the net returns of the farmers will increase from Rs. 12,400 ha^{-1} to Rs. 27,600 ha^{-1}, Rs. 43,942 ha^{-1}, Rs. 43,200 ha^{-1}, and Rs. 47,935 ha^{-1} under sole rice cropping, rice-green gram, rice-cucumber, and rice-sunflower cropping systems, respectively.

The productivity, percentage increase in yield of different crops grown close to the rubber dam installed at Chandeswar prior to the installation of rubber dam and after installation is presented in Table 2.

The yield of pumpkin, ridge gourd, cowpea, and brinjal was enhanced from 6.3 t ha^{-1}, 5.5 t ha^{-1}, 5.2 t ha^{-1}, and 4.8 t ha^{-1} during pre-project condition to 8.5 t ha^{-1}, 8.1 t ha^{-1}, 6.6 t ha^{-1}, and 6.7 t ha^{-1}, respectively, after the installation of rubber dam. The percentage yield enhancement was 34.9 %, 47.2 %, 26.9 %, and 39.6 %, respectively, for pumpkin, ridge gourd, cowpea, and brinjal. Thus, the productivity of summer vegetables at Chandeswar enhanced significantly due to assured water supply from the installed rubber dams. Similarly, 30 ha of rice fields were irrigated at critical stages through

Table 2 Yield enhancement of different crops after the installation of rubber dam at Chandeswar

Crop	Pre-project condition (productivity; t ha^{-1})	Yield (t ha^{-1}) after installation of rubber dam	% Yield enhancement
Pumpkin	6.3	8.5	34.9
Ridge gourd	5.5	8.1	47.2
Cowpea	5.2	6.6	26.9
Brinjal	4.8	6.7	39.6

stored water from rubber dams at Chandeswar during kharif 2011. The paddy grain yield recorded a jump of 23 % from 4.14 t ha^{-1} during pre-project period to 5.09 t ha^{-1} after installation of rubber dam in Chandeswar 1 and a jump of 19 % from 4.48 t ha^{-1} during pre-project period to 5.33 t ha^{-1} after installation of rubber dam in Chandeswar 2. The rubber dam at Badapokharia has been instrumental in augmenting groundwater recharge.

Conclusions

From the preliminary agricultural and hydrologic data observation, it is apparent that rubber check dams can be well utilized for achieving sustainable crop production and could be instrumental for enhancing crop and water productivity in watersheds. It does not have adverse impact on environment. It can be easily installed and operated by farmers of the watersheds. There is almost no maintenance except the running cost of filling (inflating) with water at the time of need.

References

Jena SK (2002) Development and evaluation of hydrological models for agricultural watersheds using remote sensing and GIS. Unpublished PhD thesis, Indian Institute of Technology, Kharagpur, India

Koohafkan AP (2000) Land resources potential and sustainable land management – an overview. In: Lead paper of the international conference on land resource management for food, employment and environmental security during 9–13 November 2000, New Delhi, India, pp 1–22

Singh G, Babu R, Narain P, Bhushan LS, Abrol IP (1992) Soil erosion rates in India. J Soil Water Conserv 47 (1):97–99

Tam PWM (1998) Application of inflatable dam technology-problems and countermeasures. Can J Civil Eng 25:383–388

Tam PWM, Zhang X (2002) Management of rubber dams in Hong Kong. Can J Civil Eng 29:409–420

Velayutham M (2000) Status of land resources in India. In: Lead paper of the international conference on land resource management for food, employment and environmental security during 9–13 November 2000, New Delhi, India, pp 67–83

Impact of Climatic Variability on Crop Production in Mahanadi Delta Region of Odisha

Mausumi Raychaudhuri, D.K. Panda, Ashwani Kumar, S.K. Srivastava, P.S.B. Anand, S. Raychaudhuri, and G. Kar

Abstract

The impact of climatic variability on crop production was studied in the Mahanadi delta area where upland, midland and lowland sites were selected covering Dhenkanal, Kendrapara, Nayagarh, Cuttack and Puri district. The monthly run-off (10^6 m^3) data of 19 gauging stations adequately representing the sub-basins of the Mahanadi River basin with a command area of 1,41,589 km^2 during the period 1972–2004 was analysed to investigate the run-off variability and trends of the basin. Analysis of hydroclimatic parameters using advanced statistical tools revealed that due to alteration in the hydrological cycle the severity of droughts and intensity of floods increased and the quantity of the available

M. Raychaudhuri (✉) • S. Raychaudhuri
Discipline Soil Science (Soil Chemistry/Fertility/Microbiology), Organization Directorate of Water Management, Bhubaneswar, Odisha 751023, India
e-mail: mausumiraychaudhuri@gmail.com

D.K. Panda
Discipline Agricultural Statistics, Organization Directorate of Water Management, Bhubaneswar, Odisha 751023, India

A. Kumar
Discipline Soil and Water Conservation Engineering, Organization Directorate of Water Management, Bhubaneswar, Odisha 751023, India

S.K. Srivastava
Discipline Agricultural Economics, Organization Directorate of Water Management, Bhubaneswar, Odisha 751023, India

P.S.B. Anand
Discipline Agronomy, Organization Directorate of Water Management, Bhubaneswar, Odisha 751023, India

G. Kar
Discipline Soil Science (Soil Physics/Soil and Water Conservation), Organization Directorate of Water Management, Bhubaneswar, Odisha 751023, India

A.K. Singh et al. (eds.), *Climate Change Modelling, Planning and Policy for Agriculture*,
DOI 10.1007/978-81-322-2157-9_12, © Springer India 2015

water resources also gets affected. The average annual result run-off during the period 1990–2004 exhibits an increased run-off, in comparison to the previous subseries 1972–1989 for all the stations. However, this increased run-off is also associated with an increase in the variability (standard deviation), suggesting that the wet period is more prone to uncertainty in comparison to the corresponding dry period. The rainfall status of the experimental sites on the Mahanadi delta showed occurrence of frequent drought and flood in the last 25 years (Table 1). Uneven distribution of rainfall affected the crop yield adversely. Increase in rainfall in May and September and severe drought in November and December affected the paddy yield. It has been observed that in spite of canal irrigation sources, water is unavailable during the critical stages of crop growth and moisture stress condition. To mitigate the vagaries caused by climate change, water harvesting structures for assured irrigation, microirrigation techniques to increase the water use efficiency, crop diversification and multiple use of water to improve water productivity are some of the technological interventions for resilience and sustainable identified production.

Keywords
Crop production • Climatic variability • River basin • Monsoon

Introduction

Rising temperature, drought and more intense precipitation affect the soil and water resources in an adverse way which in turn affects the crop yield. Erosion rates, soil organic carbon losses, soil moisture, root growth and function, root microbe associations and plant phenology as they are related to mineral nutrition and surface and groundwater resources in terms of quantity and quality are affected. Since 1990 both drought and floods have severely affected the agricultural prospects of Odisha that lies in the Eastern plateau and hills region and also covers the east coast plains and hills region under agroclimatic zones (ACZ) of ACZ-7 and ACZ-11. The mean annual rainfall is 1,482 mm with variation in both space and time. It was found to be very erratic over years and the state has experienced three natural calamities – flood, drought and cyclones – over the year. Floods have occurred in 1990, 1994, 1995, 2001, 2003 and 2005. Severe droughts have occurred in 1996, 2000 and 2002. Further, erratic distribution of rainfall has caused localised drought during 1997, 1998 and 1999. The agricultural economy of the state is primarily a food crop economy and more than 90 % of the cropped areas are devoted to food grains growing in the fertile region of Mahanadi basin and Mahanadi delta.

Odisha is an agriculture-dominant state whose economy is mostly agrarian based and depends on the onset of monsoon and its further behaviour. Rainfall is a random hydrological event, which is unevenly distributed in both space and time. Being the primary source of water in any region, its estimation at different probabilities is important for efficient planning of soil and water conservation programme and the optimum utilisation of water resource in various production systems. Under National Initiative on Climate Resilient Agriculture (NICRA) project an attempt has been made to study the impact of rainfall variability on paddy production in the

Mahanadi delta area where upland, midland and lowland sites were selected covering Dhenkanal (4,452 km^2), Nayagarh (3,051 km^2) and Puri (3,890 km^2) districts. Paddy is the main crop grown in this region along with black gram, sugar cane, green gram, groundnut, sweet potato and vegetables.

Experimental Details

Study Area

Mahanadi River basin is the 3rd largest basin in India with a command area of 1,42,589 km^2 along its length of 851 km (Chhattisgarh 357 km; Orissa 494 km) and diameter of 400 km. Three sites were selected, namely, Kanas and Satyabadi block in Puri (84°29′E to 86°27′E longitude and 19°28′N to 26°35′N latitude) in the coastal plains, Dhenkanal (85°58′E to 86°2′E longitude and 20°29′N to 20°11′N latitude) in the upland and Nayagarh (84°29′E to 85°27′E longitude and 19°54′N to 20°32′N latitude) in the middle land, to study the impact of rainfall variability on paddy yield.

Methodology

Monthwise rainfall of the whole basin area for the period 1901–2004 was analysed using the robust plotting technique, i.e. LOWESS. The monthly run-off (10^6 m^3) data of 19 gauging stations (Fig. 1) adequately representing the sub-basins of the Mahanadi River basin during the period 1972–2004 was analysed to investigate the run-off variability and trends of the basin. A standardised value of monthly run-off for each gauging station was calculated and then

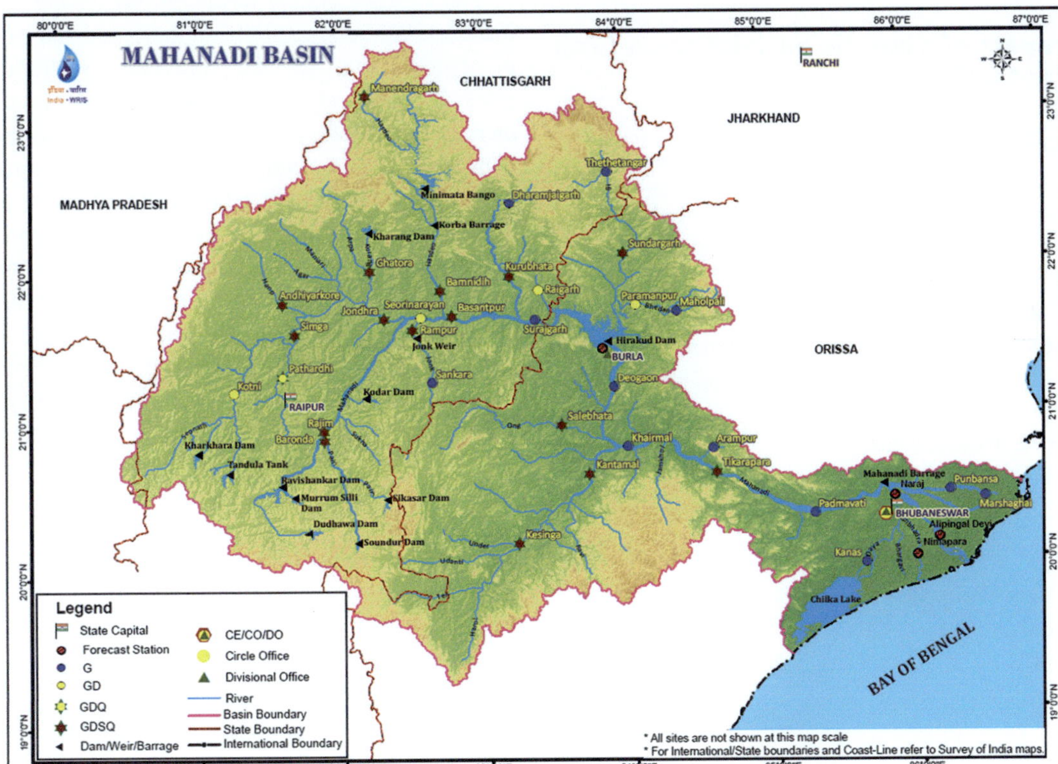

Fig. 1 Mahanadi basin area and gauging stations

arranged over the basin to get a basin-wide pattern of run-off anomalies.

The soil types of this region are saline, laterite, alluvial and red soils. For analysing the rainfall distribution pattern of the blocks of three districts, viz. Puri, Nayagarh and Dhenkanal, the time series data of 24 years (1988–2011) was collected from the respective block offices and collectorate of the district headquarter. Coefficient of variation (CV) measures the variability of rainfall and is the indication of dependability of rainfall expressed in percentage. The threshold levels for CV for any interpretation are taken as daily rainfall, <250 %; weekly rainfall, <150 %; monthly rainfall, <100 %; seasonal rainfall, <50 %; and yearly rainfall, <25 % (Veeraputhiran et al. 2003). Probability analysis was done by Weibull's method for weekly rainfall. It was calculated for annual, seasonal, monthly and weekly rainfall series for different blocks of three districts.

Impact of Climatic Variability on Crop Production

Rainfall: Run-Off Analysis

Analysis of hydroclimatic parameters using advanced statistical tools revealed that due to alteration in the hydrological cycle, the severity of droughts and intensity of floods increased and the quantity of the available water resources also gets affected. The pre-monsoon (Jan–May) rainfall shows an increasing pattern during 1971–2006, although it was declining during the pre-warming period (Fig. 2). It is important to note a significant decrease in July rainfall during the global warming era (1971 onwards) in conjunction with high variability as July rainfall provides irrigation during the critical stages of monsoon crops (Fig. 3). The August rainfall, which is equally important for crop growth, has also experienced an unstable behaviour in terms of dry and wet years (Fig. 4). The monsoon rainfall indicates that most of the drought years of the last century have occurred from 1971 onwards (Fig. 5). The annual rainfall (Fig. 6) indicates an increasing trend, although rainfall has decreased during the active monsoon months of July and August. Therefore, it is misleading to consider the annual rainfall for the success of the crop. Recent increase in the post-monsoon rainfall (Fig. 7) has contributed to the annual increases.

The average annual result run-off during the period 1990–2004 exhibits an increased run-off, in comparison to the previous subseries 1972–1989 for all the stations. However, this increased run-off is also associated with an increase in the variability (standard deviation), suggesting that the wet period is more prone to uncertainty in comparison to the corresponding dry period showing occurrence of more monsoon droughts in recent years as evident from the negative departures (Fig. 8).

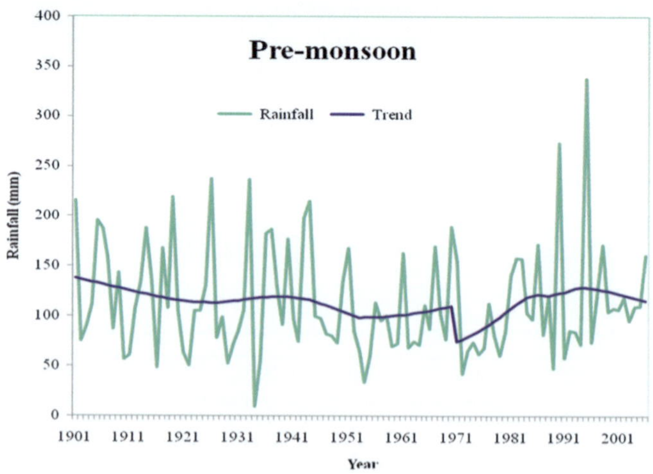

Fig. 2 Pre-monsoon (Jan–May) rainfall pattern of Mahanadi basin

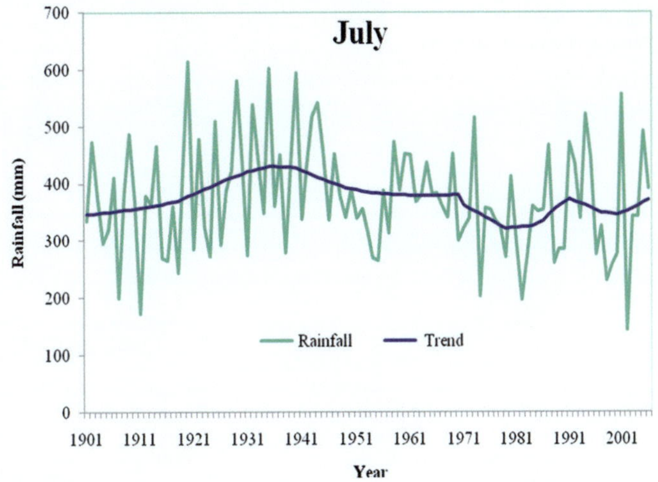

Fig. 3 July rainfall pattern of Mahanadi basin

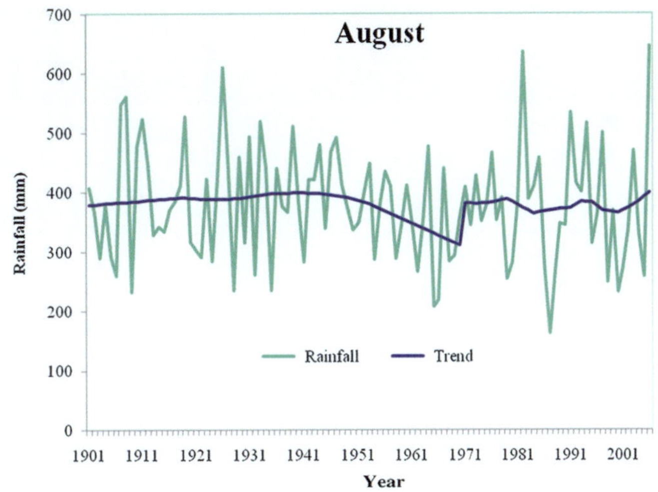

Fig. 4 August rainfall pattern of Mahanadi basin

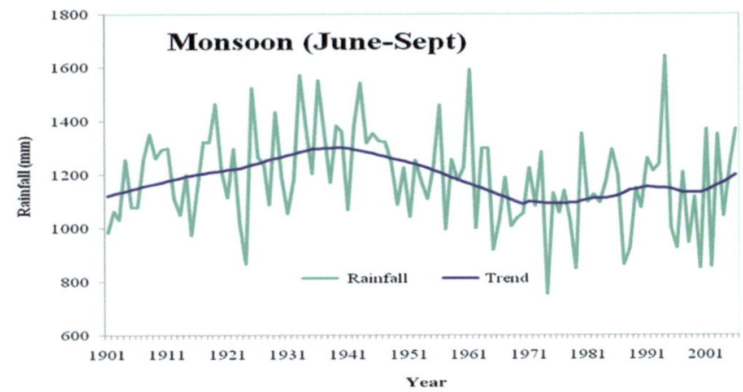

Fig. 5 Monsoon rainfall pattern of Mahanadi basin

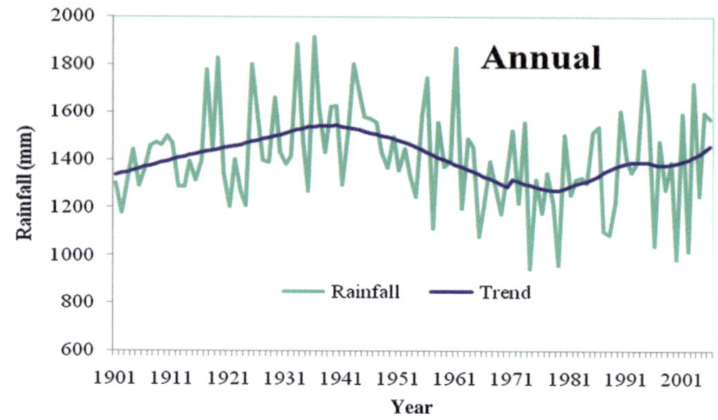

Fig. 6 Annual rainfall pattern of Mahanadi basin

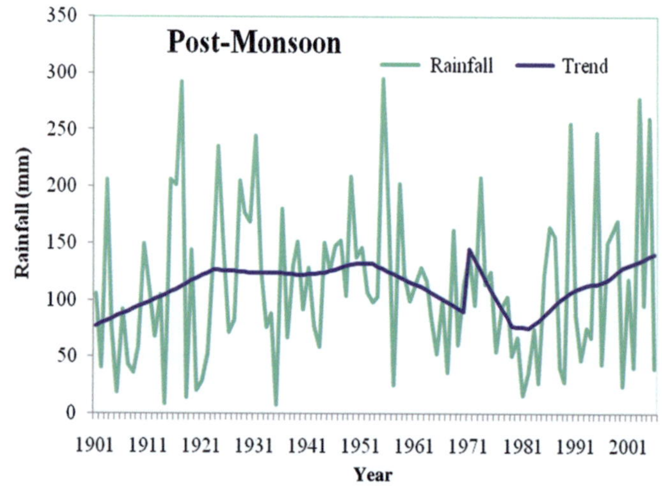

Fig. 7 Post-monsoon rainfall pattern of Mahanadi basin

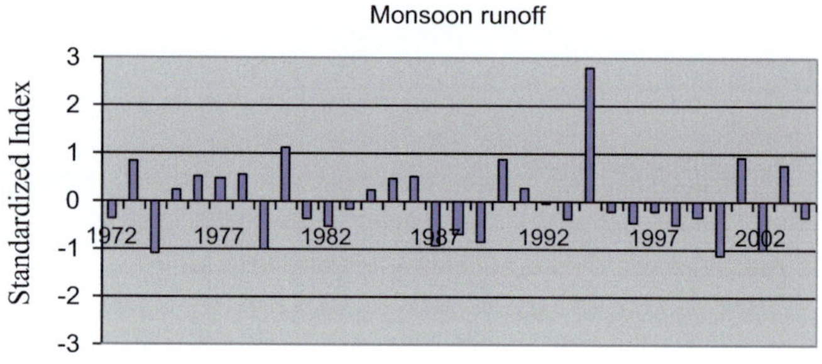

Fig. 8 Variability in the monsoon run-off of Mahanadi basin

Seasonal Variation

The rainfall analysis of the experimental sites on the Mahanadi delta revealed the occurrence of frequent drought and flood in the last 25 years (Table 1). Excess rainfall occurred maximum in Kanas block of Puri district which is a low lying area, whereas only Satyabadi block of Puri district experienced scanty rainfall during the last 25 years. Seasonal rainfall variation of Nayagarh, Puri (Kanas and Satyabadi) and Dhenkanal for the last 24 years (1988–2011) is presented in Table 2. Data showed that Nayagarh, Puri (Kanas and Satyabadi) and Dhenkanal received the mean seasonal rainfall of 1,215.1 mm, 978.57 mm, 1,138 mm and

Table 1 Rainfall status of the study site for the last 25 years

Block (district)	Annual average rainfall (mm)	Maximum annual rain fall		Minimum annual rain fall		Normal + excess (years)	Deficit + scanty (years)
		Year	Amount (mm)	Year	Amount (mm)		
Satyabadi (Puri)	1,686.71	1991	5,161	1988	508	7 + 7	8 + 3
Kanas (Puri)	1,228.25	1990	2,396.4	2000	507	11 + 9	5 + 0
Ranapur (Nayagarh)	1,533	1995	2,275	1996	841	17 + 4	4 + 0
Dhenkanal (Dhenkanal)	1,475.13	1992	2,736	2000	408.3	11 + 7	7 + 0

Table 2 Seasonal rainfall (mm) analysis for 24 years (1988–2011)

Season	Probability level												Mean				CV %			
	75 %				50 %				25 %											
	S	D	N	K	S	D	N	K	S	D	N	K	S	D	N	K	S	D	N	K
SWM (June–Sep)	605	796	1,059	584	1,008	1,018	1,283	897	1,541	1,360	1,392	1,383	1,138.8	1,064.6	1,215.1	978.6	75.2	42.3	19.7	60.0
PM (Oct–Dec)	26	37	66	22	35	126	134	75	266	224	258	407	203.0	153.8	182.1	219.0	140.4	104.8	90.8	129.8
WM (Jan–Feb)	0	30.1	58	0	0	67	107	0	0	176	151	12	17.8	121.3	149.6	13.6	295.7	121.0	119.8	229.9
SM (Mar–May)	12	0	0	5	34	0	28	25	227	23	45.3	95	176.1	21.4	33.4	99.3	191.0	178.9	95.1	208.8

S Satyabadi, *D* Dhenkanal, *N* Nayagarh, *K* Kanas

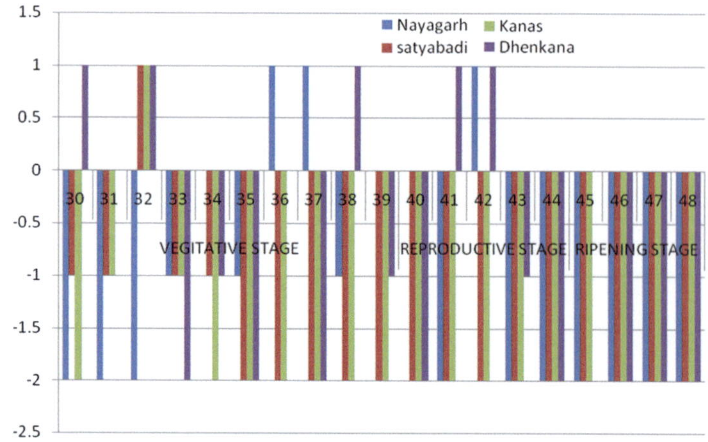

Fig. 9 Distribution of rainfall of the blocks over the growing period of paddy

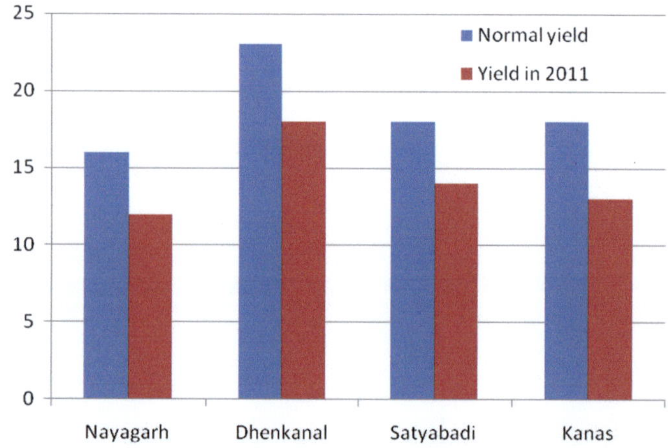

Fig. 10 Yield deviation in paddy from normal of the experimental blocks

1,064.6 mm, respectively, during the southwest monsoon and occurrence of rainfall during the southwest monsoon (SWM) period is dependable only for Nayagarh and Dhenkanal as per the respective CV (Table 2). The rest of the seasons showed high CV values indicating high variability of rainfall during post-monsoon (PM), winter monsoon (WM) and summer monsoon (SM) period. The probability of getting rainfall at 75 % probability level for Nayagarh, Puri (Kanas and Satyabadi) and Dhenkanal is 1,059, 584, 605 and 769 mm, respectively.

Impact on Paddy Yield

Monthwise rainfall trend of different blocks for the last 10 years (2001–2010) revealed an uneven distribution of rainfall that affected the crop yield adversely. Increase in rainfall in May and September and erratic distribution during the monsoon months shows advancements of monsoon. Moreover, to meet the water requirement of paddy, availability of water during the critical growth periods, viz. transplanting, vegetative growth period, panicle initiation, booting and heading, is essential. Rainfall analysis of the paddy growing period (30–48 weeks) during 2011 revealed the occurrence of severe drought in October to December (Fig. 9) in all the blocks that affected the paddy yield (Fig. 10) adversely. It has been also observed that though the experimental sites were in the canal command areas in spite of canal irrigation sources, water is unavailable during the critical stages of crop growth and moisture stress condition.

Conclusion

Rainfall pattern of the Mahanadi basin area showed uneven distribution and frequent occurrence of drought and flood. Water availability during the paddy growing period has also been scarce due to erratic distribution, shifting of monsoon trend and occurrence of severe drought during the critical growth stages. Life-saving irrigation during critical stages may help the farmers to adapt the adverse moisture stress condition and enhance the crop production. To mitigate the vagaries caused by climate change, water harvesting structures for assured irrigation, microirrigation techniques to increase the water use efficiency, crop diversification and multiple use of water to improve water productivity are some of the technological interventions for resilience and sustainable identified production.

Acknowledgement The authors are grateful to Dr. S. Ayyappan, Director General (ICAR) and Secretary (DARE), Government of India, for his constant support and encouragement to this important National Initiative on Climate Resilient Agriculture (NICRA) project under which the study was performed. The authors also express sincere gratitude to Dr. A. K. Singh, Deputy Director General (NRM), and Dr. J. C. Dagar, ADG (Agronomy and Agroforestry) for their support and guidance.

References

Manorama K, Ravichandra G, Joseph TA (2007) Rainfall analysis and crop planning for the Nilgiris. J Agrometeorol 9:209–215

Veeraputhiran R, Karthikeyan R, Geetha Lakshmi V, Selvaraju R, Sunder Singh SD, Balasubramanian TN (2003) Crop planning climate Atlas. A E Publications, Coimbatore

Website

India-WRIS wiki: http://india-wris.nrsc.gov.in/wrpinfo/index.php?title=Mahanadi#Description

Role of ICTs in Sustainable Agriculture: A Study of *e*-Sagu in Andhra Pradesh

Naveen Kumar Cheripelly and Raghava Reddy Chandri

Abstract

Indian agriculture, for long, was characterized as subsistence farming, carried out by small and marginal farmers adopting primitive techniques and thus incurring low yields. Importantly, Indian agriculture has also been negotiating the difficulties in food grain production. A series of changes in agricultural technology introduced in the 1960s, referred to as green revolution technology, helped India overcome the problems of food grain production. Green revolution technology not only offered new, high yielding seeds but also fertilizers which replaced organic manures and nitrogen-fixing crops, pesticides replacing biological, cultural, and mechanical methods for controlling pests. With the introduction of green revolution, the structure of Indian agriculture also underwent changes, bringing agriculture into the fold of capitalism.

Without having access to credible advice, the farmers have been adopting unscientific cultivation practices mostly on the advice of the dealers of pesticides and seeds. As a result, farmers have been relying on wrong, untimely, unnecessary information leading to losses. The net result is very devastating for farmers, environment, and public health. For farmers, they are facing severe crisis due to reduced crop output or crop failure. The task of making agriculture sustainable is of no less importance to India than to any other country. Emerging technologies need to be profitable and ecologically sound. Half of India's total land area is estimated to suffer from problems of degradation on account of unsustainable practices like excess use of chemical fertilizers, pesticides, herbicides, etc.

Sustainability agriculture in India is achievable when the farmers are supplied with adequate, appropriate, accurate and timely information. Agricultural information has the key role in facilitating the participation of people relating to sustainable development. Several attempts have been

N.K. Cheripelly • R.R. Chandri (✉)
Department of Sociology, University of Hyderabad,
Gachibowli, Hyderabad, Telangana, India
e-mail: craghava@gmail.com

A.K. Singh et al. (eds.), *Climate Change Modelling, Planning and Policy for Agriculture*,
DOI 10.1007/978-81-322-2157-9_13, © Springer India 2015

made in the dissemination of agricultural information to the farmers using the advances in information and communication technologies (ICTs), one such attempt is e-Sagu, which uses ICTs in providing continuous, relevant, and latest technological information to farmers through computer and the Internet in Andhra Pradesh.

Keywords
Information communication and technology • Sustainable Agriculture

Part I: Introduction

Agriculture can serve as an important engine for economic growth in developing countries, yet yields in these countries have lagged far behind those in developed countries for decades. One potential mechanism for increasing yields is the use of improved agricultural technologies, such as fertilizers, seeds, and cropping techniques. Public-sector programs have attempted to overcome information-related barriers to technological adoption by providing agricultural extension services (Aker 2011). Despite a wide range of reform initiatives in agricultural extension in India in the past decades, the coverage of, access to, and quality of information provided to marginalized and poor farmers is uneven. While such programs have been widely criticized for their limited scale, sustainability, and impact, the rapid spread of mobile phone coverage in developing countries provides a unique opportunity to facilitate technological adoption via information and communication technology (ICT)-based extension programs.

Recent developments in information and communication technology (ICT) offer exciting new opportunities for rapid and efficient information transfer including transborder data flow through electronic systems such as the Internet. The versatile application of ICT in diverse fields including agriculture is in fact an outcome of its special capabilities in terms of speed, accuracy, consistency, and storage capability. In today's fast-changing scenario, the Indian agriculturist cannot afford to ignore the use of ICT for transfer of information. In the last decade, many ICT projects in Indian agriculture have emerged, which either substitute or support extension services by enabling farmers' access to information. Recent global food price increases and high levels of inflation have provided an opportunity to increase farmers' profitability. However, to realize the benefit of higher prices, farmers need to access a wider range of information, related not only to production technologies but also to post-harvest processes, access to remunerative markets, price information, and business development (Sulaiman and Van den Ban 2003). This information could be integrated with services that support the use of the information. Information that is context-specific to farmers' local situations can have important farm outcomes.

The paper examines the challenges and constraints of each agricultural extension approach as it attempts to provide farmers with access to information that is relevant to their farm enterprises. The present paper reviews the issues associated with integration of ICT and agriculture as well as the novel developments that are evolving in agricultural information dissemination. It is identified that extension service should reach smallholder farmers for sustainable agriculture. This paper concludes that there is an increasing need to work in partnership and to share knowledge and skills in order to provide locally relevant services that meet the information needs of marginal and smallholder farmers in India.

Research Problem

The economic performance of the agricultural sector in most countries has been largely determined by the organized research and extension activity. The agricultural research system in India is a large and complex one. After the introduction of seeds-fertilizer-irrigation intensive agriculture of green revolution, information became more critical for agriculture. Due to several reasons the current agricultural extension system in India is unable to deliver the advice to the farming community in an efficient manner. In this context the farming community started relying on private seed, fertilizer and pesticide companies, and local traders for agricultural information.

More than 81 % of Indian farmers cultivate an area of 2 ha or less (India, Directorate of Economics and Statistics 2009; NSSO 2006). There is an increasing need for stronger intermediaries that can facilitate information access for diverse smallholder farmers. The 2003 National Sample Survey Organization (NSSO) survey showed that 60 % of farmers had not accessed any source of information on modern technology to assist in their farming practices in the past year. Of those who had sourced information, 16 % received it from other progressive farmers, followed by input dealers. Of those farmers who had accessed information, the major problem of extension services was found to be the practical relevance of the advice (NSSO 2005). The coverage and relevance of information provided to farmers through the agricultural extension system is therefore questionable. While this may be partly due to inadequate contact by the services, which need to reach a large and complex farming community, inappropriate or poor-quality information could also be a key hindrance to the farmers' use of extension services. In other words, the content of the information provided by agricultural extension approaches, and the information farmers actually need, may not be aligned. Advances in production technologies, particularly with relation to inputs such as seeds, fertilizers, pesticides, etc., though increased productivity failed to improve net earnings for these group of farmers. The reason for such failure is attributed to the inadequate agricultural information at the farmer level. Therefore, there is a need to reexamine the current agricultural extension approaches in India to understand where information gaps exist and determine why farmers are not accessing information through the large, well-established public-sector extension system in addition to emerging private and third-sector actors.

The diverse nature of the Indian subcontinent, with its wide variety of agroclimatic regions and broad range of socioeconomic conditions in the rural population, calls for agricultural extension approaches that are context and situation specific. Further progress in poverty and hunger reduction crucially depends on the increased productivity and profitability of these farmers, which in turn depends on the successful delivery of agricultural extension.

In this context, advances in information and communication technologies attracted the attention of researchers working in agricultural communication dissemination. It was believed that an expanding assembly of ICTs can be used to collect, store, and share information between farmers and scientists. One such attempt has been made in Andhra Pradesh, named e-Sagu, to provide ICT platform for agricultural communication between the farmers and the scientists for the exchange of information, guidance, and suggestions.

Challenges of Traditional Agricultural Extension System

The most significant shortcomings of public agricultural extension in general have been (a) unresponsiveness to the variation in farmers' needs, (b) lack of ownership by the intended beneficiaries, (c) failure to reach poor and women farmers, (d) limitations in the quality of field and technical staff, and (e) high and unsustainable public costs. Some of these problems have been eased by modifying the training and visit system, for example, by working with groups rather than individual farmers or by increasing reliance on radio and other mass media. Several emerging challenges confront Indian farmers. These include limited land and water availability, which is further exacerbated by degradation of natural resources; climate changes; changes in demand and consumption patterns, moving toward high-value agriculture; increasing population pressure; and liberalization of trade (Lele 2010).

ICTs in agriculture enable two-way communication between scientists and farmers via multiple media, particularly, the Internet. In the past, farmers used traditional knowledge and practices to optimize output and were experts in their own right, as one knew the systems well and also had an intimate knowledge of their own land, climate, and other inputs. However, with the advent of modern practices in agriculture, commercialization of agriculture, and modern technology-derived inputs such as Bt seeds, herbicides, etc., farmer's knowledge proved to be inadequate. The existing conventional models of technology transfer have also found to be of little use. There is a growing technology-knowledge gap.

Structure and Function of e-Sagu

A personalized agricultural advisory system called e-Sagu was developed to improve the performance and utilization of agricultural technology and help farmers by providing personalized information at their doorstep or farm level. It is a Web-based personalized agro-advisory system, which uses information and communication technologies to provide scientific solutions to the farmers. "Sagu" means cultivation in Telugu local language. e-Sagu is a personal agricultural advisory system in which the farmer receives specific expert advice on the agricultural problems for each and at regular intervals. It was developed in 2004 using ICTs in agricultural extension activities.

The project was started in the year 2004 by delivering agro-advisory to 1,051 cotton farms for the farmers of three villages of Andhra Pradesh. In the year 2005–2006, e-Sagu was scaled up to cover 4,894 farms in six districts of Andhra Pradesh in diverse agri-ecosystems. As part of this project, a nominal subscription fee in the form of registration fee is charged to the farmers who are interested in receiving the information. e-Sagu was developed based on a prototype experiment which was conducted for 160 fishponds in West Godavari District of Andhra Pradesh.

The expert advice is prepared by agricultural experts based on the crop status information received in the form of both digital photographs and text. Farmers trust and rely on the ideas and practices developed through scientific research. e-Sagu aims at improving the farm productivity by delivering high-quality personalized (farm specific) agro-expert advice in a timely manner to each farm at the farmers' doorstep without the farmer asking question. In e-Sagu, ICTs such as the Internet, digital photography, and database are extended to improve the performance of agricultural extension services.

Agricultural Extension Through e-Sagu

The agricultural information system of e-Sagu comprises of four key partners, namely, farmer, agricultural expert, coordinator, and computer operator. This social system is embedded into the technical system which comprises of computer, internet connectivity and digital camera. Farmers register with the system by supplying relevant information about their farms (soil data, water resources, capital availability, etc). This piece of information is collected by the coordinator and digitized and fed into the system by the computer operator at the local center.

The system functions as follows:

- The coordinator visits the farms of those farmers who are registered with the e-Sagu. The coordinator visits periodically to gather up-to-date information and take digital photographs of the crops and farm. The farmer is given prior information about the visits of the coordinator. If the farmer is not present during the visit of the coordinator, then the coordinator visits his/her home or his/her family members to take the information and also to give the advice. In case of multiple farms of a single farmer, the coordinator visits each farm separately according to schedule.

- The photographs, which were taken by the coordinator, will be given to the computer operator. The coordinator takes the photos of the farm and also takes the photos of the notes prepared by him.
- The computer operator at the local center digitizes the photos and the texts and feeds into the computer. This information is properly labeled and sent to the main lab located in Hyderabad through the Internet. If the Internet connection is not available (normally local level bandwidth is not sufficient to send the photographs through the Internet), the computer operator prepares a CD for the photographs and other information of the farmer. The CD is then sent by parcel service to the main lab.
- In the main lab by accessing specific data of each farm, crop database, and the information sent by the coordinators, the agricultural expert prepares personalized advices and stores them in the system.
- The computer operator accesses information in the system through the Internet and downloads the advices to be disseminated to the farmer of their respective regions. As the advices are text messages, even with very little bandwidth, it is possible for the computer operator to download at the local center. The advices are the text messages and are in the vernacular/English language. The computer operator passes this information to the coordinator.
- The coordinator then visits the farmer to explain the advice in the local language. The coordinator also collects feedback about the advice.
- The information gathered in the diagnostic process must be feedback into the information network if it is to have positive effect (Fig. 1).

Farmers are the first feedback source in this system of extension. The coordinator then acts as the farmer's representative and advocates by bringing information to the researchers and policy makers. The key to effective feedback is a system that values bottom-up input into continuous process of adapting itself to the needs and evolution of its clientele, i.e., farmers. The coordinator is the fundamental link of the system. He/she assures the contact with farmer not only in collecting information but also in disseminating the advices. He/she also facilitates advice appropriation by explaining them in the farmers' language.

Farmers' confidence in any extension system depends greatly on the consistency and predictability of farm visits by the concerned personnel. Any extension program can gain the confidence of the farmers only through proper interaction. The key to agricultural information transfer in e-Sagu is the regular contact with farmers. The coordinator is mandated to visit farmers as per the schedule announced in advance. The farmers' confidence in the extension system depends greatly on the consistency and predictability of farm visits (Fig. 2).

e-Sagu functions at two key levels: one, at the local level, i.e., at the village level, and another, at the nodal level located at Hyderabad. The nodal level is known as e-Sagu lab. The local center is the one where the project is

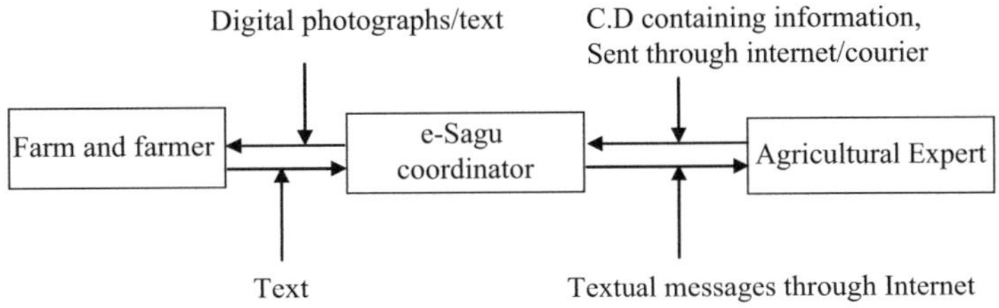

Fig. 1 Agricultural extension through e-Sagu

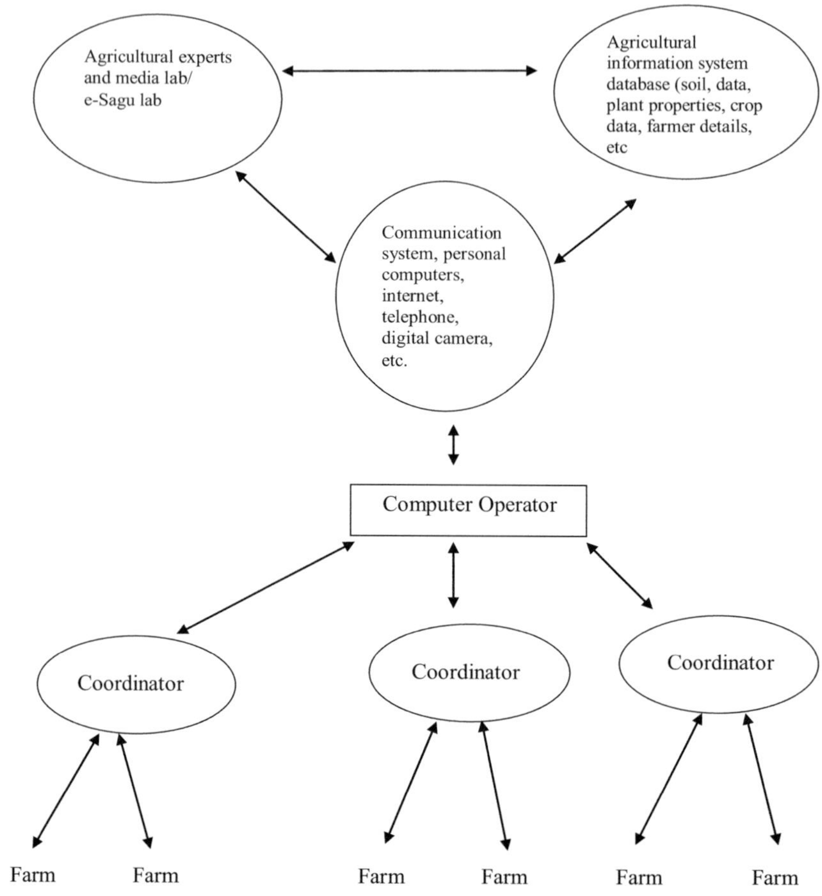

Fig. 2 Agricultural information exchange through e-Sagu

implemented. It is staffed by three members. They are the administrator, computer operator, and coordinators. Each local center is equipped with two computers and other related equipments (Fig 3 and Table 1).

With an aim to understand the functioning of e-Sagu at the village level and to know the perceptions of the registered farmers, the researcher interacted with select respondents in Malkapur village located in Warangal district. The village was selected because it was the first village in the District where e-Sagu was introduced. e-Sagu has been functioning in the village for the past 5 years. About 50 % of the farmers who have registered for the 2008–2009 season were selected as respondents. The sample was selected based on purposive sampling. The total number of farmers registered in the village in e-Sagu for the year 2008–2009 was 120. Most of them are small and marginal farmers, i.e., holding less than 5 acres of land. The researcher, using a semi-structured questionnaire, collected data from about 60 farmers.

Part II: Field Study

Caste

Caste is an important parameter in understanding the implication of any technological applications in agriculture. Out of 60 respondents, a majority

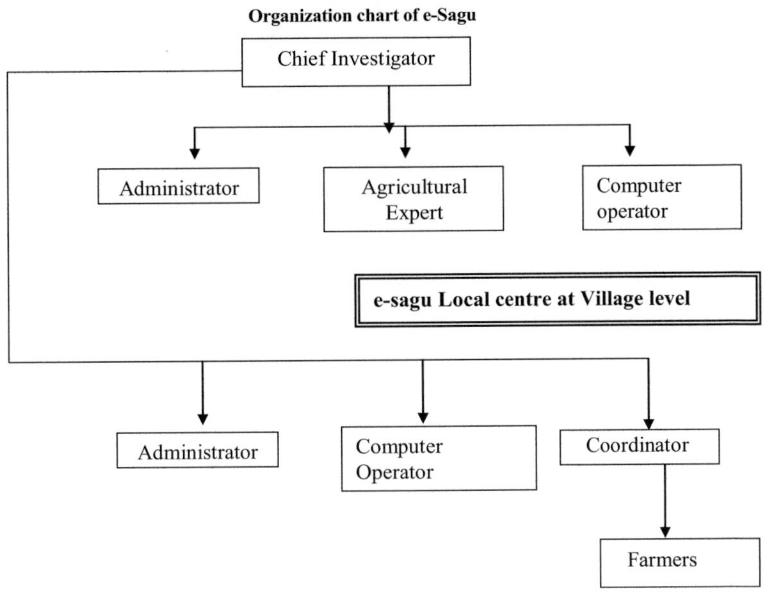

Fig. 3 Structure and functioning of e-Sagu

Table 1 A typical table showing identity of farmers maintained in the database

S. No	Name of the farmer	Farmer ID	Date of registration
1	Krishna	ap_war_gha_mal_001_jul08_cott_a	7/07/08
2	Narsaiah	ap_war_gha_mal_010_oct07_chil_a	6/09/07

Table 2 Caste-wise distribution of the respondents

S. No	Caste	Total	%
1	O.C	24	40
2	B.C	22	37
3	S.C	14	23
Total		60	100

of them (40 %) belong to upper castes. Thirty-seven percent were from backward castes and 23 % were from scheduled castes (Table 2).

Landholding Pattern

Landholding pattern among the respondents clearly indicates that a majority of them hold less than 5 acres of land. There are about 40 % of the respondents who hold less than 5 acres of land, 33 % of respondents hold 6–10 acres of land, only 17 % of respondents hold 11–15 acres of land, and only 10 % of respondents hold 15–20 acres of land.

Although caste of the respondents does not allow any significant inference on the use of ICTs in agriculture, as the difference of percentage between different castes is not much (in terms of number of respondents), particularly between the upper caste and the backward caste, the difference in the landholding pattern across different castes, however, presents a clear trend. Table 3 shows that most of the farmers in the village are small and marginal. Only a few are large farmers. If we observe the caste-wise land distribution, 100 % of the respondents who own less than 5 acres of land belong to schedule castes (14). Among the backward castes and upper castes, only 27 and 17 % of respondents owned a land between 1 and 5 acres. It shows that all the scheduled caste respondents own less than 5 acres of land. The landholding pattern among upper castes is that 38 % of the respondents

Table 3 Landholding pattern

	O.C		B.C		S.C		Total	
Acres of land	No of farmers	%	No. of farmers	%	No. of farmers	%	No. of farmers	%
1–5	04	17	06	27	14	100	24	40
6–10	09	38	11	50	–	–	20	33
11–15	08	33	02	09	–		10	17
16–20	03	12	03	14	–	–	06	10
Total	24	100	22	100	14	100	60	100

Table 4 Literacy levels among the respondents

	O.C		B.C		S.C		Total	
Education level	No. of farmers	%	No. of farmers	%	No. of farmers	%	No. of farmers	%
No formal education	2	8	6	28	10	72	18	30
Primary (1–5th)	7	30	5	23	0	–	12	20
Secondary	4	17	4	18	1	7	09	15
10 + 2	06	25	06	27	3	21	15	25
Graduation	03	12	01	4	–	–	04	7
Postgraduation	02	8	–	–	–	–	02	3
Total	24	100	22	100	14	100	60	100

owned a land up to 10 acres and 33 % owned a land up to 15 acres. Only 12 % own more than 15 acres of land. Among the backward caste respondents, a majority of them (50 %) own a land up to 10 acres, whereas no scheduled caste respondent owns a land beyond 5 acres.

Educational Background

Table **4** shows that a majority of the respondents (about 30 %) are illiterate. Only 20 % of the respondents studied up to primary level and 15 % have completed secondary education. While 25 % of the respondents have studied up to 10 + 2 level, only 7 % have completed graduation. It may be observed that only 3 % of the respondents have studied up to postgraduation.

Table **4** shows that a majority of the respondents who had no formal education (about 72 %) belong to scheduled caste. Backward caste respondents who had no formal education account to 28 %, while it is only 8 % among the upper caste respondents. Majority of the upper caste respondents (30 %) have studied up to primary level, i.e., up to the 5th standard. However, it is also significant to note that about 25, 27, and 21 % of the respondents from upper castes, backward castes, and scheduled castes, respectively, had studied up to intermediate standard, i.e., 10 + 2 level. Only among the upper caste respondents we find respondents (about 8 %) who have studied beyond graduation.

Cropping Pattern

Major crops grown in the region are paddy, maize, cotton, chili, and other cereal crops. Traditionally, paddy has been the preferred crop in farms which have irrigation facility. In the recent past, there has been a growing trend among the farmers of the region to cultivate commercial crops such as cotton and chili. Cultivation of cotton crop is not just confined to the large landholders. Even the small landholders have started cultivating cotton, as it is believed that cotton fetches good net returns when compared to other crops. The same trend is exhibited in the responses of the farmers.

Irrespective of the landholding, all the respondents cultivate cotton. If a small farmer has two acres of land, he/she has started

Table 5 Cropping pattern

Crops	No. of farmers	%
Cotton	60	100
Chili	45	75
Maize	40	67
Paddy	30	50

It is a common practice that farmers, including the small and marginal landholders, grow multiple crops. Hence, the percentage column total exceeds 100 %

Table 6 Sources of agricultural information

Sources of information	No. of farmers	%
Agriculture officer	Nil	Nil
Neighbors	28	47
Dealers	19	32
Magazines	06	10
Electronic media (TV, radio)	07	11
	60	100

cultivating cotton in at least one acre of land. Table 5 shows that all the respondents cultivate cotton crop. Next to cotton, the most preferred crop by the respondents was chili (about 75 %). The low percentage of paddy cultivation may be explained in terms of nonavailability of irrigation facilities for all farms, particularly, small farms owned by scheduled caste respondents.

Sources of Information

Agricultural information in the age of increasing commercial crop cultivation is an important input. This is because commercial crop cultivation like cotton and chili requires intensive use of external inputs like seeds, pesticides, fungicides, and fertilizers. The importance for agricultural information also increased because of the fact that even the small landholders have started commercial crop cultivation. As presented in Table 5, all the respondents, irrespective of small or large landholders, cultivate cotton. This necessitates authentic and timely agricultural information.

To know the sources from where farmers get agricultural information, the researcher collected data on the sources of information. It was revealed that a majority of the respondents get agricultural information from their neighbors who include fellow farmers and relatives. Significantly no respondent mentioned that he/she gets information from the agricultural officer. An agricultural officer is the formal source of information at the village and mandal level.

Table 6 shows that a majority of the respondents get information from their neighbors (about 47 % of the respondents) which has been proved to be inappropriate and unscientific. About 32 % of the respondents get agricultural information from an input dealer which is again unreliable and tied to commercial interests. About 21 % of the respondents get agricultural information through magazines and electronic media, but the information that they gained from the magazines and electronic media is not followed as they are not sure of its practice and suitability.

Conclusion

From the above data it may be inferred that a majority of the respondents cultivate commercial crops like cotton and chili. At the same time, a majority of the respondents' landholdings are small (less than 5 acres), particularly that of scheduled caste. It may also be inferred from the data presented above that a majority of the scheduled caste respondents had no formal education. It is also an important observation that a majority of the respondents get agricultural information from neighbors and dealers who sell the agricultural inputs. Dealers who sell the agricultural inputs often misguide or provide inappropriate information which may lead to loss of crop. There are instances reported by the respondents, like farmers spraying pesticides to control diseases or mixing two or more noncompatible pesticides to control a pest. Usage of high doses of pesticide, than what is suggested, to control a pest is also a common practice. In this context farmers were in need of appropriate, authentic, reliable, and timely information. However, the traditional extension sources (like agricultural officer) were not able to provide such information. Given these conditions ICT-based extension service, i.e., e-Sagu, has been an attempt to provide timely and scientific advice to the farmers at their doorstep. Hence, it was reported that most of the farmers in the village have registered in this project and avoided severe financial loses. It was also reported that the income gained by the farmers who registered with e-Sagu was about Rs. 3,820 per acre. This amount is a substantial amount, particularly for a small landholder. By following timely and scientific advices, farmers avoid untimely use of pesticides and overuse of fertilizers. Such types of extension services are very much useful to the small and marginal farmers and also to the farmers who have no formal education.

Important to note here is that the respondents (farmers who were registered with e-Sagu) were satisfied with the information delivered by the agricultural experts. A majority of the respondents, who are smallholders and cultivators of commercial crops, opined that they have benefited to a great extent by seeking advices through e-Sagu. The study, however, observer that, as the coordinator at the local center becomes key in the entire information flow, he/she working for personal gains and excluding the small and scheduled caste farmers may be possible. The traditional stereotypes based on caste and landholdings may often influence the functioning of the coordinator.

It is important to note that e-Sagu has been functioning out of the purview of the conventional agricultural extension system. However, the conventional agricultural extension administration is well entrenched up to the mandal level in the state. By integrating e-Sagu with the conventional agricultural extension, the benefits of ICT applications in agricultural extension may be expanded to the entire state and to all the crops. This may also result in the reduction of cost of extension service for which the farmers are paying as registration fee to e-Sagu.

References

Aker CJ (2011) Dial "A" for agriculture: a review of information and communication technologies for agricultural extension in developing countries. Int Assoc Agric Econ 42(6):631–647

Lele U (2010) Transforming agricultural research for development. Paper presented at the global conference on agricultural research for development, Montpellier, France

NSSO (National Sample Survey Organization) (2005) Situation assessment survey of farmers: access to modern technology for farming, 59th round (January–December 2003). Report No. 499(59/33/2). Ministry of Statistics and Programme Implementation, New Delhi

NSSO (National Sample Survey Organization) (2006) Some aspects of operational land holdings in India, 2002–2003. Ministry of Statistics and Programme Implementation, New Delhi

Sulaiman R, Van den Ban (2003) Agriculture extension: involvement of private sector. Occasional paper 29. Department of Economic Analysis and Research, Mumbai

Carbon Pools and Fluxes in Grassland Systems on Sodic Soils of Northern India

S.R. Gupta, R. Jangra, and J.C. Dagar

Abstract

This study analyses vegetation carbon pools and fluxes and soil carbon storage in grassland systems on sodic soils in northern India. The grassland systems on highly sodic soils show low species diversity and single-species dominance. The plant species composition and soil conditions influence above-ground and belowground carbon pools and fluxes in different grassland systems along a range of soil pH from 8.0 to 10.2. The soil organic carbon content is low ranging from 3.42 to 0.51 g kg^{-1} across soil depths. The carbon pool (Mg C ha^{-1}) in the primary producer compartment of the grassland ecosystems at Bichian was 4.945–1.721 above-ground biomass and 4.336–1.40 belowground biomass. The carbon flux through total net primary productivity ranged from 0.954 to 0.375 Mg C ha^{-1} $year^{-1}$. The organic carbon storage (up to 1-m soil depth) in soils of the natural grassland ecosystems at Bichian was 24.713–16.649 Mg C ha^{-1} over a period of 15 years of vegetation protection. By integrating trees with the naturally occurring grassland systems on highly sodic soils at Bichian, the soil organic carbon content increased by 15–57 %. After long-term protection of grassland vegetation on a sodic soil at Karnal, the soil carbon pool in 0–30-cm soil depth was 6.683 Mg C ha^{-1}(year 1982) and 13.91 Mg C ha^{-1} (year 2006). The microaggregates (250, 53 and <53 μm) formed a large fraction of soil aggregates and protected most of the soil organic carbon. In the biologically reclaimed sodic soil at Karnal, the total carbon storage in soil (SOC + SIC) up to 1-m soil depth was 89.511 Mg C ha^{-1}. Thus, the protection of native grassland vegetation on sodic soils has the potential for carbon sequestration by increasing plant biomass production and improving soil

S.R. Gupta (✉) • R. Jangra
Department of Botany, Kurukshetra University,
Kurukshetra 136119, Haryana, India
e-mail: sgupta2002@yahoo.com

J.C. Dagar
Emeritus Scientist, Central Soil Salinity Research
Institute, Karnal, India

A.K. Singh et al. (eds.), *Climate Change Modelling, Planning and Policy for Agriculture*,
DOI 10.1007/978-81-322-2157-9_14, © Springer India 2015

organic matter. Implementing practices to build up soil carbon stocks in grasslands could lead to considerable mitigation, adaptation and development benefits.

Keyword
Carbon • Grassland • Soil carbon • Microbial biomass carbon • Soil aggregates

Introduction

Soil salinity and sodicity are serious land degradation problems in arid and semiarid regions of the world. The worldwide extent of salt-affected soils is about 932.2 million hectares (Szabolcs 1989). The excessive irrigation in agriculture has mainly contributed to the increasing problems of secondary salinisation, alkalisation and waterlogging (Szabolcs 1994; Rengasamy 2006; Qadir et al. 2007). As a result of inappropriate irrigation management, salinisation and sodification of land continue to occur at an estimated rate of between 0.25 and 0.5 each year (FAO 2000). In India, 9.38 million hectares of land is salt affected, out of which 3.88 million hectares are alkali/sodic soils (IAB 2000). The maintenance of adequate levels of soil organic matter of salt lands is important to improve biological productivity and ecosystem sustainability under resource-limiting conditions. Restoration and management of grasslands on marginal soils can help in the mitigation of rising atmospheric carbon dioxide concentrations via carbon storage in the plant–soil system. Estimates of soil C storage and rates of C sequestration can be useful to estimate the potential of grasslands to help mitigate the elevated atmospheric levels of CO_2 (Lal 2004). The opportunities to benefit from grassland practices that sequester carbon can be substantial as the large populations of people who depend directly on grasslands on marginal soils tend to be poor and prone to climate variability and climate change.

In the recent past, the increase in carbon dioxide in the atmosphere has gained a lot of attention as a greenhouse gas, as it has the potential to influence the climate pattern of the world. The increase in CO_2 concentration in the atmosphere has largely been attributed to fossil fuel burning, deforestation and change in land use. For the first time in human history, the concentration of climate-warming carbon dioxide in the atmosphere has passed the milestone level of 400 parts per million (ppm) in the year 2013. The improvement of soil carbon in grasslands offers a global greenhouse gas mitigation potential of 810 Mt of CO_2 (in the period up to 2030), almost all of which would be sequestered in the soil (Conant et al. 2001; Ravindranath and Ostwald 2008). Management practices that favour carbon sequestration in grasslands also tend to enhance resilience in the face of climate variability and are thus likely to enhance longer-term adaptation to changing climates (FAO 2010). Soil carbon sequestration is an important factor in the greenhouse gas emission balance and is strongly related to site conditions (e.g. soil structure, initial soil carbon content, climate), structure of agroforestry and soil management (Montagnini and Nair 2004; Nair et al. 2009).

Soil carbon sequestration is the build-up of carbon in the entire soil profile through accumulation of organic matter from plant, fungal (and other microbial) and animal origins. Plant and microbial residues represent the major sources of carbon input into the soil, which ultimately lead to the formation of soil organic matter. Soil organic carbon storage is the most accepted method for long-term carbon sequestration in terrestrial ecosystems. In sodic soils, soil organic carbon can be rapidly lost due to disintegration of soil aggregates and desorption of carbon from clays (Wong et al. 2008). Soil aggregates and size fractions are known to affect carbon storage in soil (Six et al. 2004). Soil aggregate stability is important for soil carbon sequestration (Chivenge et al. 2011). Clay minerals strongly influence the major physical and chemical properties of soil as well as soil organic matter and its chemical nature. The soil carbon sequestration depends on clay contents and mineralogy, structural stability, moisture and temperature regimes and formation of soil aggregates (Lal 2004). Clay mineralogy is important in determining the quantity of organic carbon stored in soil, its turnover time and atmosphere–ecosystem carbon fluxes during long-term soil development (Torn et al. 1997; Laird 2001).

This chapter gives an overview of vegetation carbon pools and fluxes, soil microbial activity and diversity and soil carbon sequestration in grassland systems on salt-affected soils. It is also aimed to discuss the role of soil aggregates and clay mineralogy in soil carbon sequestration and stability.

Problem of Salt-Affected Soils

Depending upon the soil's physical and chemical nature, salt-affected soils are mainly of three types, i.e. saline, alkaline/sodic and saline/sodic soil. The saline soils have white encrustations on the surface and have high concentrations of soluble chlorides and sulphates of sodium, calcium and magnesium as dominant salts. These soils have pH below 8.2 and electrical conductivity greater than 4 dSm^{-1} at 25 °C and sodium absorption ratio of the soil solution <15. Saline soils usually remain flocculated due to the presence of excess salts and have high osmotic pressure of soil solutions which induces physiological drought and tissue injury due to direct toxic effects of individual ions and complex interactions between sodium, calcium and magnesium. A large area of canal-irrigated tracts of arid and semiarid regions has been adversely affected from waterlogging and salinity.

The alkali soils contain excess of salts capable of alkaline hydrolysis such as sodium carbonate and sodium bicarbonate ($NaHCO_3$) and sufficient exchangeable sodium on the cation exchange sites in the soil. The sodic soils are characterized by high soil pH [(saturation soil paste pH > 8.5 and often approaching 11, exchangeable sodium percentage (ESP) > 15 and varying electrical conductivity (ECe less than 4 dSm)]. The presence of high exchangeable sodium percentage in soils imparts poor physical conditions to soils, low infiltration of water and dispersion of soil organic matter. The precipitation of calcium in alkali soils causes deposition of thick $CaCO_3$ layer known as kankar pan. The adverse climatic conditions in arid and semiarid regions induce the precipitation of $CaCO_3$. High soil pH has been reported to adversely affect soil structure, physical and chemical properties and plant establishment (Qadir et al. 2002; Smith et al 2009).

The sodic soil formation includes processes like clay illuviation, deposition of pedogenic $CaCO_3$ and resultant development of subsoil sodicity (Pal et al. 2003a, b). The adverse climatic conditions induce precipitation of $CaCO_3$, thereby depriving the soils of Ca^{2+} ions and developing sodicity in the subsoils. This leads to the formation of sodic soils with exchangeable sodium percentage decreasing with depth (Pal et al. 2009). It also impairs the hydraulic conductivity of soils and eventually leads to the formation of sodium-rich soils (Pal et al. 2003a, b).

Rengasamy and Sumner (1998) have discussed various mechanisms for soil degradation under sodic conditions. Soil degradation under sodic conditions occurs due to hydration of dry aggregates, slaking and swelling of wet aggregates and dispersion of clay platelets from soil aggregates (Rengasamy and Sumner 1998; Cass and Sumner 1982).

Vegetation Composition and Species Diversity

Site characteristics of different sodic grassland systems on sodic soils in northern India are summarised in Table 1. The grassland systems located at Karnal, Kurukshetra and Bichian Saraswati Reserved Forest occur over a soil pH range of 8.8–10.03 (Fig. 1). The sodic soils support sparse growth of vegetation of salt-tolerant grasses. *Desmostachya bipinnata, Sporobolus*

Table 1 Site characteristics of different sodic grassland systems on sodic soils in northern India

Location	Year	MAT (°C)	Annual rainfall (mm)	Soil pH	Organic carbon (%)	Grassland type
Kurukshetra[a]	1975–1976	23.73	803	8.8	0.67	*Desmostachya bipinnata*
Kurukshetra[a]	1975–1976	23.73	803	9.0	0.44	Mixed grass
Karnal[b]	1981–1982	26.93	956	9.6	0.23	*Sporobolus marginatus*
Karnal[b]	1981–1982	26.93	956	9.3	0.20	*Desmostachya bipinnata*
Karnal[c]	2005–2006	24.11	506	8.24	0.65	*Desmostachya bipinnata*
Bichian[d]	2006–2007	_	362	10.03	0.26	*Sporobolus marginatus*
Bichian[d]	2006–2007	_	362	9.29	0.34	*Desmostachya bipinnata*

[a]Gupta and Singh (1982)
[b]Gupta et. al. (1990)
[c]Jangra (2009)
[d]Jangra et al. (2010)

marginatus and *Diplachne fusca* have been reported to grow in sodic soils of the Indo-Gangetic plains (Rana and Parkash 1987; Gupta and Singh 1982; Gupta et al. 1990; Neeraj et al. 2004). The sodic soils have a narrow spectrum of flora and are characterised by low species diversity and single-species dominance (Sinha et al. 1988). The vegetation of sodic soils is generally comprised of perennial grasses and the flora is highly restricted in distribution (Rana and Parkash 1987). For the different sodic grassland communities (pH ranging from 8.9 to 10.10), species richness, species diversity and species evenness have been found to decrease with increasing sodicity (Sinha et al. 1988; Neeraj et al. 2004).

The protected grassland system at the Central Soil Salinity Research Institute, Karnal, is an ecologically rehabilitated sodic soil in northwestern India. This grassland system occurring on a sodic soil has been protected since 1970 and is mainly dominated by *Desmostachya bipinnata* The grassland community of the protected grassland system at CSSRI, Karnal, during 1982 was highly sodic and characterised by low species diversity (Gupta et al. 1990). The relative density of plant species of the protected grassland during the years 1982 and 2006 based on the data of Gupta et al. (1990) and Jangra (2009) is compared in Fig. 2. The grassland system due to long-term protection of the naturally occurring vegetation on a sodic soil showed moderate species diversity; there was increase in the number of forbs, annual grasses and some shrub species in the sodic grassland of biologically reclaimed soil.

Dominance of plant species in relation to their availability of suitable niche and resource apportionment in a community has often been interpreted from the dominance–diversity curve. There was low diversity of plant species on highly sodic soils at Bichian (Fig. 3). There was a log-normal distribution of species, the curve being steeper for the *Sporobolus marginatus* community as compared to the *Desmostachya bipinnata* grassland system (Fig. 3). The values for Shannon's diversity index (H) varied from 1.107 to 1.596 and Pielou's index was 0.596–0.606 for the two grassland systems at Bichian in Saraswati Reserved Forest.

Changes in Soil Properties

The soils of the sodic grassland at Bichian were highly alkaline; a calcic hardpan at 80–120-cm depth has been found to be common on these soils resulting in impeded drainage. Soil carbon content varied from 0.10 to 0.16 % and the soil pH from 10.0 to 10.2 in surface layer of soil in the grassland systems during the year 1991 (Kaur et al. 2002a). During March 2006 to February 2007, the soil pH varied from 9.29 to 10.23 in the soil profile from 0- to 100-cm soil depth. Soil organic carbon of the two grassland sites ranged from 3.42 to 0.51 g kg^{-1} across soil depths

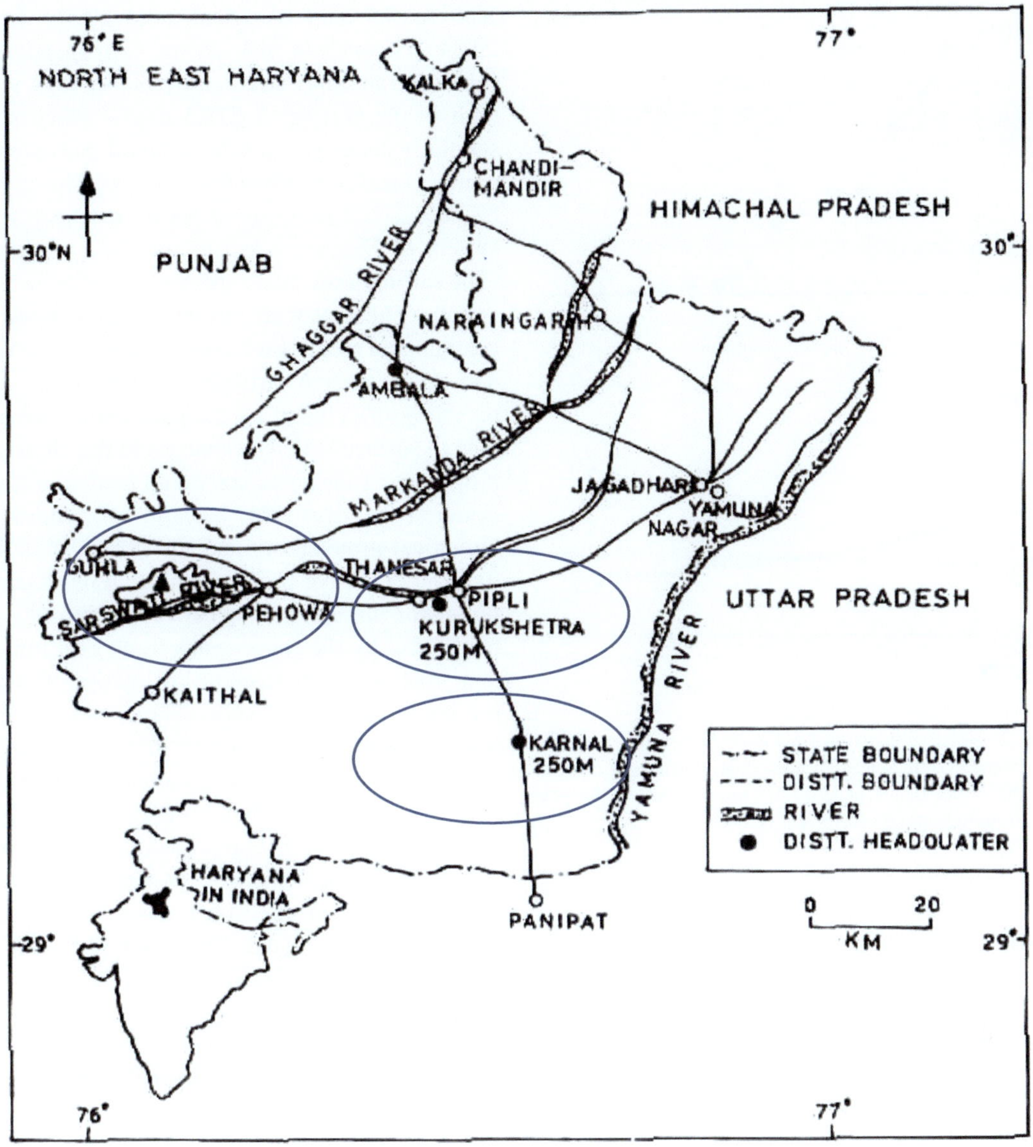

Fig. 1 Location of sodic grassland sites at Kurukshetra, Karnal and Bichian Saraswati Reserved Forest in northern Haryana

(Table 2). The protection of the grassland system over a long period of 15 years resulted in significant increase of soil carbon and biological amelioration of sodic soils; the increase in soil carbon occurred from 0.20 % (1991) to 0.44 % (2006) (Jangra et al. 2010). The carbon fixed by the plants is the primary source of organic matter input into the soil, which provides substrate for microbial processes and accumulation of soil organic matter. Thus, belowground allocation of photosynthates was found to be an important contributing factor for improving soil organic carbon.

The soil of the grassland site at Karnal is typic salonatric calciorthids characterised by deep soil profile and the soil graded from sandy loam in the surface to clay loam in the subsurface horizon (Bhumbla et al. 1970). The soil pH varied from 10.6 to 10.2 across soil depths (0–48-cm soil

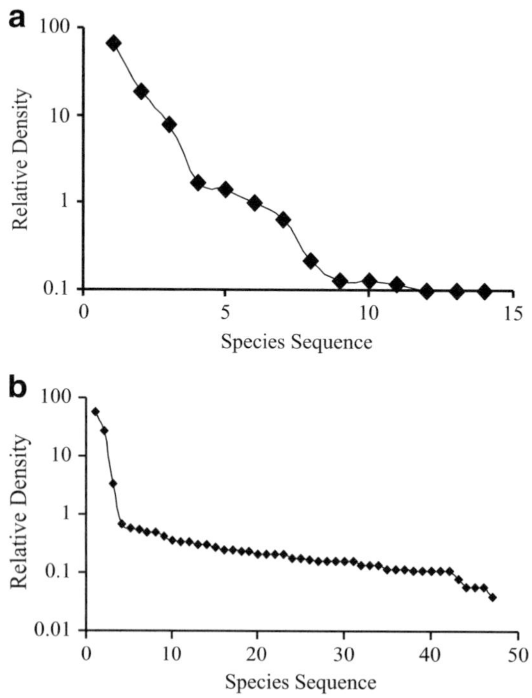

Fig. 2 Relative density of plant species in the protected grassland system at CSSRI Karnal (**a**) during 1982 (based on data from Gupta et al 1990) and (**b**) during the year 2006 (based on data from Jangra 2009)

depth) in the grassland system during the year 1970. Soil carbon content varied from 0.2 to 0.09 % (0–15-cm soil depth) during the year 1981 (Gupta et al. 1990). During March 2005 to February 2007, the soil pH of the grassland system varied from 7.72 to 8.82 in the soil profile from 0- to 100-cm soil depth. Soil organic carbon of the grassland soil ranged from 1.00 to 4.45 g kg^{-1} across soil depths (Table 3).

It is interesting to note long-term changes in soil properties due to protection of natural grassland vegetation on sodic soil at CSSRI, Karnal, from the year 1970 to 2007 (Fig. 4). The changes in soil characteristics during 1970 and 2006–2007 in the surface soil layer (0–10 cm) and subsurface soil layer (0–20 cm and 20–39 cm) of the grassland system are compared in the Fig. 3. Protection of natural vegetation on highly sodic soil for almost four decades showed that the soil pH and electrical conductivity of the grassland soil have been found to decrease from 10.6 to 7.7 and electrical conductivity from 22.34 dS m^{-1} to 0.4 dS m^{-1}, respectively. Initially, the concentration of sodium was very high in the soil (81.9–278.3 mq/l), which substantially decreased to 1.2–2.23 mq/l due to long-term protection of natural vegetation. This could be because of removal of Na by the deep root system of *Desmostachya bipinnata,* a predominant salt-tolerant grass species in the grassland system. There was an increase in exchangeable calcium and magnesium concentration in the soil due to extensive growth of vegetation and recycling of calcium by the plants from subsurface to surface soil. The changes in the chemical properties of the sodic soil could be due to amelioration of soil pH because of improved biological productivity and soil microbial activity. The extensive root system of *Desmostachya bipinnata* through biological production of carbonic acid by the roots seems to play an important role in solubilisation of native $CaCO_3$ present in the sodic soils. There was marked improvement in soil carbon in the year 2006 (0.44 %), which is almost double as compared to the soil organic carbon observed during 1982 (0.20 %). After long-term protection of grassland vegetation on a sodic soil at Karnal, the soil carbon pool in 0–30-cm soil depth was 6.683 Mg C ha^{-1}(year 1982) and 13.91 Mg C ha^{-1} (year 2006) (Fig. 5).

Vegetation Carbon Pool and Flows

The carbon fixed by plants is the primary source of organic matter inputs into the soil both from above-ground and belowground parts of plants. Carbon flow through the organic matter of soil, an important constituent of the lithosphere, is essential to the functioning of terrestrial ecosystems. A large fraction of the terrestrial above-ground net primary production finds its way to the soil surface in the form of litter fall. Vegetation acts as the source and sink of atmospheric CO_2. The plants fix atmospheric CO_2 through the process of photosynthesis in the form of organic compounds and stored in different above-ground and belowground plant

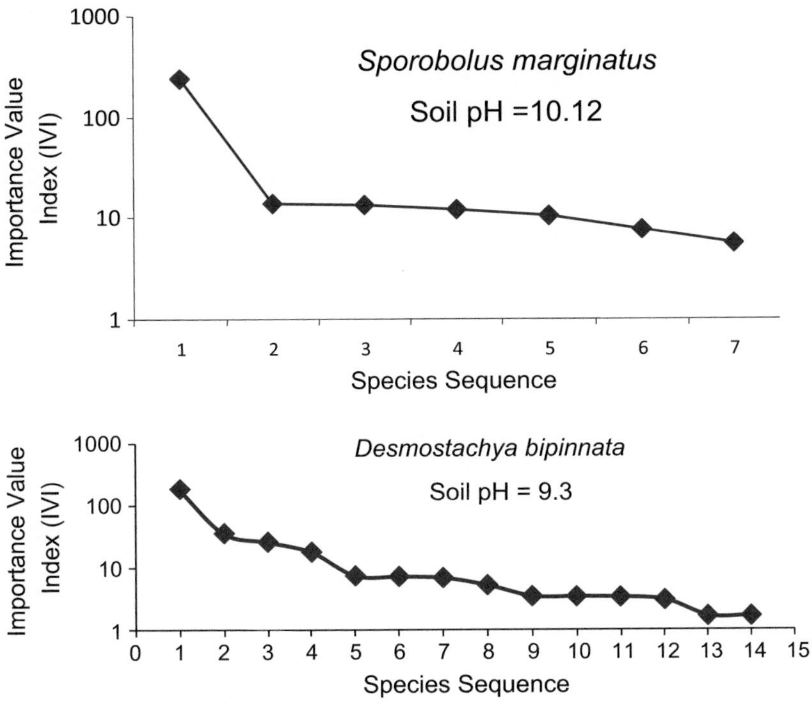

Fig. 3 Dominance–diversity curves of plant species in grassland systems on sodic soils at Bichian, northern India (based on data from Jangra 2009)

Table 2 Some physical and chemical soil characteristics at different soil depths in sodic grasslands at Bichian in Saraswati Reserved Forest

Sites/soil depth	Soil pH	Organic C (g kg^{-1})	Inorganic C (g kg^{-1})	Bulk density (g cm^{-3})
***Sporobolus marginatus* grassland**				
0–15 cm	9.60 ± 0.04	2.55 ± 0.10	1.0 ± 0.09	1.55 ± 0.02
15–30 cm	10.03 ± 0.04	1.2 ± 0.16	1.1 ± 0.13	1.64 ± 0.01
30–45 cm	10.13 ± 0.04	1.05 ± 0.06	2.15 ± 0.14	1.58 ± 0.02
45–60 cm	10.19 ± 0.04	0.75 ± 0.06	5.55 ± 0.23	1.62 ± 0.021
60–100 cm	10.23 ± 0.05	0.51 ± 0.04	9.5 ± 0.13	1.74 ± 0.01
***Desmostachya bipinnata* grassland**				
0–15 cm	9.29 ± 0.05	3.42 ± 0.11	0.87 ± 0.04	1.53 ± 0.01
15–30 cm	10.01 ± 0.05	1.72 ± 0.11	0.95 ± 0.06	1.61 ± 0.02
30–45 cm	10.05 ± 0.07	1.55 ± 0.06	1.42 ± 0.08	1.56 ± 0.02
45–60 cm	10.08 ± 0.02	1.15 ± 0.06	4.87 ± 0.08	1.56 ± 0.02
60–100 cm	10.12 ± 0.03	0.72 ± 0.08	7.5 ± 0.20	1.73 ± 0.01

Source: Jangra et al. (2010)
± standard error

components. Carbon flow refers to the input of carbon through net primary productivity into the system and its subsequent transfer to the soil through litter and root turnover.

The species composition of the grassland communities on sodic soils affected carbon pool in plant biomass (Table 4). The average live shoot biomass in different seasons ranged from 966.66 to

Table 3 Some physical and chemical soil characteristics at different soil depths in *Desmostachya bipinnata* grassland ecosystem at CSSRI, Karnal

Soil depth	Soil pH	Organic C (g kg^{-1})	Inorganic C (g kg^{-1})	Bulk density (g cm^{-3})
0–15 cm	7.72 ± 0.04	4.45 ± 0.03	–	1.46 ± 0.03
15–30 cm	8.12 ± 0.02	1.92 ± 0.11	–	1.50 ± 0.02
30–45 cm	8.20 ± 0.04	1.85 ± 0.14	1.00 ± 0.18	1.56 ± 0.014
45–60 cm	8.30 ± 0.04	1.60 ± 0.17	3.92 ± 0.27	1.61 ± 0.01
60–100 cm	8.82 ± 0.05	1.00 ± 0.09	7.25 ± 0.24	1.70 ± 0.006

Source: Jangra (2009)
± standard error

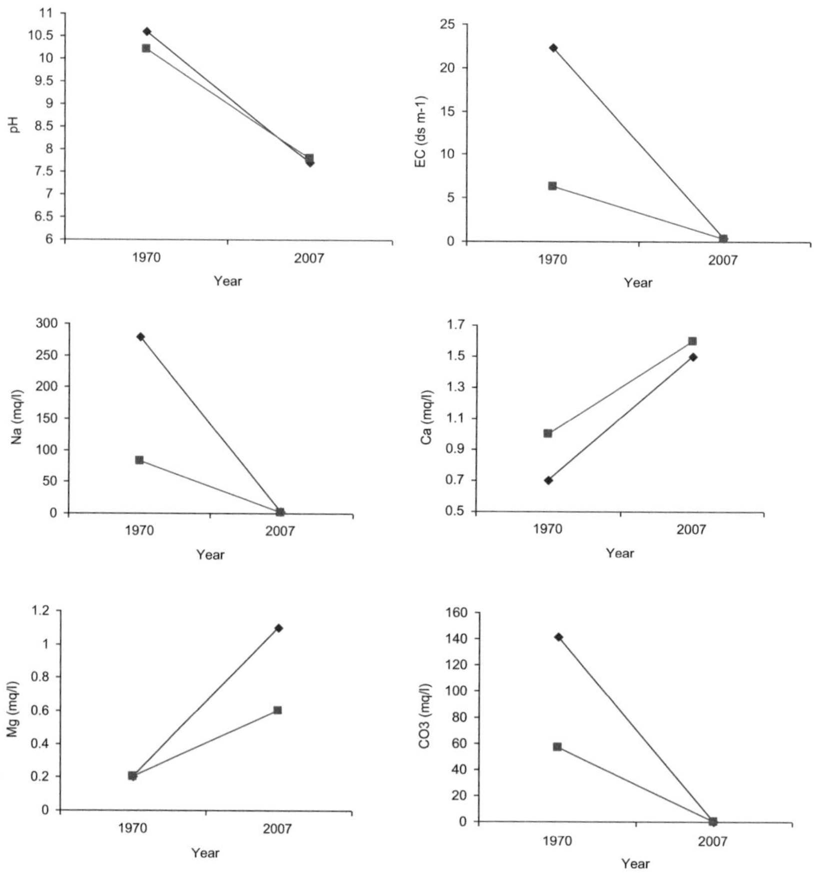

Fig. 4 Changes in soil properties due to protection of natural grassland during the year 1970 (soil depth = 0–10 cm, 10–48 cm) to 2007 (soil depth = 0–20 cm, 20–39 cm) (based on data from CSSRI Karnal, Bhumbla et al. 1970 and Jangra 2009)

2032.5 kg ha^{-1} for the *Sporobolus marginatus* grassland system and from 2,250.0 to 8,510.12 kg ha^{-1} for the *Desmostachya bipinnata* grassland system on sodic soils at Bichian (Jangra et al. 2010). Total above-ground net production was 1,227 to 607 kg ha^{-1} $year^{-1}$. The carbon flux through total net primary productivity was 0.954–0.375 Mg C ha^{-1} $year^{-1}$ (Jangra et al. 2010).

Afforestation and agroforestry systems improved biological production by increasing soil organic matter contents and availability of

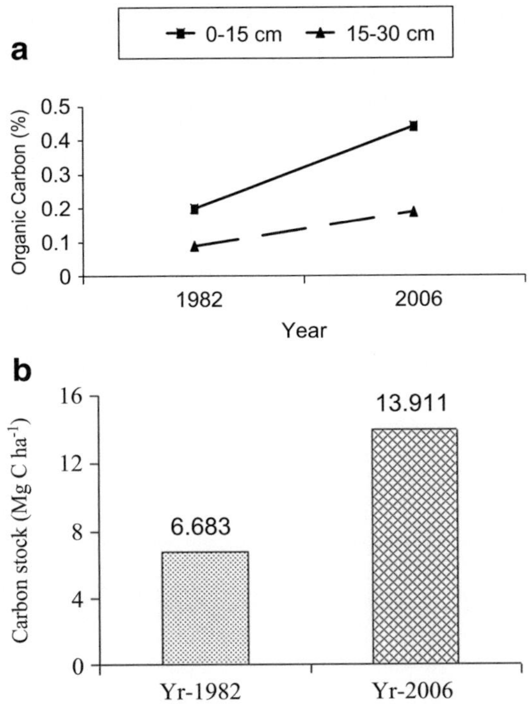

Fig. 5 (**a**) Soil organic carbon (%) and (**b**) carbon stock (Mg C ha^{-1}) in soil (0–30-cm soil depth) of the protected natural grassland system at CSSRI Karnal during the year 1982 (based on data from Gupta et al 1990) and during the year 2006 (based on data from Jangra 2009)

soil inorganic nitrogen in *Prosopis*-, *Acacia*-, *Eucalyptus*- and *Populus*-based agroforestry system (Singh 1995; Singh et al. 1997). Kaur et al. (2002a) reported an increase in total net production in different silvopastoral systems of *Prosopis juliflora*, *Dalbergia sissoo* and *Acacia nilotica* when combined with salt-tolerant grasses like *Desmostachya bipinnata* and *Sporobolus marginatus* on sodic soils. In these systems, increased input of plant residue into the soil played a significant role to improve soil properties and fertility of highly sodic soils. The silvopastoral agroforestry systems raised on sodic soils have been found to improve soil carbon and microbial activity through input of organic matter from above-ground and below-ground parts of the plants (Kaur et al. 2002a, b). There is a large potential of sequestering carbon in soil and vegetation by protecting native vegetation and adopting silvopastoral agroforestry systems on salt-affected soils (Gupta et al. 1990; Kaur et al. 2002a). Carbon partitioning in grassland and grass + tree systems in the plant–soil system of sodic grassland and silvopastoral systems is compared in Fig. 6. In silvopastoral agroforestry systems, the total carbon storage in the grass + tree systems was 1.18–18.55 Mg C ha^{-1}, and carbon input in net primary production varied between 0.98 and 6.50 Mg C ha^{-1} $year^{-1}$ (Kaur et al. 2002a). Carbon flux in net primary productivity increased significantly due to integration of *Prosopis* and *Dalbergia* with salt-tolerant grasses. In silvopastoral agroforestry systems, soil organic matter, biological productivity and carbon storage were greater than grass-only systems (Kaur et al. 2002a).

The carbon concentration in different primary producer compartments of *Desmostachya bipinnata* in the grassland system at Karnal was 31.82 % live shoots, 33.42 % dead shoots, 36.51 % litter and 34.11 % root and rhizomes. There was large pool of carbon in the grassland vegetation. In the protected grassland system, the plant carbon pool (Mg C ha^{-1}) in various primary producer compartments was live shoots 1.628, dead shoots 1.793, litter 0.977 and roots 4.163. Total carbon pool in the plant biomass was 8.561 Mg C ha^{-1} in the grassland system (Table 4). Total carbon input in net primary productivity was 0.887 Mg C ha^{-1} $year^{-1}$ in the grassland system (Table 4). About 56.59 % of total carbon flux was channelised to the belowground components.

Soil Carbon Sequestration

Plant biomass and soil organic matter constitute the major pool of carbon in terrestrial ecosystems. The biotic pool in vegetation stores about 610 Pg C at any given time (Amundson 2001). The total amount of carbon in the world's soil organic matter is estimated to be 1,500 to 1,580 Pg C (Schlesinger 1991; Amundson 2001; Lal 2004). The global soil carbon pool of 2,500 Gt C includes about 1,550 Gt of soil organic carbon and 950 Gt of soil inorganic carbon (Lal 2004). Globally, the terrestrial heterotrophic respiration is responsible

Table 4 Vegetation carbon pools and fluxes in the *Desmostachya bipinnata* and *Sporobolus marginatus* grassland systems at Bichian and *D. bipinnata* grassland at Karnal in northern India

Grassland system/site			
	Sporobolus marginatus Bichian	*Desmostachya bipinnata* Bichian	*Desmostachya bipinnata* CSSRI, Karnal
Carbon pool (Mg C ha^{-1})			
Live shoots	0.522	1.871	1.628
Standing dead	0.677	2.020	1.793
Litter	0.522	1.054	0.977
Roots	1.400	4.336	4.163
Total	*3.121*	*9.281*	*8.561*
Carbon flux (Mg C ha^{-1} $year^{-1}$)			
ANP	0.230	0.436	0.385
BNP	0.145	0.518	0.502
TNP	*0.375*	*0.954*	*0.887*

Source: Jangra (2009), Jangra et al. (2010)

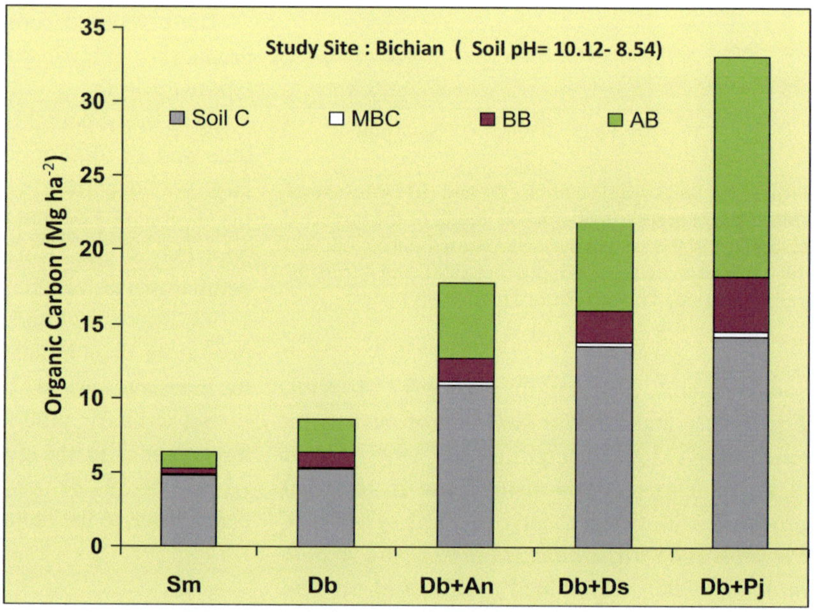

Fig. 6 Carbon partitioning in grassland and grass + tree systems (Sm *Sporobolus marginatus*, Db *Desmostachya bipinnata*, Db + Ds *Desmostachya bipinnata+ Dalbergia sissoo*, Db + An *Desmostachya bipinnata+ Acacia nilotica*, Db + Pj *Desmostachya bipinnata+ Prosopis juliflora*). *Soil C* (soil organic carbon), *MBC* (microbial biomass carbon), *BB* (belowground biomass), *AB* (aboveground biomass) (based on data from Kaur et al. 2002a, b)

for returning about 60 Pg C $year^{-1}$ to the atmosphere (Schlesinger 1991). Soil organic carbon includes plants, animals and microbial residues in all stages of decomposition. A large SIC stock has the potential to help in the establishment of vegetation as well as sequestration of organic carbon in the soils (Bhattacharyya et al. 2004).

In the *Sporobolus marginatus* and *Desmostachya bipinnata* grassland systems on sodic soils, soil organic carbon showed marked decrease with increase in soil depth. In the two grassland system, the soil organic carbon stock (Mg C ha^{-1}) at different soil depths was 5.812–7.803 (0–15-cm soil depth), 2.952–5.416 (15–30-cm soil depth), 2.495–3.584 (30–45-cm soil depth), 1.824–2.902 (45–60-cm soil depth) and 3.566–5.008 (60–100-cm soil depth). The carbon stock at 0–30-cm soil depth was 52–53 % of the total organic carbon stock up to 100-cm soil depth.

Largest increase in soil inorganic carbon was observed at 60–100-cm soil depth in both the grassland systems and found to be reflected in a substantial carbon stock of 69.578–57.953 Mg C ha^{-1} in soil, The total carbon stock (Mg C ha^{-1})

Table 5 Soil carbon pool (Mg C ha^{-1}) in grassland system up to 1 m soil depth in the grassland system at CSSRI, Karnal in northern India

Soil depth	Soil carbon pool (Mg C ha^{-1}) Soil organic carbon	Soil inorganic carbon	Total soil carbon
0–15[a] cm	9.636 ± 518.56	–	9.636 ± 518.56
15–30[a] cm	4.275 ± 206.32	–	4.275 ± 206.32
30–45[a] cm	4.196 ± 293.89	2.340 ± 458.29	6.536 ± 738.72
45–60[a] cm	3.867 ± 436.42	9.419 ± 615.53	13.286 ± 394.27
60–100[b] cm	6.818 ± 614.81	48.960 ± 1,952.70	55.778 ± 1,398.02

Source: Jangra (2009)
[a]The carbon stock represents an increment of 15 cm along the soil depth
[b]The carbon stock represents an increment of 40 cm along the soil depth

in the *Sporobolus marginatus* and the *Desmostachya bipinnata* grassland system was 8.137–9.799 (0–15-cm soil depth), 5.658–6.279 (15–30-cm soil depth), 7.604 to 6.883 (30–45-cm soil depth), 15.320–15.179 (45–60-cm soil depth) and 69.578–57.953 (60–100-cm soil depth). Bhattacharyya et al. (2008) have reported that there is a large potential of sequestration of atmospheric CO_2 in the form of soil inorganic carbon, i.e. pedogenic carbonate (Pal et al. 2009).

In the grassland system at Karnal, soil organic carbon showed marked decrease with increase in soil depth (Table 5). The soil organic carbon stock (Mg C ha^{-1}) at different soil depths was 9.636 (0–15-cm soil depth), 4.275 (15–30-cm soil depth), 4.196 (30–45-cm soil depth), 3.867 (45–60-cm soil depth) and 6.818 (60–100-cm soil depth). The carbon stock at 0–30-cm soil depth was 48 % of the total organic carbon stock up to 100-cm soil depth. The organic carbon up to 1-m soil depth was 28.792 Mg C ha^{-1} (Table 5). In contrast to soil organic carbon distribution, most of soil inorganic carbon was found concentrated below 30-cm soil depth.

There was significant increase in soil inorganic carbon with increase in soil depth ($P < 0.01$). The soil inorganic carbon increased from 0.1 % at 30–45-cm soil depth to 0.72 % at 60–100-cm soil depth. The soil inorganic carbon stock (Mg C ha^{-1}) in grassland system was 2.340 (30–45-cm soil depth), 9.419 (45–60-cm soil depth) and 48.960 (60–100-cm soil depth). Largest increase in soil inorganic carbon was observed at 60–100-cm soil depth in the grassland system, which contained a substantial stock of 48.960 Mg C ha^{-1} (Table 5). Largest increase in soil inorganic carbon was observed at 60–100-cm soil depth in the grassland system and was found to be reflected in a substantial carbon stock of 55.778 Mg C ha^{-1} in soil. The total carbon stock (Mg C ha^{-1}) in the grassland system was 9.636 (0–15-cm soil depth), 4.275 (15–30-cm soil depth), 6.535 (30–45-cm soil depth), 13.286 (45–60-cm soil depth) and 55.778 (60–100-cm soil depth) (Table 5).

The largest amount of soil inorganic carbon (SIC) is present in the form of soil carbonates (Schlesinger 2002) and accounts for one-third of the total carbon in soil (Ming 2002). Calcite is the most common carbonate in soil. Carbonates may be inherited form the soil parent material (lithogenic) or newly formed as a result of soil processes (pedogenic) (Ming 2002). Soil inorganic carbon pool comprised of primary or lithogenic carbonates is derived through dissolution of CO_2 in soil air to form carbonic acid and its reprecipitation with calcium or magnesium (Lal 2006). Total SIC pools of India are estimated about 196 Pg up to 1-m soil depths. It comprises about 27 % of the world total SIC pool (722 Pg up to 1-m soil depth) (Lal 2004; Rasmussen 2006). Pedogenic inorganic carbon is formed from non-carbonate material and is a sink for carbon.

Soil Biological Activity and Sodicity Amelioration

Soil respiration is one of the main components of ecosystem respiration and small changes in soil respiration may strongly affect soil carbon

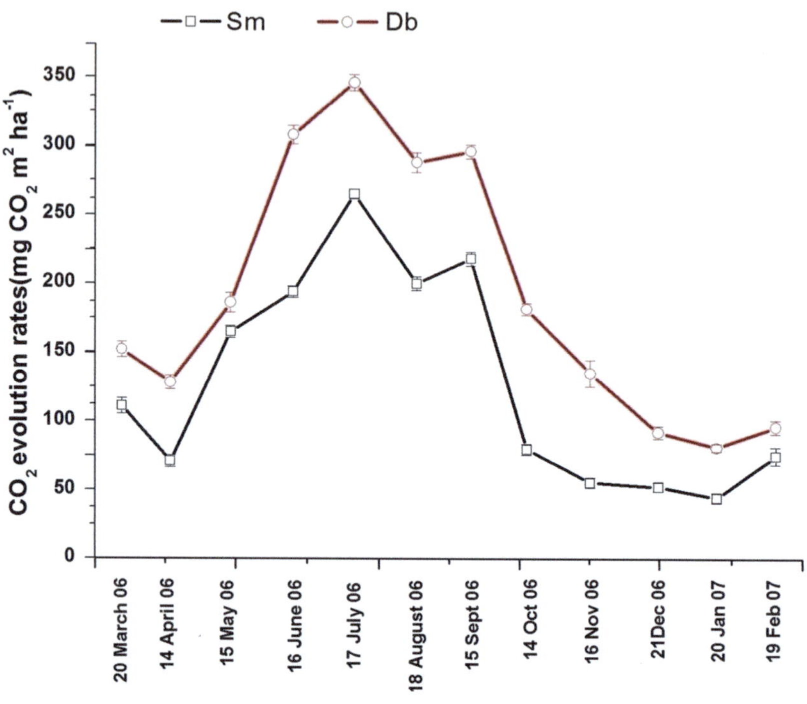

Fig. 7 Seasonal variations in CO_2 evolution rates from *Sporobolus marginatus* (Sm) and *Desmostachya bipinnata* (Db) grassland ecosystems on sodic soil under field conditions during 20 March 2006 to 19 February 2007 (based on data from Jangra et al. 2011)

sequestration (Raich and Schlesinger 1992). Soil respiration serves as a useful index of flow of carbon from soil surface to the atmosphere (Singh and Gupta 1977). The magnitude of soil respiration is primarily controlled by roots and microbial activities and by environmental factors. Seasonal changes in temperature and water affect the productivity in terrestrial ecosystems and the decomposition rates of soil organic matter, thereby driving the temporal variations of soil respiration (Gupta and Singh 1981; Yan et al. 2006). In the grassland at Kurukshetra on moderate alkaline soil, soil respiration rates were highest in rainy season, moderate in summer and least during winter (Gupta and Singh 1981). The contribution of root respiration was 42 % to total carbon dioxide evolution from the soil respiration (Gupta and Singh 1981).

The soil respiration rates (mg CO_2 m^{-2} h^{-1}) in the sodic grassland systems at Bichian over the annual cycle were 81.0–346.4 for *Desmostachya bipinnata* grassland and 44.5–265.2 for *Sporobolus marginatus* grassland (Fig. 7). A positive significant relationship was found between soil respiration rates and soil water content (r^2 = 0.89–0.87) and ambient temperature (r^2 = 0.65–0.73).

The soil microbial activity includes measures of the respiratory activity of soil organisms (Singh and Gupta 1977), soil microbial biomass (Vance et al. 1987) and microbial respiration (Ingram et al. 2005). The active fraction of soil organic matter consists mainly of microbial biomass and its metabolites and comprises 1–3 % of total soil organic matter (Jenkinson and Ladd 1981). It is a source and sink of the plant nutrients (Singh et al. 1989) which regulates the functioning of the soil system. Microbial populations can provide an early indication of changes in soil long before it can be measured by change in soil organic matter (Powlson et al. 1987). Soil microbial biomass is an important component of soil organic matter comprising 1–3 % of total soil organic matter (Jenkinson and Ladd 1981). It is a labile fraction of soil organic matter and conserves nutrients for plant growth (Singh et al. 1989). The microbial biomass primarily depends on rates of nutrient fluxes and has been used as an index of soil fertility.

Table 6 Microbial biomass carbon and soil carbon at 0–15-cm soil depth in grassland systems of salt-affected soils in northern India

Grassland systems	Microbial C (g m^{-2})	Soil carbon (g m^{-2})
Sporobolus marginatus[1]	8.5	481.6
Desmostachya bipinnata[1]	11.2	526.5
Dalbergia sissoo + Desmostachya bipinnata[1]	32.5	1,357.2
Acacia nilotica + Desmostachya bipinnata[1]	22.5	1,090.2
Prosopis juliflora + Desmostachya bipinnata[1]	34.8	1,421.1
Mixed grassland[3]	14.8	773.2
Desmostachya bipinnata[3]	40.8	1,544.3
Vetiveria zizanioides[3]	47.5	1,708.8
Desmostachya bipinnata[2]	34.7	1,394.9

Source: Kaur et al. (2002b), Neeraj et al. (2004)
Location of sites = Bichhian[1], Karnal[2], Kurukshetra[3]

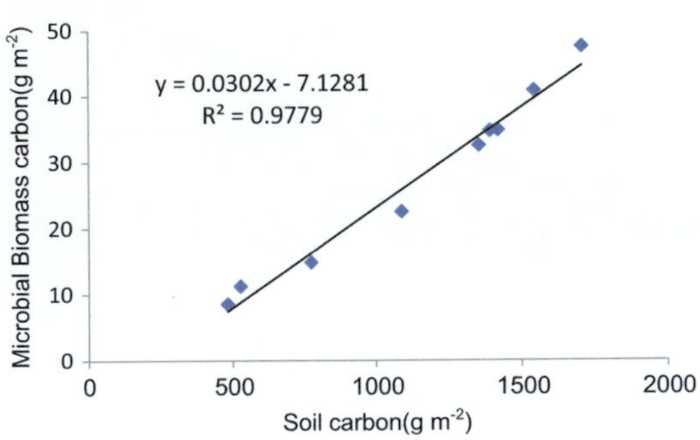

Fig. 8 Microbial biomass carbon and soil carbon at 0–15-cm soil depth in grassland systems of salt-affected soils in northern India

Plant cover through its effects on quantity and quality of organic matter influences the levels of soil microbial biomass. In salt-affected soils, the size and dynamics of soil labile carbon pool have been shown to vary with land-use type (Kaur et al. 2000) and plant species composition (Kaur et al. 2002b), as shown in Table 6. The soil microbial biomass carbon and soil carbon at 0–15-cm soil depth in the grassland systems of salt-affected soils were positively related to soil carbon (R^2 = 0.997), as shown in Fig. 8.

The soil microbial biomass was greater in the *Desmostachya bipinnata* grassland as compared to the *Sporobolus marginatus* grassland. Averaged across the sampling dates, microbial biomass carbon (MBC) was 17.00–48.49 μg C g^{-1} soil (*Sporobolus marginatus*) and 27.33–93.00 μg C g^{-1} soil (*Desmostachya bipinnata*). The biomass specific respiratory activity was 38.06–273.88 μg CO_2–C produced mg^{-1} MBC day^{-1} (Jangra et al. 2011). A significant positive relationship was found between carbon dioxide emission from the soil and soil organic carbon using the pooled data across soil depths and the grassland systems (R^2 = 0.81, d.f. =22, $p < 0.01$),as shown in Fig. 9. The variation in soil organic carbon explained 81 % variability in carbon dioxide emission from the soil on the basis of pooled data of the two grassland systems across soil depths (Fig. 9). It is interesting to find that soil organic carbon was linearly related to soil microbial carbon in grassland system using the pooled data across soil depths and the grassland systems (R^2 = 0.85, d. f. = 22, $p < 0.01$), as shown in Fig. 9.

Several workers have shown that short-term carbon mineralisation of soil after drying at 40–60 °C and then rewetting could be a rapid index of carbon mineralisation potential of soils under different soil management practices (Franzluebbers et al. 2000; Ingram et al. 2005). The soil microbial respiration rates and the microbial biomass carbon decreased significantly with increase in soil depth ($p < 0.05$) in the sodic grassland systems (Jangra et al. 2011). The soil microbial respiration rates (mg CO_2–C kg^{-1} soil) were 3-d ($C_{min0\text{-}3d}$) = 5.22–42.66 and 21-d ($C_{min0\text{-}21d}$) = 6.84–76.13. The soil microbial respiration and soil microbial biomass have been found to be good indicators of improved soil conditions of sodic soils due to long-term grassland vegetation protection. There is need to

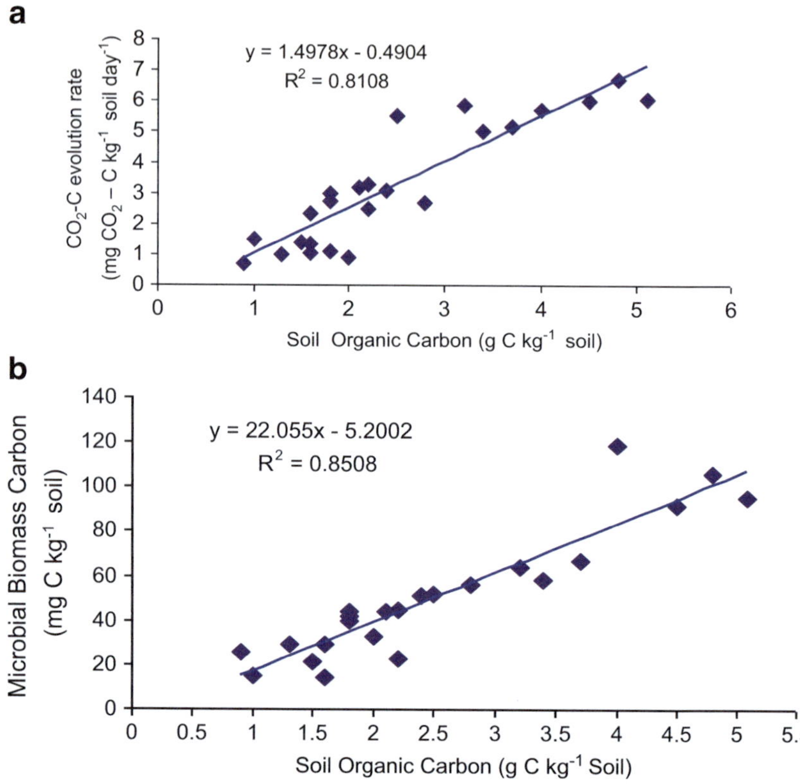

Fig. 9 A relationship between (**a**) soil organic carbon and carbon dioxide evolution rates, (**b**) soil organic carbon and soil microbial biomass in grassland systems at Bichian in Saraswati Reserved Forest

provide additional insight into microbial function and processes that affect C sequestration under normal and changing environmental conditions.

AM Mycorrhizal Fungi and Soil Carbon Storage

The arbuscular mycorrhizal fungi (AMF) are a key, integral component of the stability, sustainability and functioning of terrestrial ecosystems. In terrestrial ecosystem, arbuscular mycorrhizal fungi characterise a delicate balance between plant, fungus and the soil (Mosse 1986) and are present in about 80 % of plant species, including most agricultural crops (Trappe 1987; Smith and Read 1997).

There were significant variations in the diversity of AM fungal spores in the sodic grassland systems at Bichian (Jangra et al. 2011). A total of 27 species of AM fungal spore belonging to six genera, i.e. *Acaulospora*, *Entrophospora*, *Gigaspora*, *Glomus*, *Sclerocystis* and *Scutellospora* were identified in soils. A total of 18 AM fungal spore species belonging to four genera, i.e. *Acaulospora*, *Entrophospora*, *Gigaspora* and *Glomus* were recorded from the *Desmostachya bipinnata* grassland system. In the *Sporobolus marginatus* grassland, 10 species of *Acaulospora*, three species of *Entrophospora*, three species of *Gigaspora*, 19 species of *Glomus* and one species each of *Scutellospora* and *Sclerocystis* were recorded. The arbuscular mycorrhizal species of *Glomus* and *Acaulospora* dominated the AM fungal communities. The density of arbuscular mycorrhizal (AM) fungal spores in soil of the sodic grassland systems was, at 0–15-cm soil depth, 2.28–6.08/g soil and, at 15–30-cm soil depth, 0.96–1.84/g soil. The *Glomus* spp. and *Acaulospora* spp. were the predominant AM fungal species in the soils. The AM fungi associated with salt-adapted grasses could play an important role in bioamelioration and soil carbon storage.

The root colonisation of *Desmostachya bipinnata* showed the presence of abundant

Table 7 Seasonal variations in AM fungal root colonisation of *Sporobolus marginatus* and *Desmostachya bipinnata* grassland systems at Bichian in Saraswati Reserved Forest

	AM fungal root colonisation (%)	
Sampling dates	*Sporobolus marginatus*	*Desmostachya bipinnata*
15 March 2007	51.86 ± 4.43	28.84 ± 2.34
24 September 2007	81.52 ± 4.89	50.54 ± 3.25
25 November 2007	66.5 ± 6.31	42.74 ± 4.52
LSD ($p < 0.05$)	18.17	12.01

Source: Jangra et al. (2011)
± standard error

fungal mycelium; the AM fungal infection consisted of both fine and coarse hyphae with distinct vesicles and arbuscules. In the *Sporobolus marginatus* grassland ecosystem, the mycorrhizal fungal colonisation of the roots varied from 68 to 80 % in different seasons with abundant mycelia, arbuscules and vesicles in the cortical cells (Table 7). In the roots of *Sporobolus marginatus*, there was abundance of vesicles at peak growth of plants during September (Fig. 10). Roots of *Sporobolus marginatus* growing in sodic soils were highly colonised by AM fungi. The capability of becoming densely colonised by AM fungi is an important trait of sodicity-tolerant plants. The mycorrhizal root colonisation was observed in different forms such as mycelium (H and Y types) and formation of arbuscules and vesicles (elliptical, globose and round types). The profuse vesicle formation in *Sporobolus marginatus* reflects the completion of life cycle so as to form more propagules and support the regrowth of intercellular hyphae in this bunch-forming grass.

Mycorrhizal fungi could play an important role in the sequestration of C in soil under elevated CO_2 and N deposition (Treseder and Allen 2000). This group, which symbiotically colonises plant roots, forms associations with 80 % of plant species and is found in nearly every habitat in the world (Smith and Read 1997). A substantial amount of C allotted to mycorrhizal tissues could be long lived in the soil. For example, chitin, which is not readily decomposed (Gooday 1994), can constitute up to 60 % of fungal cell walls (Muzzarelli 1977). Arbuscular mycorrhizal (AM) fungi are also the sole producers of glomalin, a potentially recalcitrant glycoprotein (Wright and Upadhyaya 1996, 1999). In many systems exposed to elevated CO_2, mycorrhizal fungi might sequester increased amounts of C in living, dead and residual hyphal biomass in the soil (see Treseder and Allen 2000).

Soil Aggregate Carbon Storage and Clay Mineralogy

Soil texture, soil mineralogy and soil organic matter play a key role in ecosystem functioning and maintaining soil productivity. Soil organic matter forms a highly heterogeneous mixture of organic materials in the soil along a continuum from freshly fallen litter to highly decomposed organic materials and regulates a range of physical, chemical and biological properties of the soil. Clay mineralogy controls soil structure, porosity and stability through formation of microaggregates (Tisdall and Oades 1982). Physical fractionation techniques have also been used to separate soil organic matter pools into primary particles (sand, silt and clay), microaggregates (53–250 μm) and macroaggregates (>250 μm). Primary soil particles (sand, silt and clay) are associated with organic matter to form microaggregates (<250 μm) and macroaggregates (>250 μm) in the soil (Jastrow and Miller 1998; Tisdall and Oades 1982) and play an important role in protection of soil organic matter.

In the grassland systems at Bichian, Jangra et al. (2010) reported that only a small amount of macroaggregates (2 mm–250 μm) were recovered from soils up to 0–30-cm soil depth (Table 8).

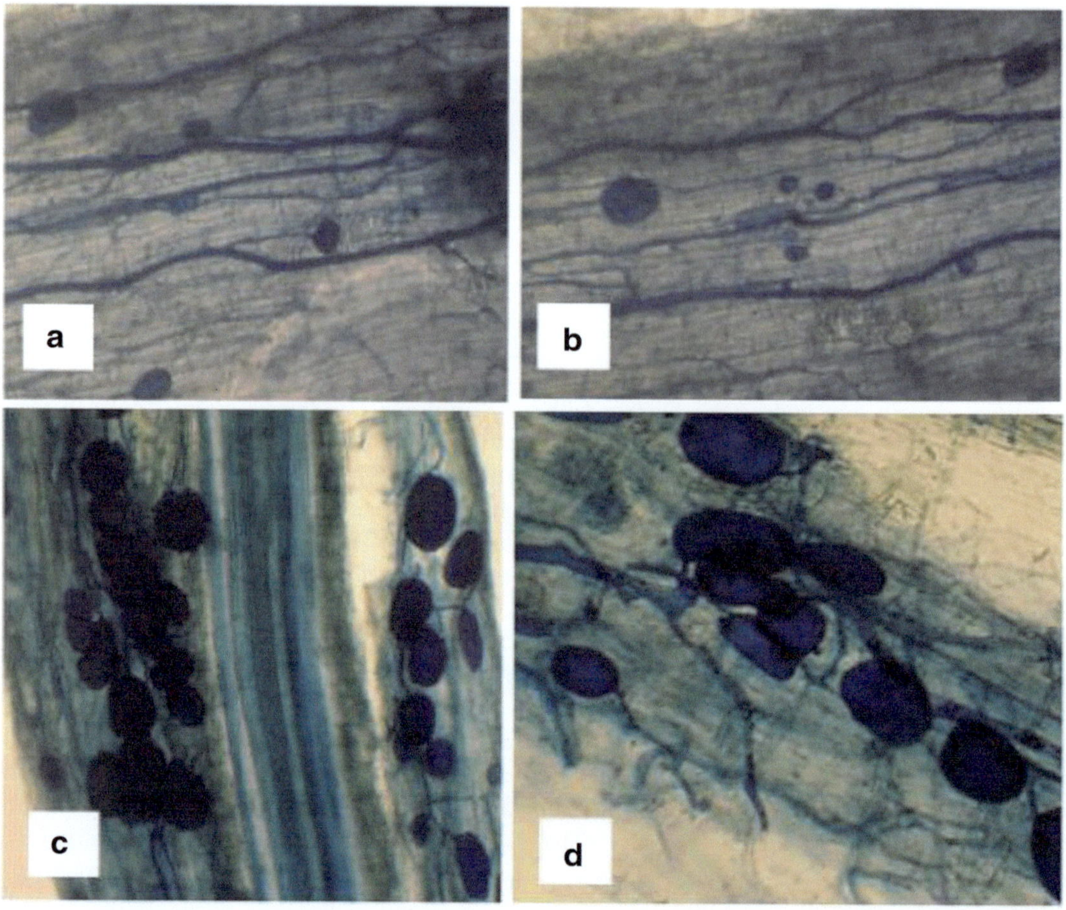

Fig. 10 (**a–b**) Arbuscular mycorrhizal fungal infection in roots of *Sporobolus marginatus* with presence of hyphae in cortical cells, (**c–d**) showing presence of hyphae and vesicles. The AM fungi associated with salt-adapted grasses could play an important role in bioamelioration and soil carbon storage. The *Glomus* spp. and *Acaulospora* spp. were the predominant AM fungal species in the soils (based on data from Jangra 2009)

The proportion of small macroaggregates (2 mm–250 μm) was higher in *Desmostachya bipinnata* grassland (4.42–5.45 %) compared to that in the *Sporobolus marginatus* grassland system (3.10–3.79 %). Microaggregates (250 μm–53 μm) varied from 28.12 to 38.34 % in the *Sporobolus marginatus* and from 56.41 to 61.56 % in *Desmostachya bipinnata* grassland. The clay- and silt-associated aggregates (<53 μm) formed a large fraction of the soil, the values being 58.00–68.09 % for the *Sporobolus marginatus* grassland system and 34.02–38.14 % for the *Desmostachya bipinnata* grassland system. The microaggregates (250 μm, 53 μm and <53 μm) formed a large fraction of soil aggregates and protected most of the soil organic carbon (Jangra et al. 2010). Organic carbon distribution in aggregate size classes in soils of grassland system at various soil depths at Bichian in Saraswati Reserved Forest is given in Table 9.

Distribution of carbon within aggregate size fractions in grassland soils at Kurukshetra is shown in Fig. 11. The proportion of small macroaggregates (2 mm–250 μm) and the clay- and silt-associated aggregates (<53 μm) formed a large fraction of the soil (Fig. 11). The rapid turnover of macroaggregates reduces the formation of microaggregates within the

Table 8 Percent soil weight distribution in aggregate size classes in the grassland system across soil depths at Bichian in Saraswati Reserved Forest (±standard error)

	Percent soil weight in soil aggregates		
Soil depth (cm)	2 mm–250 μm	250 μm–53 μm	<53 μm
Sporobolus marginatus			
0–5	3.79 ± 0.77	28.12 ± 1.85	68.09 ± 0.42
5–15	3.10 ± 0.43	32.78 ± 1.98	64.12 ± 1.23
15–30	3.66 ± 0.58	38.34 ± 1.60	58.00 ± 2.85
LSD ($p < 0.05$)	1.36	5.87	3.79
CV (%)	23.76	9.91	3.40
Desmostachya bipinnata			
0–5	5.45 ± 0.32	56.41 ± 1.26	38.14 ± 1.60
5–15	4.79 ± 0.77	59.18 ± 2.84	36.03 ± 0.81
15–30	4.42 ± 0.74	61.56 ± 1.09	34.02 ± 1.05
LSD ($p < 0.05$)	1.36	4.42	4.30
CV (%)	15.89	4.25	6.83

Source: Jangra et al. (2010)

Table 9 Organic carbon distribution in aggregate size classes in soils of grassland system at various soil depths at Bichian in Saraswati Reserved Forest

	Organic carbon (%)			
Soil depth (cm)	Soil	2 mm–250 μm	250 μm–53 μm	<53 μm
Sporobolus marginatus				
0–5	0.32 ± 0.01	0.41 ± 0.01	0.25 ± 0.02	0.21 ± 0.01
5–15	0.19 ± 0.01	0.31 ± 0.02	0.18 ± 0.01	0.16 ± 0.01
15–30	0.12 ± 0.01	0.19 ± 0.01	0.13 ± 0.01	0.09 ± 0.01
LSD ($p < 0.05$)	0.062	0.067	0.059	0.034
CV (%)	18.08	12.63	17.88	13.20
Desmostachya bipinnata				
0–5	0.46 ± 0.01	0.58 ± 0.03	0.33 ± 0.01	0.20 ± 0.01
5–15	0.23 ± 0.01	0.33 ± 0.02	0.22 ± 0.01	0.16 ± 0.01
15–30	0.17 ± 0.01	0.28 ± 0.01	0.16 ± 0.02	0.09 ± 0.01
LSD ($p < 0.05$)	0.062	0.059	0.045	0.034
CV (%)	13.36	9.15	10.52	11.07

Source: Jangra et al. (2010)
± standard error

macroaggregates and favours the stabilisation of carbon within the microaggregates (Six et al. 2004). On the basis of several reports on soil organic carbon stabilisation, Six et al. (2002) reported that silt-plus-clay fraction in soil plays a key role in the protection of soil organic matter.

Clay minerals strongly influence the major physical and chemical properties of soil as well as soil organic matter and its chemical nature. Clay mineralogy varies spatially as a function of climate and parent material and temporally as a function of soil development (Torn et al. 1997). Clay mineralogy plays an important role in the stabilisation of soil organic carbon (Laird 2001; Ramson et al. 1998). The largest changes in the quantity and turnover of soil organic carbon across landscapes and over long time scales may be due to variation in passive (mineral-stabilised) carbon in the soil (Torn et al. 1997).

In the sodic grassland systems, the illite and montmorillonite (a member of the smectite

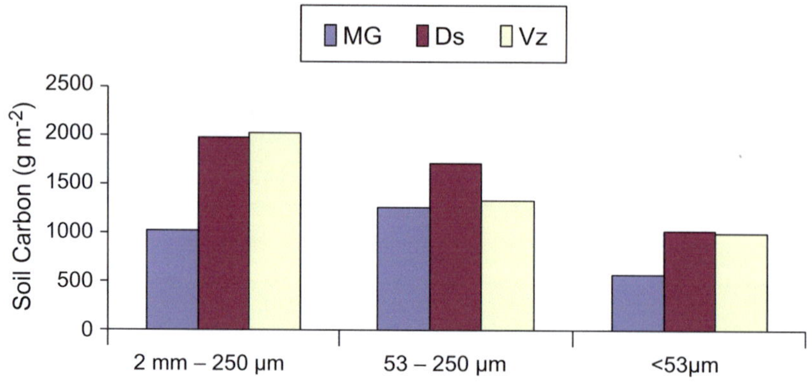

Fig. 11 Distribution of carbon within aggregate size fractions in grassland soils at Kurukshetra. *MG* mixed grass, Db *Desmostachya bipinnata*, Vz *Vetiveria zizanioides* (based on data from Neeraj et al. 2004)

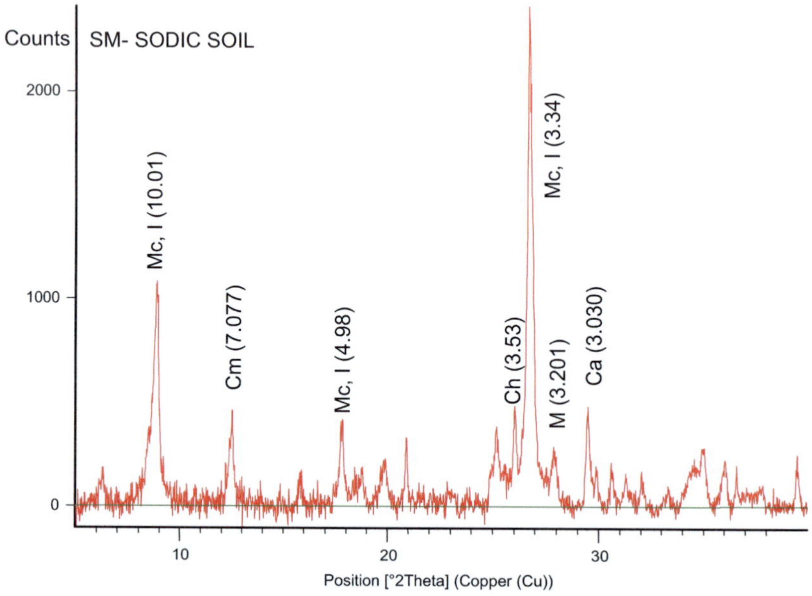

Fig. 12 X-ray diffraction pattern of untreated clay from highly sodic soil of the *Sporobolus marginatus* grassland at Bichian (*I* illite, *Mc* mica, *Ch* chlorite, *M* montmorillonite, *Ca* calcite, *Cm* chamosite) (based on data from Jangra 2009)

family, 2:1 clay) were predominant in the soil (Figs. 12 and 13). The stabilisation of organic material by soil matrix is a function of chemical nature of mineral fraction and its surfaces capable of adsorbing the organic material (Baldock and Skjemstad 2000). Illite is the main precursor mineral for the formation of smectite in salt-affected soils. In the surface layer of salt-affected soils in Alberta, Canada, Kohut and Dudas (1994) reported highly diffuse smectite diffraction maxima because of mineralogical interactions with organic matter. The distribution of newly formed humic materials into mineralogically distinct clay size fractions on a silt loam soil showed that new humic materials are preferentially accumulated on smectite surfaces (Gonzalez and Laird 2003). The predominance of illite and montmorillonite in the clay could be related to soil carbon stability. The association of organic matter in soil with minerals is a

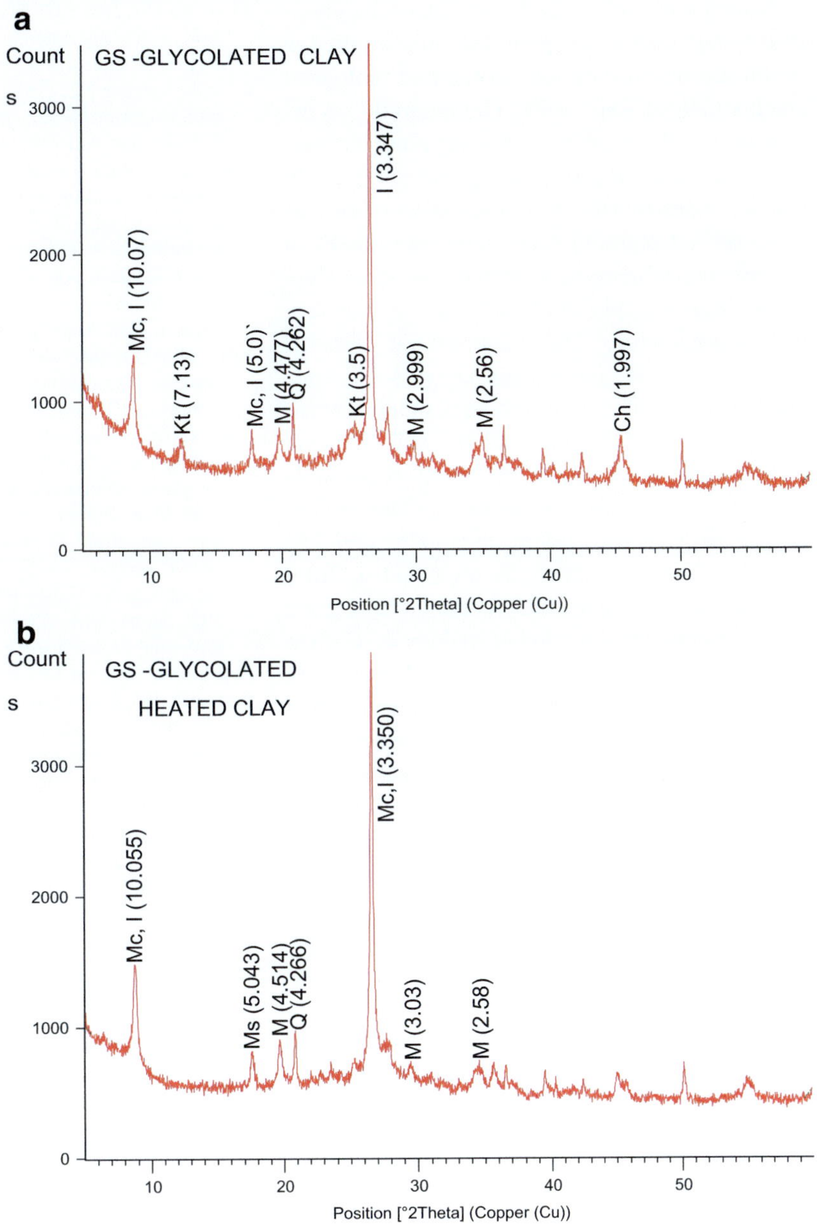

Fig. 13 X-ray diffraction pattern of clay fraction from a grassland soil at CSSRI Karnal. (**a**) Glycolated. (**b**) Glycolated Heated Clay (*I* illite, *Kt* kaolinite, *M* montmorillonite, *Mc* mica, *Ch* chlorite, *Q* quartz) (based on data from Jangra 2009)

controlling factor of organic carbon storage in soil (Kohut and Dudas 1994). The soil organic matter protection through intimate association with clay particles can provide long-term stability for carbon sequestration (Qualls 2004; Ratnayke et al. 2008). Montmorillonite, chlorite, illite, kaolinite and vermiculite were found to be the main clay minerals in the sodic soil (Jangra et al. 2010).

Conclusions

The native grassland vegetation on salt-affected soils has the potential for carbon sequestration by increasing plant biomass production and improving soil organic matter. The differences in soil inorganic carbon stock were found to be one of the major determinants of total soil carbon in the reclaimed sodic soils. The protection of the grassland system over a long period of time resulted in significant increase of soil carbon and biological amelioration of sodic soils. The naturally occurring grassland system, by increasing plant biomass production and soil carbon pool, has been found to play an important role in ecological restoration of sodic soils. Integrating trees with the grasses in silvopastoral systems have been found to be effective to improve soil fertility and increase soil carbon sequestration. The silvopastoral systems greatly improved soil properties of sodic soils due to increase of carbon storage in plant biomass and organic matter input into the soil. Organic carbon content increased by 24–62 % in soils under the silvopastoral systems as compared to that in the grassland system on a sodic soil. The microbial biomass carbon, as regulated by litter and root carbon input, was found to be a good indicator of bioamelioration of sodic soils. Carbon sequestration also provided associated ecosystem co-benefits such as increased soil water holding capacity, better soil structure, improved soil quality and nutrient cycling and reduced soil erosion. Soil carbon sequestration in salt-affected soils is critical to formulate management strategies for their ecological restoration and for improving soil productivity so as to meet increasing demands of human society for food, fodder, biomass energy and industrial products. Implementing appropriate management practices to build up soil carbon stocks in grasslands could lead to considerable mitigation, adaptation and development benefits.

References

Amundson R (2001) The carbon budget in soils. Annu Rev Earth Planet Sci 29:535–562

Baldock JA, Skjemstad JO (2000) Role of the soil matrix and minerals in protecting natural organic materials against biological attack. Org Geochem 31:697–710

Bhattacharyya T, Pal DK, Chandran P, Ray SK, Gupta RK, Gajbhiye KS (2004) Managing soil carbon stocks in the Indo-Gangetic plains, India. Rice-Wheat Consortium for the Indo-Gangetic plains, New Delhi, 44 pp

Bhattacharyya T, Pal DK, Chandran P, Ray SK, Mandal C, Telpande B (2008) Soil carbon storage capacity as a tool to prioritize areas for carbon sequestration. Curr Sci 95:482–493

Bhumbla DR, Abrol IP, Bhargava GP, Singhla SK (1970) Soil of the experimental farm. Bulletin No.1, Division of Soil Science and Agronomy, CSSRI, Karnal, 43 pp

Cass A, Sumner ME (1982) Soil pore structural stability and irrigation water quality. Sodium stability data. Soil Sci Soc Am J 46:507–512

Chivenge P, Vanlauwe B, Gentile R, Six J (2011) Organic resource quality influences short-term aggregate dynamics and soil organic carbon and nitrogen accumulation. Soil Biol Biochem 43:657–666

Conant RT, Paustian K, Elliott ET (2001) Grassland management and conversion into grassland: effects on soil carbon. Ecol Appl 11:343–355

FAO (2000) Crops and drops: making the best use of water for agriculture. Food and Agriculture Organization of the United Nations, Rome

FAO (2010) Challenges and opportunities for carbon sequestration in grassland systems: a technical report on grassland management and climate change mitigation. Plant Production and Protection Division Food and Agriculture Organization of the United Nations (FAO), Rome.

Franzluebbers AJ, Haney RL, Honeycutt CW, Schomberg HH, Hons FM (2000) Flush of CO_2 following rewetting of dried soil relates to active organic pools. Soil Sci Soc Am J 64:613–623

Gonzalez JM, Laird DA (2003) Carbon sequestration in clay mineral fractions from ^{14}C-labeled plant residues. Soil Sci Soc Am J 67:1715–1720

Gooday GW (1994) Physiology of microbial degradation of chitin and chitosan. In: Ratledge C (ed) Biochemistry of microbial degradation. Kluwer, Dordrecht, pp 279–312

Gupta SR, Singh JS (1981) Soil respiration in a tropical grassland. Soil Biol Biochem 13:261–268

Gupta SR, Singh JS (1982) Influence of floristic composition on the net primary production and dry matter turnover in a tropical grassland. Aust J Ecol 7:363–374

Gupta SR, Sinha A, Rana RS (1990) Biomass dynamics and nutrient cycling in sodic grassland. Int J Ecol Environ Sci 16:57–70

IAB (2000) Indian agriculture in brief, 27th edn. Directorate of Economics and Statistics, Department of Agriculture and Cooperation, Ministry of Agriculture, Government of India, New Delhi

Ingram LJ, Schuman GE, Stahl PD, Spackman LK (2005) Microbial respiration and organic carbon indicate nutrient cycling recovery in reclaimed soils. Soil Sci Soc Am J 69:1737–1745

Jangra R (2009) Carbon dynamics in land-use systems on a reclaimed sodic soil. PhD thesis, Department of Botany, KUK

Jangra R, Bhalla E, Gaur A, Gupta SR (2010) Carbon sequestration in sodic grassland ecosystems in north western India. Am Euras J Agric Environ Sci 9(1):27–35

Jangra R, Gupta SR, Kumar R, Bhalla E (2011) Soil respiration, microbial biomass, and mycorrhizal diversity in sodic grassland ecosystems in Northwestern India. Am Euras J Agric Environ Sci 10:863–875

Jastrow JD, Miller RM (1998) Soil aggregate stabilization and carbon sequestration: feedbacks through organomineral associations. In: Lal R, Kimble JM, Follett RF, Stewart BA (eds) Soil processes and the carbon cycle. CRC Press LLC, Boca Raton, pp 207–223

Jenkinson DS, Ladd JN (1981) Microbial biomass in soil: measurement and turnover. In: Paul EA, Ladd JN (eds) Soil biochemistry. Marcel Dekker, New York, pp 415–471

Kaur B, Gupta SR, Singh G (2000) Soil carbon, microbial activity and nitrogen availability in agroforestry systems on moderately alkaline soils in Northern India. Appl Soil Ecol 15:283–294

Kaur B, Gupta SR, Singh G (2002a) Carbon storage and nitrogen cycling in silvopastoral systems on sodic soil in Northwestern India. Agrofor Syst 54:21–29

Kaur B, Gupta SR, Singh G (2002b) Bioamelioration of a sodic soil by silvopastoral system in north western India. Agrofor Syst 54:13–20

Kohut CK, Dudas MJ (1994) Characteristics of clay minerals in saline alkaline soils in Alberta, Canada. Soil Sci Soc Am J 58:1260–1269

Laird DA (2001) Nature of clay–humic complexes in an agricultural soil: II. Scanning electron microscopy analysis. Soil Sci Soc Am J 65:1419–1425

Lal R (2004) Soil carbon sequestration impacts on global climate change and food security. Science 304:1623–1627

Lal R (2006) Carbon management in agricultural soils. Mitig Adapt Strateg Glob Change Biol 12:303–322

Ming DW (2002) Carbonates. In: Lal R (ed) Encyclopedia of soil science. Marcel Dekker, New York, pp 139–141

Montagnini F, Nair PKR (2004) Carbon sequestration: an underexploited environmental benefit of agroforestry systems. Agrofor Syst 61:281–298

Mosse B (1986) Mycorrhiza in a sustainable agriculture. In: Lopez-Real JM, Hodges RH (eds) Role of microorganisms in a sustainable agriculture. Academic, London, pp 105–123

Muzzarelli RAA (1977) Chitin. Pergamon, New York

Nair PKR, Kumar BM, Nair VD (2009) Agroforestry as a strategy for carbon sequestration. J Plant Nutr Soil Sci 172:10–23

Neeraj, Gupta SR, Malik V, Kaur B, Neelam (2004) Plant diversity, carbon dynamics and soil biological activity in tropical successional grassland systems at Kurukshetra. Int J Ecol Environ Sci 30:285–298

Pal DK, Srivastava P, Bhattacharyya T (2003a) Clay illuviation in calcareous soils of the semi-arid part of the Indo-Gangetic plains, India. Geoderma 115:177–192

Pal DK, Srivastava P, Durge SL, Bhattacharayya T (2003b) Role of microtopography in the formation of sodic soils in the semi-arid part of the indo-gangetic plains, India. Catena 51:3–31

Pal DK, Bhattacharyya T, Sarastava P, Chandran P, Ray SK (2009) Soils of the Indo-gangetic plains; their historical perspective and management. Curr Sci 96:1193–1202

Powlson DS, Brookes PC, Christensen BT (1987) Measurement of microbial biomass provides an early indication of changes in total soil organic matter due to the straw incorporation. Soil Biol Biochem 19:159–164

Qadir M, Qureshi RH, Ahmad N (2002) Amelioration of calcareous saline sodic soils through phytoremediation and chemical strategies. Soil Use Manag 18:381–385

Qadir M, Oster JD, Schubert S, Noble AD, Sahrawat KL (2007) Phytoremediation of sodic and saline-sodic soils. Adv Agron 96:197–247

Qualls RG (2004) Biodegradability of humic substances and other fractions of decomposing leaf litter. Soil Sci Soc Am J 68:1705–1712

Raich JW, Schlesinger WH (1992) The global carbon dioxide flux in soil respiration and its relationship to vegetation and climate. Tellus B 44:81–99

Ramson B, Dongseon K, Kastner M, Wainwright S (1998) Organic matter preservation on continental slopes: importance of mineralogy and surface area. Geochim Cosmochim Acta 62:1329–1345

Rana RS, Parkash V (1987) Floristic characterization of alkali soils in northwestern India. Plant Soil 99:447–451

Rasmussen C (2006) Distribution of soil organic and inorganic carbon pools by biome and soil taxa in Arizona. Soil Sci Soc Am J 70:256–265

Ratnayke RR, Seneviratne G, Kulasooriya SA (2008) Characterization of clay bound organic matter using activation energy calculated by weight loss on ignition method. Curr Sci 95(6):763–766

Ravindranath NH, Ostwald M (2008) Carbon inventory methods. Handbook for greenhouse gas inventory, carbon mitigation and roundwood production. Springer, Dordrecht

Rengasamy P (2006) World salinization with emphasis on Australia. J Exp Bot 57:1017–1023

Rengasamy P, Sumner ME (1998) Processes involved on sodic behaviour. In: Sumner ME, Naidu R (eds) Sodic soil: "distribution, management and environmental consequences". Oxford University Press, New York, pp 35–50

Schlesinger WH (1991) Biogeochemistry: an analysis of global change. Academic Press, San Diego, p 443

Schlesinger WH (2002) Inorganic carbon and the global carbon cycle. In: Lal R (ed) Encyclopedia of soil science. Marcel Dekker, New York

Singh G (1995) An agroforestry practice for the development of salt lands using *Prosopis juliflora* and *Leptochloa fusca*. Agrofor Syst 29(1):61–75

Singh JS, Gupta SR (1977) Plant decomposition and soil respiration in terrestrial ecosystems. Bot Rev 43:449–527

Singh JS, Raghubanshi AS, Singh RS, Srivastava SC (1989) Microbial biomass acts as a source of plant nutrients in dry tropical forest and savanna. Nature 338:499–500

Singh G, Singh NT, Dagar JC, Singh H, Sharma VP (1997) An evaluation of agriculture, forestry and agroforestry practices in a moderately alkali soil in northwestern India. Agrofor Syst 37:279–295

Sinha A, Rana RS, Gupta SR (1988) Phytosociological analysis of some natural grassland communities of sodic soils. Trop Ecol 29:136–145

Six J, Conant RT, Paul EA, Paustian K (2002) Stabilization mechanisms of soil organic matter: implication for C-saturation of soils. Plant Soil 241:155–176

Six J, Bossuyt H, Degryze S, Denef K (2004) A history of research on the link between (micro) aggregates, soil biota, and soil organic matter dynamics. Soil Tillage Res 79:7–31

Smith SE, Read DJ (1997) Mycorrhizal symbiosis. Academic Press, San Diego, p 928

Smith AP, Chen D, Chalk PM (2009) N_2 fixation by faba bean (*Vicia faba* L.) in a gypsum-amended sodic soil. Biol Fertil Soils 45:329–333

Szabolcs I (1989) Salt affected soils. CRC Press, Boca Raton, p 274

Szabolcs I (1994) Soil and salinization. In: Pessarakli M (ed) Handbook of plant and crop stress. Marcel Dekker, New York, pp 3–11

Tisdall JM, Oades JM (1982) Organic matter and water stable aggregates in soil. J Soil Sci 33:141–163

Torn MS, Trumbore SE, Chadwick OA, Vitousek PM, Hendricks DM (1997) Mineral control over soil carbon storage and turnover. Nature 389:170–173

Trappe JM (1987) Phylogenetic and ecological aspects of mycotrophy in the angiosperms from an evolutionary standpoint. In: Safir GR (ed) Ecophysiology of VA mycorrhizal plants. CRC press, Boca Raton, pp 5–25

Treseder KK, Allen MF (2000) Mycorrhizal fungi have a potential role in soil carbon storage under elevated CO_2 and nitrogen deposition. New Phytol 147:189–200

Vance ED, Brookes PC, Jenkinson DS (1987) An extraction method for measuring soil microbial biomass C. Soil Biol Biochem 19:703–707

Wong VNL, Dalal RC, Greene RSB (2008) Salinity and sodicity effects on respiration and microbial biomass of soil. Biol Fertil Soils 44:943–953

Wright SF, Upadhyaya A (1996) Extraction of an abundant and unusual protein from soil and comparison with hyphal protein of arbuscular mycorrhizal fungi. Soil Sci 161:575–586

Wright SF, Upadhyaya A (1999) Quantification of arbuscular mycorrhizal fungi activity by the glomalin concentration on hyphal traps. Mycorrhiza 8:283–285

Yan JH, Wang YP, Zhou GY, Zhang DQ (2006) Estimate of soil respiration and net primary production of three forests at different succession stages in South China. Glob Chang Biol 12:810–821

Achieve and Sustain Climate Resilient Agriculture Through Focused Involvement of the Farming Community: A Bottom-Up Public Leadership Perspective in Support of the Missions of the Government of India

Kirit N. Shelat and R. Gopichandran

Abstract

The paper focuses on three important dimensions of climate resilient agriculture. (1) The various missions launched by the Government of India have set the context to mutually reinforce climate resilient agriculture. At this juncture, it is important to prepare the farming community across the country to understand the benefits of preventive practices including conservation and help implement appropriate measures. A framework for capacity building of communities is presented duly considering the various initiatives launched by the State Agriculture Universities and the Department of Agriculture. This framework clearly states the synergies across programmes and helps avoid duplication of efforts. (2) It is also well known that farming community across the country has periodically shown that it is feasible to develop and implement locally relevant and innovative nutrient, bio resources and soil and water conservation strategies aligned with the goals of the missions. It is however important to sustain these initiatives to upscale them and help adapt them suitably for the benefit of a larger number of communities. It is equally important to define policies and plans to foster them far beyond the tenure of State/national initiatives so that positive results can be maintained. The architecture of some policies to meet this objective is presented for the consideration of the national missions. Communities and their leadership therefore are important in this context. Some institutional mechanisms to involve elected and non-elected leaders are also indicated. (3) The framework for capacity building and the architecture proposed highlight the integrated benefit of agriculture to tackle impacts of climate change. The photosynthetic ability of crops and associated vegetation has to be recognized and further strengthened by recruiting larger tracts of land.

K.N. Shelat (✉)
ISPL &The NCCSD, Ahmedabad, Gujarat, India
e-mail: drkiritshelat@gmail.com

R. Gopichandran
GERMI, Gandhinagar, Gujarat, India

A.K. Singh et al. (eds.), *Climate Change Modelling, Planning and Policy for Agriculture*,
DOI 10.1007/978-81-322-2157-9_15, © Springer India 2015

This will also expand livelihood opportunities and meet emerging food and nutritional security related challenges. These are critical aspects of sustainability and have to be addressed on a priority basis to complement the national missions through a bottom-up intervention. The spread and depth of specific and cross-cutting interventions are discussed in the paper.

Keywords

Agriculture • Photosynthetic ability • Mitigation and adaptation • Communities • Public leadership • Capacity building • Sustainable development

The Context of India's Plans and Missions

The Government of India has launched preparations towards the 12th five-year plan. This is with special reference to aspects of agriculture that have to recognize and tackle multiple challenges that determine economic and environmental sustainability. It is quite appropriate that the national missions on the management of impacts related to climate change have recognized the cross-cutting aspects of policy, plans, programmes and projects across sectors and the convergence needed to avoid duplication of efforts. Deliberations on preventive environmental strategies as part of the missions take note of the importance of resource quality and the need for appropriate remedial and augmentation interventions to sustain productivity.

A Snap Shot of the History and Impacts of Some Interventions by the Government at the National Level

The Government of India introduced a special programme for assistance of small and marginal farmers – SFDA programme (1975–1980). This was followed by Integrated Rural Development Programme (IRDP) (1980). The National Extension Programme also strengthened the focus was on individual poor family, particularly farmers and artisans, and the VLW was responsible for providing technical assistance to poor farmers. Special subsidy was provided to obtain productive assets and inputs with back-up bank credit. The programme achieved great success through which poverty was on the decline and the small and marginal farmers benefited. However, a good proportion of the farmers was left out and needed continuous attention. In the 1990s, the focus of IRDP programme shifted to non-farm activities. Poor farmers no longer received special individual attention. The focus shifted to farmers who can obtain assistance on their own. However, a positive step taken by Government of India was to write-off farmers debts about Rs.50,000 crores. The role of women farmers cannot be over emphasized in this context (Reid et al. 2010).

Lack of control on quality seeds and basic guidance for selection of crop to be grown in rain-fed areas based on soil health further compounded the context. Poor farmers imitated wealthy farmers and lost a great deal on all fronts and tended to resort to action that was socially disruptive. Since several of these externalities have not been resolved, it is only logical to expect that issues of climate change will only accentuate impacts. Alternating floods and droughts have made it impossible for reservoirs to capture enough drinking water in many parts of our country while hill areas are often washed out by storm surges and mudslides. Changing temperatures will have a profound effect on the plants and animals. Crops that flourished will have to adapt to the new one and will increasingly loose of yields and milch cattle, especially cross-breed cows will lose on milk yield. It may be useful to revisit the Small & Marginal farmers development programme (SFDA) of the late-seventies. It is therefore essential to mainstream appropriate mitigation and adaptation strategies into local development planning (Rattana and Krawanchid 2012; Community leadership and climate change) and derive inspiration from comparable initiatives world over (IREX invites Community Leaders from Select Countries to attend the Community Solutions Program in the U.S 2011; Adaptation Learning Programme in Ghana Women take lead in tackling climate change 2011).

Traditional extension services involved VLWs who most often were involved in diverse multifarious activities. They however predominantly provided knowledge only about production. But the need is to enable choosing the appropriate crop based on soil moisture and health analysis and market prices; help understand the impacts of climate change and the use of eco-agriculture of biodiversity. Farmers need extension support prior to Khariff. Scientists along with the extension team at the district level (VLW extension Officers, APMC representatives and Bank officials) could interact with guide them suitably. These can be followed by interactions on TV/Internet/Telecommunication system during the rest of the season. A similar exercise needs to be undertaken prior to Rabi crop.

The recent MGNREGA, the Integrated Watershed Management Programme influencing livelihoods of people in the watershed, the NAPCC and the NICRA that integrate the aspects of agriculture and allied sectors can convergence on farm-level mobilization – of knowledge, technologies, expertise and market information to overcome stresses, synergies amongst ATMA-KVK-CBO linkages and their roles in different districts and adapting watershed – level plans to individual farmer plans through innovations in institutional arrangements, policy support and other technologies.

Some Important Considerations

Agriculture has exhibited only about 2–3 % growth compared to about 8 % in the other sectors. There is significant scope for productivity enhancement and the need to reduce the spread and depth of natural and artificially induced perturbations and harmonize productivity gains for all sections of the farming community. This is based on the fact that only some have been able to overcome challenges through diligence. The others have to be assisted with appropriate action-oriented information support and technical capacity building on preventive, remedial and adaptive strategies for applied resource conservation and optimize yields. There is a significant gap between leaders and communities in so far as perceptions of issues, delivery of solutions and involvement in decision making and sustainability action are concerned. This gap has to be bridged through focused community leadership enhancement initiatives that focus on "Awareness for Action & Leadership that enables Participatory Sustainable Development". This calls for an integrated approach with appropriate planning and policy framework supported by effective programmes. The central feature involves the improved management of land, water and bio resources to strengthen productivity and enhance livelihood improvements with focus on poor farmers.

India has achieved self-sufficiency to meet the requirement of food and also exports food. But the micro-level growth in agriculture is far behind the service and industry sector. This is so over the last two decades. This has a direct impact on continuously increasing the rural–urban divide and attraction to leave farming and move to urban centers. There is a wide disparity between states in the growth of agriculture. Farmers find farming increasingly unattractive in almost all states. According to the NSSO report (2005), 60 % of farmers do not prefer farming. Within rural areas the disparity is much more. There are some (perhaps a few) farmers who make money, achieve higher yields and are prosperous. The majority of small farmers (80 % of total farmers) lag behind with no or minimal grow that about 1 % as indicated above. Those with negative growth land in the debt trap. Some amongst them leave farming and migrate to urban centers. With same land and water resources, one farmer makes profit but others commit suicide. The young and educated are attracted to Naxalism and their number is growing.

Agriculture is under threat. Increasingly area under agriculture is being reduced due to rapid urbanization, indiscriminate growth of industrial townships, mining including illegal mining and infrastructure projects such as roads and highways, railways, ports etc. The worst threat is due to the speculative value of land – land developers go on buying lands outside urban townships for investment. The net result is even now retail investors and wealthy farmers are attracted to this. The greener areas are getting increasingly reduced at an alarming rate all over country. This is further aggravated by the climate change. Its adverse effect is well known and is being experienced by one and all. Heavy rains episodes with fewer rainy days, untimely or delayed rains and variation in temperature or frost as inclement phenomena have already affected the productivity of farm lands, milch cattle and fish catch. The worst affected continue to be small and marginal farmers. The overall impact of above – if not taken care of properly, would have disastrous consequences:

On the other hand, India has rich and varied experience of facing adverse climatic conditions across the various agroclimatic and biogeographic regions. With local initiatives of government officials and Ngos, our farmers have converted drought-prone areas into green-ones in the states of Gujarat, Rajasthan, Andhra Pradesh, Karnataka and Maharashtra. Similar initiative has, for instance, turned around local level economics of areas affected due to the earthquake (Kutch-Guj) and Tsunami (Tamil Nadu & Kerala).

The country has significant knowledge, successful research, well-developed technology and empirical experiences. It has the capacity to meet the challenges of lopsided development referred above and not only stop reduction in green areas but enhance the green cover. Although all states are not growing equally, it is possible to have enhanced growth in all states with determined and consistent efforts by its Public Leadership and that it is feasible to achieve growth rates of 6–8 %, as already achieved by some states, by the entire country as the recent example of the Gujarat shows. It has overcome obstacles to growth in past and can do it so again.

As indicated above, food security can not only be achieved; the country can supply food to other parts of the world where problems are already worse. But for achieving this it is necessary to recognize that the agriculture vegetation is the main tool to meet the problem of poverty and that of global warming – the enhanced release of CO_2 in atmosphere. Agriculture – The green cover is the nature's factory to absorb CO_2 from atmosphere and release oxygen through photosynthesis. There is no other comparable technology to reduce CO_2. It has to be used optimally to reduce the imbalances in the atmosphere. This can be achieved by enhancing land areas under agriculture by development of waste lands, wetlands, grasslands and compulsory green cover in open areas of urban centers and townships which are increasingly reducing vegetative cover. Suitable landscaping and natural resources management interventions have to be dovetailed with crop mixes and schedules for optimal benefits. Small farmers (who represent about 80 % of the community) in particular have to be enabled to overcome related challenges through greater diligence than seen presently.

There is a felt need also for community leadership enhancement initiatives that focus on awareness for action and leadership that enables participatory sustainable development. Awareness has to be created in all stakeholders right from the highest level, say the Chief Minister and Chief Secretary to lowest level viz. Sarpanches of villages and village level workers (VLW). For instance, the Government of India may invite Chief Ministers and Chief Secretaries for review of progress of small farmers and initiatives that enhance the form and function of agriculture, related landscapes and ecosystems services.

It will be useful to devise an integrated framework on "Thrust areas for Research and Initiatives through Public Leadership" (TRIPLE) for expedited and focused delivery of value-added services through a two-way process. This could help mainstream knowledge resources of communities to tackle location-specific and dominant challenges and exert rights to strengthen agriculture through locally adapted action. One of the important perspectives in this context is about the positive role of agriculture to twin benefits of mitigation and adaptation. It is well known that the photosynthetic ability of crops is central to carbon mobilization. This aspect mutually reinforces its role in integrated protection with implications for efficient management of other natural resources. However, the most important question we need to ask is about the extent to which the twinning benefits of agriculture is recognized and mainstreamed in policies for integrated management. This perspective is different from the often discussed role of agriculture as an emitter of greenhouse gases. Answers to this question will further strengthen the case for promoting locally adapted agriculture with positive land and water management spin-offs. This aspect has not been adequately addressed in our country context.

A complementing aspect is of the role of all stakeholders associated with agriculture. Of special relevance is the preparedness of the farming community to exert its influence on the right choice of inputs that can sustain productivity. The vagaries exerted by monsoon, the quality and quantity of land and water resources compound issues and call for location-specific interventions that have to simultaneously deliver on two fronts. The first front is about preventing further deterioration of resources and remediation. The second front is about capacity building of communities to assess changes in the profiles of resources, report and exert rights through leadership mechanisms and well-informed action to ensure sustainability. This gravitates towards an "awareness to action" paradigm. Continuous interactions with decision makers is essential for mid-course corrections/modifications of policies and plans and hence the need to foster leadership qualities in communities. The role of elected and non-elected leaders cannot be overemphasized in this context. The present paper therefore calls for immediate attention on these two aspects. Importantly this is the first attempt in our country to highlight the relevance of these areas from a public policy perspective. The paper also takes note of eight important emerging interventions and insights from several parts of the world. These are stated as part of the references. The role and impact of initiatives already in progress are important for a dynamic convergence. The value-added dimensions of sustainable agriculture in this context include the following:

- Mainstreaming agriculture has the important integrated mitigation and adaptation option to tackle challenges posed by climate change through appropriate policy and plan provisions supported by well-designed programmes and projects to deliver the desired impact of integration
- Build capacities of leadership (elected and non-elected individuals and institutions) to assess information and technical capacity needs of communities; provide appropriate information based on the assessments and build capacities in a timely manner so that communities can develop and implement locally relevant and feasible mitigation and adaptation practices. Both these dimensions are interrelated and are proposed in response to a felt need in our country to improve preparedness of communities to reduce impacts of climate change especially at the microclimate and local levels.

More importantly, the present framework highlights the need to recognize the potential in communities to coalesce and meet the collective goals of productivity and sustainable agriculture. This submission is a forerunner of the concept of integrated mitigation and adaptation-centered agriculture in our country, with a special focus on community leadership to deliver. This is to enable a much needed bottom-up approach based on ground realities of resource quality and community action through ownership to overcome challenges. It will also enable the direct involvement of the affected stakeholders in decision making for the development and implementation of locally relevant solutions expeditiously.

In order to make this a reality it is however important to suitably complement and adapt existing systems of agriculture extension, policy and infrastructure support and optimize benefits.

This will be possible if we have a clear understanding of the dynamics of some important aspects raised in the following. The important questions that have to be answered on a priority basis are as follows:

- Are our institutional mechanisms, especially at the grass-root level (Panchayati Raj Institutions and others) adequately interactive with communities on aspects of biodiversity conservation and management? This is essential to recognize the spread and depth of knowledge and locally adapted practices communities hold and suitably augment them for sustainability. It is well known that communities have been able to mobilize resources on their own to physically modify landscapes, conserve bio-resources and use natural systems to augment external inputs; enhance productivity and most importantly make well-informed choices. It is essential to build on these insights to develop a portfolio of locally relevant and sustainable solutions and ensure optimal community participation. The important groups of elected leaders include Sarpanch, Taluka Pramukhs, District Pramukhs, MLAs, MPs, trade and industry associations/Cooperatives and the like. The set of non-elected leaders includes entrepreneurs, heads of industry, NGOs, civil servants, educational leaders and development agencies.
- Are communities aware of the signals of climate change impacts and the need to adopt mitigation and adaptation strategies including conservation of bio-resources directly and indirectly associated with agriculture? This calls for an understanding of the linkages between soil, water and bio-resources management with implications for sustainable livelihoods. Integrated planning for resources management, mitigation and adaptation include aspects of agro-ecology and micro irrigation plans that dove tail with other management initiatives. Agro-ecology focuses on enhancing carbon sinks through soil organic matter and above-ground biomass; minimizing carbon dioxide or other greenhouses gas emissions from farms by reducing direct and indirect energy use. Introduction of village level micro irrigation plans for contour bunding, gully plugging, check dams and village ponds and farm ponds based data from satellite imagery are equally important.
- If, programmes and projects have delivered locally adaptable solutions to manage impacts of changes on local resource quality and quantity (including those of climate change at the micro-level), are they designed to sustain delivery of benefits across longer time periods? Very often the positive impacts of the initiatives appear to be co-terminus with the period of implementation of the project. This in turn appears to be often related to the period of funding of the project. Is it therefore possible to device sustainability mechanisms that can go beyond the short and medium terms?
- Community action can be stimulated and sustained with appropriate policy incentives for mitigation and adaptation measures. These policy incentives can target conservation of resources, sustainable practices of extraction and replenishment of the resource, payments for ecosystem services, use of appropriate technologies and avoid the use of environmentally harmful tools, techniques and material inputs. Are our systems adequately sensitive to these specific requirements of integrated and sustainable action?
- Can communities exert their right to a safer and productive environment through appropriate grass-root institutional mechanisms guided by well-informed public leadership? This aspect calls for focused capacity building of communities and institutions to understand the relevance of information for local action and assess the specific information and capacity building needs of communities; as essential aspects of public leadership for action. This also signifies a paradigm shift from environmental awareness to environmental action.
- What is the extent to which agriculture is mainstreamed as a mitigation option considering its photosynthetic potential in our

agronomy strategies? Normally agriculture is only seen as an emitter of greenhouse gases. On the other hand, it is extremely important to take note of the carbon sequestration potential of agricultural crops. This aspect of agriculture has to be mainstreamed into policies on agronomy to highlight multiple benefits of promoting land and water conservation, choice of appropriate climate resilient crops and their sustainable production. This has enormous implications for widening livelihood options for communities, based on improved management of land and water resources. It is also possible to recruit waste lands and less productive lands through appropriate remediation and value addition. Agriculture therefore can become an integrated solution to the problems of greenhouse gas emissions; further implying improved natural resources management and widened livelihood options. Based on the above, it is clear that the scope for integrated mitigation and adaptation interventions is quite significant.

The TRIPLE framework for sustainable and climate resilient agriculture therefore proposes three well-defined goals and approaches and allows the use of a wide variety of tools and techniques to deliver outputs that can be verified with indicators.

- Provide guidance to local authorities to strengthen community leadership to enable for well-informed action. The government has significant experience assisting communities overcome challenges due to adverse impacts of policies and programmes including access to materials for survival and livelihoods. These experiences can be useful to assess the economic impacts of adaptation, especially in communities that highly dependent on agriculture. The extension approach for food security supported by skill enhancement wherein knowledge economy tools to facilitate community action and leadership through institutions including Panchayati Raj can be used efficiently.
- Interface with other link departments including Department of Forests & Environment, Irrigation, Rural Development, Department of Earth Sciences, Space and Remote Sensing, DST and micro-level planning for preventive measures understand limits and limitations of technology and practices to sustain positive impacts with respect to biodiversity, soil management, post harvest management, water, river and other water basins, grasslands, waste lands, wetlands, nutrients, bio-char, organic agriculture, agriculture and urban environment, fisheries and weather forecasting. Augment and increase the diversity of independent income streams to enhance ownership of alternatives. Reward adaptation and mitigation initiatives as community enterprises and scale up and suitably adapt depending on location specific needs to establish and further build resilience.
- Assess the purpose of existing programmes, externalities they generate, the diversity of compensatory interventions needed to overcome barriers and nature and quantum of assistance to tackle issues. This assessment will help minimize anomalies in policies and plans that determine resource allocation and harmonize initiatives for optimal results.
- Differentiate between impacts on account of natural climate change and impacts generated through artificially induced variations. The latter can pertain to unsustainable practices and cross-sectoral externalities. These impacts could be location and time-specific and have to be duly recognized and addressed. It is essential to gather significant insights on these three aspects through reality checks and articulate policy and plan prescriptions that can be suitably adapted across the country.

Useful Lessons from Some Pilots that Demonstrate Success of Public–Private Partnership

- The case of Rukumavati River Basin – Kutch Dist, Gujarat – is a proof of this possibility. This has helped understand potential of various natural resources like water, land, vegetation etc. within the basin and educate the stakeholders about the situation and generate multiple benefits pertaining to river basin management, prevention of salinity ingress, socio-economic status, decentralized and people centered institutional resource management mechanisms, micro climate in the basin area, drought-resistant agriculture and horticulture crops, impacts of high velocity of water or flooding of field and indigenous animal breeds.
- A watershed approach was adopted in about 2,000 acres in two villages of Lakhpat Taluka. The Chuger village farmers generated about 3–4 lakh income and employment of about 15,970 man-days. Soil erosion reduced and improved productivity. The water table and quality of water improved in the wells of surrounding area. Animals from surrounding five villages got water for drinking and land areas were taken up under joint forest management.
- The transformation of Dahod – Gujarat is a case in point. Dahod is a remote tribal district in Gujarat with a difficult terrain. Prior to 1974 it was designated the poorest district in the country; drought prone area with highly unpredictable monsoon. These circumstances were responsible for migration. By 2010 the same district achieved food security, housing conditions improved, school enrolment and attendance increased manifolds, 68,000 ha of land were brought under irrigation. 17,000 wells were recharged and the irrigation coverage is around 30 %. 700 community water resources were developed and managed by 325 village level irrigation cooperative societies. 2,700 village institutions – users groups manage their affairs and assets; 65 rivers and rivulets made perennial through series of structures connected to lift irrigation system. Migration diminished with better livelihood options through appropriate cropping patterns of horticulture and floriculture. This socio-economic and ecological impact was achieved by initiative taken by Jagavats of Sadguru Development Foundation that commenced in the early seventies. Their efforts converged with the Government programmes and promoted local level initiatives and leaders to manage their affairs.
- The Community biogas project – Chhota Udepur generated significant benefits for the community tackling agro wastes and dung implicated in methane generation. The community biogas plan set up with the involvement of the village community produced 3 Ton slurry per day and network of gas pipeline for village household. The liquid slurry was converted into vermicompost and transported to urban areas too.

The Gujarat Experience in Sustainable Agricultural Development

Diverse in its topography, it boasts of a 1,600 km coast line and is home to the largest desert in the country the Rann of Kutch. The state has all possible handicaps faced by agriculture wherein 70 % is rain-fed, droughts recur; with untimely/irregular rainfall and some areas receiving rain only three to 4 days in a year. Gujarat's agriculture suffered heavily whenever there were droughts. The growth rate of agriculture used to be negative during such years. In a normal year, the agricultural growth rate used to be 2–3 %. Agriculture was not sustainable in many parts of the state due to recurrent crop failures. However, this is a story of the past. In the new millennium, Gujarat, with determination and persistent efforts, changed the agriculture scenario. From 2004 agriculture witnessed a major turnaround with a growth of 11 % per year. The state became a front-runner in agricultural production in the country. This turnaround became possible due to certain effective experiments and steps taken by people on the basis of the experience of Pujya

Kaka mentioned earlier, government's experience and that of agricultural universities. Such successful experiences did not remain specimen or model projects, but also became a base to launch an overall initiative in all 18,000 villages in the state, known as "Krishi Mahotsav".

The key to this success was public leadership both elected and non-elected, which introduced effective soil and water management and proper land use by using mass communication approach-based, micro level management model. On the water front, more than 100,000 check dams got constructed, and the Mahi and Sabarmati were interlinked. These rivers, in turn, were linked with Narmada and Mahi canals. In its rain-starved areas, such as North Gujarat and Kutch, a special scheme for irrigation known as "Sujalam-Suflam" was introduced. Scientific agriculture was introduced by distributing Soil Health Card to every farmer. From 2004 onwards, 50 farmers from each village were given such cards every year along with soil moisture analysis and past 5 year's average market price of the crop grown in their area. This helped them make informed choices about crops that gave higher returns and were sustainable in the soil.

A direct door-to-door extension programme for guiding the farmers at village level was introduced under a pre-Kharif programme, known as "Kirishi Mahotsav". Every village was visited by a development team comprising agri-scientists and officers from the veterinary department, co-operative, irrigation department, rural development department and local banks etc. High-yield crops were identified. The farmers were guided about using certified seeds and looking at price of APMCs before selling their farm produce. BISAG, a institute set up in collaboration with Space Application Centre (SAC), Ahmedabad, by the Government of Gujarat, prepared a micro-level plan for land use by identifying sites for check dams and village ponds for every village. ICT material was made available to the farmers in their mother tongue for crop management, including the use of fertilizers and pesticides. Free telephonic helplines were introduced to answer the queries of farmers. All these initiatives were backed by total involvement of public leadership, both elected and non-elected. The major transformation occurred in the state's most difficult areas of Saurashtra, Kutch and North Gujarat. These insights can help replicate initiatives with comparable impacts across our country. The turnaround in Gujarat was due to effective public leadership.

The Way Forward

It will be useful to carry out a reality check on the dynamics of policy implementation and community preparedness in at least six states across India that represent diverse challenges. Simultaneously assessments can be carried out in all the agro-climatic zones in Gujarat to derive a micro perspective. It will be possible to synthesize learnings from these across – states and within the State contexts with respect to decision making and enabling mechanisms. Lessons from this reality check will be useful to develop action – based delivery – centered interventions in the states including Gujarat that culminate in remedial and preventive action. This will also set the pace and the context for appropriate integration in other States across the country as the logical way forward. The links between policies and plans are critical to ensure local relevance and acceptance by communities. Climate-smart agriculture practices have to be robust and tackle multiple cross-cutting challenges (Policy Oriented research/Scientific Support to Policies Call FP6-2005-SSP-5A 2008; Kelsey and Wall 2003; FAO's success stories on climate smart agriculture) that modulate livelihood profiles.

References

Adaptation Learning Programme in Ghana Women take lead in tackling climate change (2011) 2p. http://www.careclimatechange.org/files/adaptation/ALP_Ghana_Saamini_Nov2011.pdf

Community leadership and climate change (2001) Guidance for local authorities 2001, 12p. http://www.idea.gov.uk/idk/aio/396534

FAO's success stories on climate smart agriculture, 28p. http://www.fao.org/3/a-i3817e.pdf
IREX invites Community Leaders from Select Countries to attend the Community Solutions Program in the U.S. (2011) http://www.fundsforngos.org/developing-countries-2/cambodia/irex-invites-community-leaders-select-countries-attend-community-solutions-program/#ixzz1iwEJ5Mo0
Kelsey KD, Wall LJ (2003) Do agricultural leadership programs produce community leaders? A case study of the impact of an agricultural leadership program on participant's community involvement. http://pubs.aged.tamu.edu/jae/pdf/Vol44/44-04-35.pdf
Policy Oriented research/Scientific Support to Policies Call FP6-2005-SSP-5A. Deliverable D6 synthesis report 2008. Policy incentives for climate change mitigation agricultural techniques. http://climatechangeintelligence.baastel.be/piccmat/files/D6_CS%20Synthesis%20report_final.pdf
Rattana K, Krawanchid D (2012) Main-streaming adaptation into local development planning: a case study in Chainat, Thailand. Adaptation Knowledge Platform, 32p. http://www.climateadapt.asia/upload/publications/files/51091ce0b11f8Chainat_Mainstreming_adaptation_web.pdf
Reid H, Huq S, Laurel M (2010) Community champions: adapting to climate challenges, 112p. http://pubs.iied.org/pdfs/10028IIED.pdf

Climate Change and Indian Agriculture: Impacts, Solutions, and Adaptation

Hari Ram Prajapati

Abstract

This paper has examined the impact of change in climate variables on the productivity of wheat and rice, with the help of agricultural dataset, spanning 1971–2005, for agricultural yield at the district level. The expected impact of productivity of wheat and rice due to climate variables has been turned to declined wheat production and increased rice production in some regions, while the magnitude of impacts is varied. Any change in climate variables directly affected inputs such as water for irrigation, amounts of solar radiation that affect plant growth, as well as the prevalence of pests. The study finds that impact of climate change is not identical; it varied in all the geographical regions because agricultural activities in India mainly depend on climate variables: monsoon, rainfalls, temperature, growing degree days, etc. Any changes in these variables projected to have adverse effects on agricultural productivity, water resources, coastal ecosystems, and biodiversity.

Keywords

Temperature • Wheat • Rice • Climatic variability • Ecosystems • Biodiversity

An Overview of Climate Change and Indian Agriculture

An alarming change has been seen in the climatic variables of the current decade such as global warming, rise in temperature, and variations in rainfall. The impact of these variables is visible at the local and national level. Various studies have noticed the change in climatic variables and its impact on agriculture in Indian economy. There is strong evidence that the annual mean temperature of India has remained consistently above from normal temperature since 1993–2011 (except 1997). The wheat crop has been adversely affected by these changes; its output remained depressed for several years till 2011 due to early onset of summer. Other crops and its allied fields will be definitely affected in the long run. The annual climate summary (2007)

H.R. Prajapati (✉)
Center for Studies of Economics and Planning, School of Social Sciences, Central University of Gujarat, Gandhinagar, India
e-mail: harry.cug@gmail.com

A.K. Singh et al. (eds.), *Climate Change Modelling, Planning and Policy for Agriculture*,
DOI 10.1007/978-81-322-2157-9_16, © Springer India 2015

released by the Indian Metrological Department (IMD) in April 2008 contend several amazing facts, providing evidence of the global warming on the Indian climate; 2007 was the fourth warmest year on record since 1901. The annual mean temperature of India as a whole was recorded at 0.55 °C above from the normal temperature. According to the IMD future projection, monsoon rainfall is expected to increase by 15–30 % by the end of the twenty-first century. The mean annual temperature is predicted to rise by 3–6 °C, but the northern India will be warmer than the other part of the country.

Recently, a study conducted by the Indian Agriculture Research Institute (IARI) finds that around 4–5 million tons of wheat production is loss with every 1 °C rise in temperature throughout the growing season, despite some above probable positive effect on yields due to likely increase in atmospheric carbon dioxide content. An actual loss in productivity may be higher than expected, because the above study is based on the assumption that irrigation availability would continue at the present level, which may be actually less or more in future. Climate change will also affect water supplies in India, where high-intensity floods and droughts are expected in the near future. Increasing water storage is a key adaptation strategy, and the experience of irrigation tanks illustrates both the potential and challenges of this adaptation response. Although there are over 208,000 water tank in india, (See ISEC, Bangalore, 2006 report No E15590AP0) in India, irrigating about 2.3 million hectares in 2000–2001, the net area irrigated by tanks declined by 29 % between 1990–1991 and 2000–2001 and by 32 % between 2001 and 2008.

The Intergovernmental Panel on Climate Change (IPCC) reported in its 4th assessment report that the earth's climate system has made remarkable change at both regional and global scale with compared to the preindustrial era. Further, the report also pointed out that the global mean temperature may increase somewhere between 1.4 and 5.8 °C by 2100. This drastic increase is expected to have brutal impacts on various aspects of climate system including changes in the global hydrological system, increase in the sea level, and changes in crop production. It is now widely recognized that developing countries are more vulnerable as compared to the developed countries with the impact of climate change and variability. It happens because in developing countries, the ecological environments are fragile and the vulnerability of these economies is high due to low-income level and less ability to cope with the climate change. The initial circumstances of each country in terms of its climatic conditions, socioeconomic setting, and growth prospects will also partly determine the scale of the social, economic, and environmental impacts of climate change (Stern 2007). Climate change is likely to impact all the natural ecosystems as well as socioeconomic systems as marked by the National Communication Report of India to the United Nations Framework Convention on Climate Change (UNFCCC). Different sectors like water resources, forests, agriculture, and coastal zones are projected to have several potential impacts, which will bring changes in hydrological cycles, in rainfall, as well as in the magnitude and timing of its runoff. However, the distribution of the incidences of climate change will also vary within the geography boundary of nations.

Purpose of the Paper

The main purpose of this paper is to identify the probable impact of climate change on Indian agriculture, highlighting the extent of India's vulnerability to different regions. The impact of climate change on Indian agro-system has been examined with the help of certain variables like productivity, especially wheat and rice; changes in rainfall pattern and temperature during crops' growing days; and increases in emission of greenhouse gases and temperature by human and nonhuman activities. Finally, it discusses the policy-related reforms for forthcoming problem that may provide an appropriate solution, applicable in rural as well as in urban areas.

Review of Literature

Most of the previous studies have followed two methodologies for the analysis of economic effects of climate change: one is *the production function approach* and other is the *Ricardian approach*. The production function approach is also known as the *Crop Modeling Approach*, based on controlled agricultural experiments, where specific crops are exposed to varying climates in laboratory-type settings such as greenhouses and yields and then compared across climates. This approach has an advantage of careful control and randomized application of environmental conditions. However, these laboratory-style outcomes may not reflect the adaptive behavior of optimizing farmers. Some adaptation is modeled, but how this will correspond to actual farmer behavior is uncertain. If farmer actual practices are more adaptive, the production function approach is likely to produce estimates with a negative bias. On the other hand, if the presumed adaptation misses constraints on farmers, then it does not consider adjustment costs into account. These estimates could be overoptimistic.

The Ricardian approach, pioneered by Mendelsohn et al. (1994), attempts to allow for the full range of compensatory or mitigating behaviors by performing *cross-sectional regressions* of land prices on county-level climate variables, plus other controls. If markets are functioning well, land prices will reflect the expected present discounted value of profits from all, fully adapted uses of land, so, in principle, this approach can account for both the direct impact of climate on specific crops and farmers adjustment of production techniques, substitutions of different crops, and even exit from agriculture. However, the success of the Ricardian approach depends on being able to account fully for all factors correlated with climate and influencing agricultural productivity. The omitted variables, such as dicey weather, rainfall, and soil quality, could lead to bias of unknown sign and magnitude.

More recently, economists have turned to a *panel data approach*, using presumably random year-to-year fluctuations in realized weather across US counties to estimate the effect of weather on agricultural output and profits (Deschênes and Greenstone (2007), Schlenker and Roberts (2006), Schlenker and Roberts (2006, 2009), Deschenes and Greenstone (2007), and Krishnamurthy (2010)). This approach overcomes the omitted variable-type issues encountered in the Ricardian approach and permits controlling for unobserved individual heterogeneity (via the use of county fixed effects), such as farmer/soil quality, while at the same time allowing for some farmer adaptation, providing presumably better estimates than the production function approach. A major drawback with this approach is an inability to reflect changes in technology and choices (such as switching of crops, exit from agriculture, and change in cropping seasons). Using this approach for India, along with gridded reanalysis weather data, Guiteras (2008) explores the impact of climate change on annual profits at the district level and estimates moderate losses up to 9 % for the medium term and significant losses up to 25 % for the long term, under a variety of climate scenarios.

Methodology and Data Sources

The present study explored the impact of climate change on Indian agro-system: past, present, and future scenario with the help of certain climate variables. The study has classified variables into two categories: *controlled* and *uncontrolled variables*. The controlled variables have been considered as those which are controlled by the government through policy regulation, such as agricultural price policy, input subsidy policy, environment policy, and irrigation policy, whereas uncontrolled variables are those which cannot be controlled by policy instruments such as monsoon, temperature, and rainfall. This paper is concerned about both controlled and uncontrolled variables, based on the secondary data sources collected from the Ministry of Agriculture, annual reports, Census of India, Indian Metrological Department, published and unpublished government reports, and IPCC report, World Report on Climate Change, annual report of IACR, books, journals, and electronic media.

The novel *panel data approach* has been applied for linking observed weather outcomes and agricultural outcomes at the district level chosen quantiles (like talukas). Under this framework, crop yield, at each quantile, is modeled as being a (potentially nonlinear in covariates) function of weather variables, primarily growing degree days and seasonal (monthly) rainfall and district-specific fixed effects. The use of district-specific fixed effects allowed unobserved determinants and time-invariant effects (such as soil type and topography) to be controlled for the estimation of result. This approach regressed yield productivity to with observed weather and temperature outcomes, at district-fixed effects, for each quantile of interest.

The analysis is based on the findings of Krishnamurthy (2010) based on recent available dataset of gridded temperature and rainfall data for India, from the Indian Meteorological Department, spanning 1969–1999, and an agricultural dataset, spanning 1971–2005, for agricultural yield at the district level, for estimation Rice (218 districts) and Wheat (222 districts) are grown in different seasons, with very different characteristics, and projections of climate changes are very different for these two seasons. It is therefore important to account for considerations of separate growing seasons and hence very different.

It is important to account for considerations of separate growing seasons and different crops in the different growing seasons. Prior estimation of climate change effect on agriculture productivity in India suffers from two drawbacks: first, given that there are two agricultural seasons, it is not clear that ignoring completely weather (climate) in this season while including agricultural outcomes for this season as part of the response (as in Sanghi et al. (1998), Kumar and Parikh (2001) and Guiteras (2008)) provides sensible estimate of climate change. This issue aside, it is not clear how to interpret the monetary equivalent of crop yield when summed over crops grown in different growing seasons (i.e., $\sum_k y_{itk}$ p_{ik} where k denotes the crop, y_{itk} denotes the yield of crop k in year t and district i, and prices P_{ik} are fixed in time) at the district level (Guiteras (2008)). For any interpretation of this variable as potential productivity, one must also control for all inputs, which is not possible in this case (due to unavailability of data). A final issue is the possible confounding of price impact from productivity; i.e., results are likely sensitive to impacts of the specific price used. The approach outlined in this paper avoids the issues raised above altogether by working directly with changes in physical crop yield, addressing crops grown in different seasons separately, and accounting for substantial changes in seasonal weather directly.

The plan of this chapter has followed the process: first introduce a brief outline of agronomic and economic basis of the model, then outline the advantages of quantile regression (QR) in the current setting, indicate the benefits and interpretation of fixed effects in quantile regressions, then proceed toward the econometric model adopted for the purpose, and, finally, report results of the estimation, including impacts of climate change on yields.

Theoretical Framework

For the estimation of the impact of weather on crop yields, it has used the latest available data weather set and also used a newly developed framework for quantile regression for panel data with fixed effects. The approach outlined here allows estimation of the potentially *nonlinear relationship between temperature and rainfall on crop yields*, at different quantiles.

Temperature and Crop Yield

The statistical model has been proposed here; with selected climate variables, it is imperative to obtain a seasonally aggregate metric of temperature relevant for crop growth (given that aggregate metrics such as "average temperature" yield almost no relevant information). Following the approach outlined in Schlenker and Roberts (2009), Guiteras (2008) and Krishnamurthy (2010) assumed a temperature limit within which heat is beneficial to plants, in a linear and time-separable manner. The upper and lower limits of temperature will vary between 8° and 34 °C for the summer growing season. Following Schlenker et al. (2006), who illustrate the superiority of the agronomic measure of growing degree days (GDD), we compute this measure using two different methods. In the first case, which we term *linear degree days*, the computation is straightforward. Given daily average temperature, T, compute

$$D(T) = \begin{cases} 0 & T \leq T(\text{low}) \\ T - T(\text{low}) & T(\text{low}) \leq T \leq T(\text{upper}) \\ T(\text{upper}) - T(\text{low}) & T \geq T(\text{upper}) \end{cases} \quad (1)$$

where the upper and lower threshold limits have been set between 8 and 34 °C for the summer growing season. The linear GDD metric was computed for each day, summed over the growing season (kharif and rabi seasons) to obtain a seasonal sum of GDD. The fixing of the growing season has less to do with the endogeneity of farmers' sowing and harvesting decisions (which are, for the rainfed areas, dependent on expected onset of the summer monsoon) and more to do with biophysical conditions, including availability of water. In order to account for potentially harmful effects of temperatures beyond 34 °C upon plant growth, we also compute "harmful degree days" as the difference between 34 °C and the daily average temperature.

The choice of temperature limits for the rabi season was made more difficult by having little India-specific literature on the issue, and much of the literature used only a lower bound and does not use any specific upper bound. Given, however, the significant temperature constraints on wheat, it was decided to choose 6 °C as the lower threshold (temperatures only rarely fall below this limit over much of the wheat-growing regions) and an upper threshold of 28 °C. Choosing slightly different upper thresholds (27° and 29 °C) did not appreciably alter the results. Harmful degree days were computed as in the case of the kharif season. Two points regarding this formulation are worth emphasizing: (a) linearity of plant growth as a function of temperature during the growing season and (b) the complete independence between temperature and precipitation, when used in the regression framework with yield (not its transformations) as the dependent variable. For instance, in the regressions in Guiteras (2008), this implies the complete separation of the effects of temperature and precipitation upon yield, which is somewhat unrealistic in the case of the summer growing season.

Precipitation and Crop Yield

Most of the econometric research focused on estimation of climate change impacts due to temperature change on plant growth. Agronomic and climate literature for India, however, has tended to focus, for rainfed agriculture, on the significance of summer monsoon characteristics. However, in the case of India, even without climate change, climatic variability, especially the variability in monsoon rainfall, has been of major interest for agriculture. This is due to (1) the season (the earlier part of the summer growing season from May to Sept is also the hottest months of the year, and temperatures generally reduce only with rainfall) and (2) lack of water availability, seasonal characteristics, and the largely rainfed nature of agriculture in India (current estimates indicate that only about 40–50 % of agriculture land is fully irrigated, which means that sowing operations are critically dependent on the arrival of rainfall).

The significance of monsoon rainfall for Indian agriculture is well documented in both agricultural and development economics (Rosenzweig and Binswanger' 1993) and in the agriculture and climate literature Krishnamurthy et al. (2000). In all four regions, it is well known that the major constraint is availability of water during the critical sowing period, May to June, and therefore, onset of the monsoon is of some importance. Further, contrary to popular perception of the monsoon as marching southwest to northwest, the monsoon has what are known as *break periods*, during which there are long dry spells (up to a fortnight or more in duration) over large parts of India. There is some speculation of harmful effects on these periods, but little actual evidence about its importance for agriculture. When place of if identification is based on weather variations from district mean weather, we argue that it is important to (a) account for rainfall variations which are not captured by monthly totals (b) condition monthly and seasonal total rainfall on onset and dry spells (Guiteras 2008). In summary, the following are the variables for summer monsoon season measures of rainfall: seasonal (monthly) total rainfall and onset day, dry spells. For the winter season, use, in addition to seasonal rainfall, a bimonthly rainfall sum.

Econometric Model

Let us begin with a brief interpretation of linear quantile regression as an extension of linear regression to the heterogeneous effects of covariates case and then turn to indicating the major ideas that underlie an extension of quantile regression to the fixed effect setting.

Quantile Regression and Heterogeneous Covariate Effects

Consider the so-called location-scale model:

$$Y_{it} = E_{it}\beta + \left(X_{ity}\right)\sigma_{it} \tag{2}$$

where σ_{it} is an error term and $\gamma = 0$ represents "heterogeneity" (observed). It is evident that the condition $\gamma = 0$ is sufficient for the conditional quantiles to be parallel to one another, and this is generally a maintained assumption in mean regressions, parametric or nonparametric. In other words, most mean regression frameworks assume that the conditional quantiles are all parallel to one another and that the covariates exert an effect only on the mean of the distribution. Evidently, this assumption is restrictive. In this context, quantile regression may be viewed as an extension of regression to the *location-scale* models; i.e., covariates also affect the scale parameter (variance, for the normal distribution) of the conditional distribution of the response, due to which the quantiles are no longer parallel to one another. For instance, in the current application, it is quite plausible that weather covariates have differential impacts on distinct quantiles of the yield; further, if the effect of climate change for India is likely manifested in increases in variance along with changes in mean, then impacts computed based on mean regression are likely to understate the true magnitude. Similarly, if there is a change in seasonal distribution of rainfall, computing the mean impact will tend to smooth out variations in yield at different quantiles.

Quantile Regression for Fixed Effect Panel Data Models

There is an extensive literature in economics, particularly in fields such as labor economics, on the usage of quantile regressions (Fitzenberger et al. 2002). Effectively, linear quantile regression is the simplest parametric framework for a scenario in which covariates are expected to exert differential impacts on the distribution of the response.

Panel data with fixed effects allow applied researchers to base identification purely "within" variation in the covariates. Further, this approach allows arbitrary correlation between the covariates and the fixed effects. There has been, however, little research on the intersection of the two methods. Two approaches have recently been developed to deal with the fixed effect quantile regression (FE-QR) case, differing primarily (from an applied perspective) in how the fixed effects are viewed (Krishnamurthy 2010). Consider the following two regression frameworks:

$$E(Y_{it}|\alpha_i, X_{it}) = \alpha_i + X_{it}\beta \quad (3)$$

$$QY_{it}(\tau|\alpha_i, X_{it}) = \alpha_i(\tau) + X_{it}\beta(\tau) \quad (4)$$

where QY_{it} $(\tau|X_{it}, \alpha)$ is the τth conditional quantile of Y_{it}. Model (3) may be seen to be obtained from

$$Y_{it} = \alpha_i + X_{it}\beta + €_{it} \quad (5)$$

By imposing the following orthogonality condition, $E\left(€_{it}|X_{it}, \alpha_i\right) = 0$, while model (4) might appear, at a first glance, to be obtained by imposing the following restriction on the model in (5): $Q_{€it}\left(\tau|X_{it}, \alpha_i\right) = 0$ (along with the separability assumption on α_i and $€_{it}$). However, the two models are not equivalent, except under special conditions, as indicated in detail in Rosen (2009). In both cases, the α_i are (a) unobservable and (b) potentially correlated with the X_{it} and with the residuals of the model (accounting for omitted variables effects), and we maintain the assumption that X are (conditionally) exogenous, i.e., $X|\alpha \perp €$. The $\{\alpha_i\}$ in Eq. (3) represent "unobserved" heterogeneity, while such an interpretation in the case of Eq. (4) is somewhat misleading, since even in the absence of $\{\alpha_i\}$, there is considerable heterogeneity allowed by the model. In fact, the $\{\alpha_i\}$ are not "fixed" in the true sense, since they vary across quantiles and provide additional flexibility in modeling unobserved heterogeneity (since they vary over the distribution of Y), due to which Harding and Lamarche (2009) term them *quantile fixed effects* (QFE).

Identification and Interpretation

There are three important issues to note with the framework above. First, following Powell (2010), it is important to note that the quantiles are defined, in the approaches outlined above, based on each individual unit's residual, i.e., based on $€_{i,t} = Y_{i,t} - \alpha_i - X_{it}\beta$ (the alternative definition being $(\alpha_i + €i)$) as a result of which one "loses" the position of an observation in the distribution, compared to the cross-section distribution. However, given that the effects identified here (weather on yield) require a panel, and do not make sense in a cross-section, it is evident that the definition used here is the most sensible one. This, however, leads to the interpretation of the coefficients being somewhat different from those in the purely cross-sectional case. Second, if year fixed effects are not included in the above framework, then, as pointed out in Powell (2010), the resulting τth quantile corresponds to different parts of the distribution for different years, a possibly undesirable artifact in applications. Third, unlike in the case of the linear panel data models (where identification and consistency follows from a simple orthogonality condition between X and €), conditions for identification and consistency of β involve high-level conditions on the marginal density of the residuals evaluated at zero and on moment conditions on X_i Koto et al. (2010), due to which they have no interpretable or intuitive content.

Covariance Matrix Estimation

For the un-penalized case, Koenker (2004) derives the asymptotic variance of the estimation and points out that bootstrap inference maybe more accurate (another cause for concern is that the asymptotic variance requires T to grow "fast enough," compared to N; in particular, Na $T \rightarrow 0$ for some $a > 0$). Koto et al. (2010), for the same un-penalized model, also point out that asymptotic and bootstraps inference require, in their case, T to grown faster than N_2, a clearly restrictive condition, and they recommend the cross-sectional bootstrap as being the better approach, after comparing both cross-sectional and time bootstraps for their coverage accuracy. We note, however, that their comparisons are for coverage for individual coefficients only.

There is a vast literature on the use of the bootstrap for confidence interval estimation in a cross-section QR setting (see, for instance, Buchinsky (1995), Fitzenberger (1998), Fitzenberger et al. (2002), Krishnamurthy (2010), and others). For the panel data case, there are three possible basic version of the bootstrap, which, following Kapetanios (2008), are termed *time* bootstrap, with resampling across the time dimension; *cross section*, with resampling across the *individual unit* dimension; and a combination. It is noted that all three versions correspond to the "paired" or the "x-y" bootstrap, involving bootstrapping the data matrices rather than the residuals. Thus normal time bootstrap uses for inference, which accounts unknown spatial dependence but not for time dependence.

Extension to Semi-parametric Quantile Regression

Let us outline an extension of the above linear conditional QR-FE estimation strategy to semi-parametric settings. Consider the linear FE-QR estimation detailed above:

$$QY_{it}\left(\tau\middle|\alpha_i, X_{it}\right) = \alpha_i(\tau) + X_{it}\beta(\tau) \quad (6)$$

It is observed that, once a particular estimation framework is settled upon, the problem above is in essence a straightforward linear QR problem (with involved conditions for consistency). Consider instead the more general semi-parametric conditional quantile estimation problem:

$$QY_{it}\left(\tau\middle|\alpha_i, X_{it}\right) = \alpha_i(\tau) + X_{1i}, t\gamma + g(Z_{it}) \quad (7)$$

where (Z_{it}) is a univariate variable to be modeled nonparametrically. There exist two possible approaches to nonparametric modeling of quantiles (Koenker (2005, Chapter 7) for a survey), the local-polynomial-based one and a penalized spline version. We note, however, that even without these options, the function $g\ (Z_{it})$ may always be approximated by an orthogonal polynomial of a suitable order, and the problem is one of the estimations of the order of the polynomial. Inference is even more problematic in this setting; however, since consistency is likely guaranteed (under possibly additional conditions), a bootstrap approach may be easily used in this case.

The alternative approach, of penalized splines, is a different variant of the same approach, in that the scalar penalty term must be chosen, generally through means of AIC-like measures (see Koenker (2010) for details and an application); we briefly pursue this approach here, more as a robustness check, but note that in most nonparametric QR cases, inference is generally problematic and computationally expensive, and this is especially true in the FE case.

Model Specification

The basic econometric model followed Krishnamurthy's (2010) production frontier for chosen crop:

$$Y_{it} = \alpha_i + \beta_1 GDD_{it} + \beta_2 GDD^2{}_{it} + \beta_3 \text{Precip}_{it} + \beta_4 \text{Precip}^2{}_{it} + \sum\nolimits_{k=1}^{K} \times \sum\nolimits_{j=0}^{J} \beta_{5+k,j} \text{Trend}^k * R_j \quad (8)$$

where y_{it} is the log of yield, R_j is a region indicator (we use 5 regions, North, West, South, East, and Central, with "North" being the omitted category), and $K = 3$ is the order of the time trend. A second possible specification where we use, in addition, monsoon onset at the district level and the number of dry spells for the summer growing season is

$$Y_{it} = \alpha_i + \beta_1 GDD_{it} + \beta_2 GDD^2{}_{it} + \beta_3 \text{Precip}_{it} + \beta_4 \text{Precip}^2{}_{it} + \sum\nolimits_{k=1}^{K} \sum\nolimits_{j=0}^{J} \beta_{5+k,j} \text{Trend}^k * R_j + \beta_{5+J+1} \text{Onset}_{it} + \beta_{5+J+2} \text{dryspells}_{it} \quad (9)$$

The third specification is that study used monthly precipitation data instead of using seasonal totals as above.

$$Y_{it} = \sum\nolimits_{j=1}^{K} \text{precip}_{kit}\beta_k + GDD_{it}\beta_{K+1} + GDD^2{}_{it}\beta_{K+2} + \sum\nolimits_{j=1}^{K} \text{precip}^2{}_{kit}\delta_j + \sum\nolimits_{j}^{J} \sum\nolimits_{k}^{K} \gamma_{k,\,j} \text{Trend} * R_j + \nu \text{Onset}_{it} + \zeta \text{dryspells}_{it} \quad (10)$$

with $K_1 = 6$ being the number of months in the winter (rabi) growing season and $K_1 = 4$ the number of months in the summer growing season (kharif), with onset and dry spells included for the summer growing season only. Another covariate added to many regressions is the ratio of irrigated area to un-irrigated area at the district level. Finally, for the winter season which spans 6 months, instead of including rainfall for all 6 months separately, in a specification, we aggregate them into three separate periods, early, middle, and late season rainfall.

Summary Statistics

Summary statistics of the key variables are presented in Tables 1 and 2 for wheat and Tables 3 and 4 for rice by regions on the basis of Krishnamurthy's (2010)[1] findings. Both tables presented descriptive statistics of regressed sample for *wheat/rice* (rabi/kharif seasons) and mean coefficients reported, with standard deviations in parenthesis. Three points are noteworthy in Table 1: the predominance of the Northern region in wheat production, its high yields, and the very high irrigation ratios. For rice, it is evident that the Eastern region dominates in rice production, followed by the Southern and Northern regions; in addition, there is a significant difference in rainfall between the Eastern regions and the rest of India. Finally, the relatively late onset of the monsoon is evident in the Northern and Western regions.

Table 1 Summary statistics of percentage change in yield of *wheat* for different regions of India under different scenarios at the district level[a]

Tau	Min	Median	Max	Min	Median	Max
	South			**North**		
Scenario 1						
0.25	−11.5	−10.6	−0.88	−11.25	−3.16	1.63
0.50	−8.89	−8.1	0.38	−8.91	−1.59	2.14
0.75	−5.96	−5.12	0.81	−6.26	−0.96	2.21
0.90	−7.10	−5.43	−0.55	−8.225	−1.89	1.27
Scenario 2						
0.25	−21.99	−21.10	−2.79	−22.47	−7.44	3.19
0.50	−17.12	−16.00	−0.10	−17.79	−4.16	4.12
0.75	−11.74	−9.63	0.88	−12.44	−2.72	4.38
0.90	−14.04	10.50	−1.44	−16.47	−4.07	2.49
	Central			**West**		
Scenario 1						
0.25	−11.23	−5.99	1.64	−9.96	−6.26	1.64
0.50	−8.86	−4.10	2.14	−7.58	−4.3	2.14
0.75	−6.25	−3.15	2.22	−5.42	−3.19	2.21
0.90	−8.61	−3.68	1.27	−5.08	−3.22	1.27
Scenario 2						
0.25	−22.37	−13.00	3.20	−20.22	−13.5	3.18
0.50	−17.64	−9.17	4.21	−15.36	−9.45	4.21
0.75	−12.36	−7.05	4.38	−10.71	−7.00	4.38
0.90	−16.23	−7.55	2.52	−10.08	−6.51	2.49

Source: Krishnamurthy (2010)

[a]Scenario 1 and 2 correspond to uniform increase in daily temperatures over the growing season by 0.5 and 1C, respectively. Median impacts are reported along with (time) bootstrapped confidence intervals

Impacts on Wheat Production

Results of estimation provided the log-linear nature of the regression; it interpreted the coefficients as approximately corresponding percentage changes. Turning first to effects of temperature, two conclusions are evident. First, it appears to be an inverted U-shape relationship between yield and temperature, indicating that temperature up to a certain point is beneficial and harmful thereafter. The results were obtained by Schlenker and Roberts (2009) and Guiteras (2008), using very different methods. A second, and more surprising, one is that the difference across quantiles of yield of the inflection point is minimal, indicating a significant degree of uniformity across differing crops and ecology types is found (controlling for irrigation and rainfall). Rainfall also appears to have a similar relationship to yield as does temperature, and interestingly, the coefficients across different quantiles are very different (for instance, in absolute terms, the coefficients in many instances differ by up to 50 %), reflecting the high spatiotemporal variability in winter season rainfall (Table 1). When accounting for a richer interaction

[1] For detail, see his (2010) extended work (Krishnamurthy (2012)) The Distributional Impacts of Climate Change on Indian Agriculture : A Quantile Regression Approach, Working Paper no. 69/2012, MSE.

Table 2 Descriptive statistics of regressed sample for *wheat* (rabi season)

	Central	North	South	West	Total
Yield (kg/hectare)	1,050	2056.1	786.1	1313.0	1551.2
	(506.5)	(829.5)	(438.5)	(592.0)	(849.6)
Production (000's of ton)	80.24	335.40	7.09	84.51	193.50
	(63.28)	(297.1)	(15.09)	(106.3)	(251.9)
Seasonal rainfall (mm)	88.13	105.0	185.0	67.69	102.7
	(68.31)	(73.48)	(101.9)	(65.68)	(82.44)
Growing degree days (°C)	2808.2	2444.1	3383.1	2845.5	2708.1
	(217.9)	(311.5)	(164.4)	(315.3)	(417.3)
Harmful degree days (°C)	8.50	4.15	32.16	15.04	10.43
	(8.90)	(5.70)	(23.50)	(14.92)	(14.70)
Average seasonal temp (°C)	21.44	19.41	24..74	21.68	20.90
	(1.23)	(1.73)	(0.99)	(1.99)	(2.35)
Irrigation ratio	0.417	0.849	0.596	0.673	0.692
	(0.293)	(0.180)	(0.334)	(0.274)	(0.301)
Number of observations	911	2,053	486	893	4,343
Districts	45	97	28	53	222

Source: Krishnamurthy (2010)

Table 3 Summary statistics of percentage change in yield of *rice* for different regions under different scenarios at the district level[a]

Tau	Min	Median	Max	Min	Median	Max
	South			**East**		
Scenario 1						
0.25	−10.86	4.60	11.40	−9.84	8.51	11.55
0.50	−11.73	2.36	5.60	−8.77	4.70	5.50
0.75	−10.98	−0.15	2.22	−8.55	1.68	2.20
0.90	−12.22	−0.59	1.71	−8.24	1.29	1.66
Scenario 2						
0.25	−20.28	12.70	22.19	−15.84	19.40	22.15
0.50	−22.37	4.40	10.65	−13.56	9.22	10.41
0.75	−20.65	−0.72	3.94	−14.24	3.11	3.89
0.90	−23.87	−1.80	2.93	−13.33	1.71	2.96
	North			**West**		
Scenario 1						
0.25	−8.57	9.56	11.26	−9.81	7.29	11.51
0.50	−7.71	4.24	5.45	−10.04	3.90	5.59
0.75	−7.40	1.46	2.17	−9.06	1.09	2.21
0.90	−9.62	0.55	1.68	−11-51	0.97	1.69
Scenario 2						
0.25	−12.40	18.30	21.64	−15.48	16.60	22.19
0.50	−17.60	6.56	10.47	−19.81	8.41	10.64
0.75	−16.04	1.36	3.90	−17.51	2.61	3.95
0.90	−20.41	−1.16	2.91	−21.57	1.81	2.94

Source: Krishnamurthy (2010)

[a]Scenario 1 and 2 correspond to uniform increase in daily temperatures over the growing season by 1 and 2 °C, along with a 10 % increase in seasonal rainfall, respectively. Median impacts are reported along with (time) bootstrapped confidence intervals

Table 4 Descriptive statistics of regressed sample for *rice* (kharif season)

	East	North	South	West	Total
Yield (kg/hectare)	1206.4	1593.6	1961.8	1098.0	1523.4
	(462.8)	(863.6)	(645.0)	(562.7)	(757.6)
Production (000's of ton)	310.0	161.0	197.6	39.46	170.8
	(248.5)	(174.4)	(217.5)	(75.34)	(207.2)
Seasonal rainfall (mm)	1346.2	782.4	844.7	801.4	904.6
	(546.9)	(297.7)	(707.0)	(451.2)	(562.6)
Growing degree days (°C)	2460.8	2625.7	2320.5	2477.4	2472.1
	(147.7)	(120.9)	(293.9)	(167.5)	(233.0)
GDD lower threshold	2456.3	3598.7	2315.3	2462.3	2458.3
	(145.3)	(107.4)	(289.2)	(155.3)	(222.7)
Harmful GDD (°C)	0.754	11.01	1.383	5.968	5.194
	(2.716)	(13.37)	(4.285)	(11.14)	(10.32)
HGDD lower threshold	5.161	38.03	6.654	21.06	21.06
	(9.274)	(29.74)	(12.74)	(28.11)	(26.37)
Days of monsoon on set (from 1 April)	29.41	72.20	47.97	76.63	58.43
	(23.57)	(21.41)	(31.92)	(19.13)	(30.56)
Dry spells (no. of days)	3.964	5.067	6.000	5.014	5.139
	(1.786)	(1.575)	(1.858)	(1.536)	(1.826)
Average seasonal temp (°C)	28.18	29.61	27.04	28.37	28.31
	(1.22)	(1.06)	(2.42)	(1.43)	(1.95)
Observations	880	1,497	1,490	1,127	4,994
Number of districts	36	66	60	56	218

Source: Krishnamurthy (2010)

between temperature and rainfall in scenario 1, the additively separable nature of harmful degree days yields little additional information. Irrigation, unsurprisingly, is highly significant and positive at all quantiles, with a possibly smaller coefficient at the 90th percentile. The median range losses from 2 to 5 % in the relatively benign 0.5 °C temperature increase scenario to a substantial 5–10 % in the 1 °C warming scenario, with many districts losing as much as 22 % of yield.

Table 1 provides a visual description of the magnitude of losses: for scenario 2, losses rise to as high as 20 % for the lower percentiles, indicating significant losses for the already lower productive districts. Western and Central regions, where large poverty and substantial wheat production already exist, also lose significantly, up to 6 % of yield with a uniform increase of 0.5 °C and 9 %, with an increase of 1 °C. Losses in the Northern region are smaller, at about 2–5 %, whose impacts, however, can translate into large welfare losses due to the very high production losses embodied. The loss of wheat production is varied in the entire four regions at different quantiles, but the maximum loss is similar in some regions.

Impacts on Rice Production

There is extensive work on climate change and its impact on various crops (Kumar and Parikh (2001), Mall and Aggarwal (2002), Attri and Rathore (2003), and Mall et al. (2006a, b)); results varied, but quite clear, most of the studies reported minimal direct impact of major kharif crops, due to minor increase in daily temperature.

The Eastern region has a very large share in area and production of rice, while most of the losses are driven by it. The impact of climate change on rice cultivation in eastern India has been investigated, with some studies indicating possible increases in yield. Further, the study by

Welch et al. (2010) already cited reports a similar increase in south-eastern India. The impact of climate change, as simulated here, on rice appears at a first glance in Table 3. The situation is diametrically opposite, with modest to substantial increases in yield, of 2–18 %, except at the higher quantiles in all the regions. However, two features are worth noting: median changes are substantial only at the first two quartiles, under both scenarios, and changes at the highest quantile are close to zero or even modestly negative. The uniform climate change scenarios for India indicate a significant negative impact on wheat yields, of up to 11 %, primarily in Southern and Central India, with more moderate losses in the more important Northern region. Further, these impacts were seen to be most negative on the most productive districts, indicating losses in production which are likely significant.

The results of the regional analysis help inform the pathways to possible gains. The impact of climate change on rice cultivation in eastern India has been investigated, with some studies indicating possible increases in yield. Further, the study by Welch et al. (2010) already cited reports a similar increase in south-eastern India. This study therefore adds to the weight of evidence on this important issue. Most studies report significant negative impacts on crops grown in the rabi (winter) season, primarily due to increase in nighttime temperature. Our results also indicate, for the moderate scenarios considered here, substantially negative impacts on wheat yields, mostly in the southern and central regions, which are consistent with the results obtained in many crop modeling efforts, and modest increases in rice yield, driven for the most part by increases in the Eastern and Northern regions.

Rainfall and temperature has the inverted "U"-shaped relationship with yield of rice and wheat; i.e., changes in temperature and rainfall are beneficial but up to a certain extent; after that yield curve starts declining as we can see in Fig. 1. In Fig. 1, x-axis shows temperature and rainfall and y-axis shows yield of food grains. First, due to increase of minor temperature and rainfall, production increase then starts to fall. e^* is the turning point at that point yield will be maximum and then start to fall.

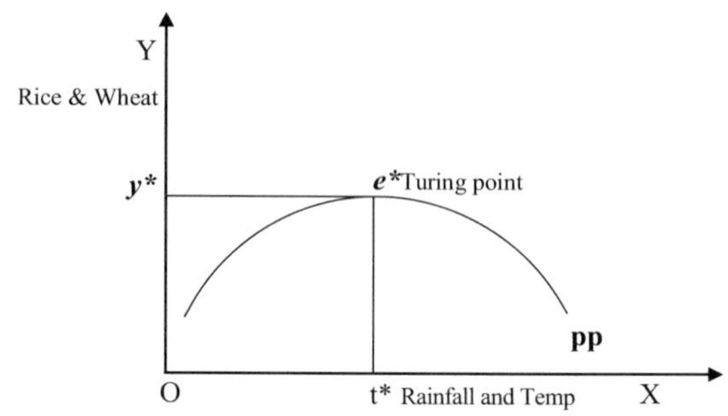

Fig. 1 Relation between yield production and climate factors

Possible Solutions

From the above analysis, it is clear that climate change will affect various sectors, but from all of them, agriculture is more vulnerable. Because production is also a function of climate variable input along with other input variables that includes seeds, fertile, temperature, rainfall, etc. Meanwhile, most of the small and medium farmers in India are using traditional equipments for cultivation and their knowledge is negligible relating to climate change and change in soil. For reducing the impact of climate change, strategies are in initial stage that can meet future food production needs and environmental challenges if deployed simultaneously. Adding them jointly, they will increase food availability and meet projected demands while lowering greenhouse gas emissions, biodiversity losses, and water use. However, all four strategies are needed to meet our food production and environmental goals; no single strategy is sufficient.

A general approach has been described here for solving national agricultural challenges, but much work remains, which needed to transform into action. Specific land use and agricultural and food system policy must be developed. Fortunately, many such strategies already exist, including precision agriculture, drip irrigation, organic soil remedies, buffer strips and wetland restoration, new crop varieties that reduce the needs for water and fertilizer, perennial grains and tree-cropping systems, and paying farmers for environmental services. However, these actions are needed to be implemented effectively, in all the states. For example, reforming trade policies includes eliminating price-distorting subsidies and tariffs and will be vital for achieving our strategies. In developing improved land use and agricultural practices, Foley et al. (2011) recommend the following four strategies.

Solutions should focus on critical biophysical and economic "leverage points" in agricultural sector: *First* is to develop a system, under which food production or environmental performance may be achieved with the least effort and cost. *Second*, new practices must increase the resilience of the food system. High-efficiency, industrialized agriculture has many benefits, but it is vulnerable to disasters, including climatic disturbances, new diseases, and economic calamities. *Third*, agricultural activities have many costs and benefits, but methods of evaluating the trade-offs are still poorly developed. There is the need of better data and decision support tools to improve management decisions, productivity, and environmental stewardship.

Fourth, there is a need to develop an appropriate low-cost technology that does not affect the soil productivity. These multiple efforts would increase the production and handle the problem of food security and environmental performance of agriculture. We should not be locked into a single approach, a main concern, whether it may be conventional agriculture, genetic modification, or organic farming.

Climate Adaptation Policy in India

It is well proved that climate change has significantly adverse impacts on agriculture productivity. It is more vulnerable in countries like India because the agriculture activities are more dependent on monsoon than other nations, so any change in climatic variables more affects the agriculture productivity. As Aggarwal (2009) pointed out, 1 °C increase in temperature may reduce yields of wheat, soya bean, mustard, groundnut, and potato by 3–7 %. Kumar and Parikh (2001) and Sanghi and Mendelsohn (1998) have estimated that in a moderate climate change scenario, there could be about 9 % decline in farm-level net revenues in India. More adverse impacts are predicted in high productive agricultural regions such as Punjab, Haryana, and Uttar Pradesh and dry regions Gujarat and Rajasthan. But the states like Bihar and West Bengal could be marginally benefited.

Kumar (2010) favored controlling spatial effects of climate change on farmers. Along with other things, strong flow of information among farmers could also contribute for being able to adapt to the climate change more efficiently. His information-based proposal deals largely with private adaptation measures undertaken by farmers. Research priorities in this context include exploring the factors that facilitate information diffusion in agriculture.

There is no formal policy detailing India's response to climate change; however, some document has inter alia made reference to climate change. The National Environment Policy (NEP) 2006 had addressed to some multilateral efforts for climate change. This constitutes adherence to the principle of common but differentiated to the respective capabilities of different countries regarding both mitigation of greenhouse gases (GHGs) and adoption. However, it trusts on multilateral approaches that opposes bilateral and unilateral approaches. Thus, all countries should be entitled to equal per capita of the global environment resources. Moreover, over emphasizing the priority for the right to development is identified as the one of the key vulnerabilities for the climate change in the areas like water resources, forest, coastal areas, and agriculture. Not only there is need for adaptation to future climate change policy, but also there should be the scope of incorporating this in the relevant programs such as watershed management, coastal-zone planning and regulation, forestry management, agricultural technological practices, and health programs.

The government of India has formulated detailed plans for dealing with contingencies arising in the wake of natural calamities, including drought, flood, and cyclones. The Ministry of Home Affairs and United Nations Development Program (UNDP) have introduced a program that aims to build community level awareness and strengthen capacity for such disasters. A high-level National Disaster Management Authority has been also set up to coordinate actions by all concerned agencies far taking up of emigration measures and management of the disasters, both natural and man-made. Other programs, including initiatives in agriculture sector for crop diversification, promoting zero tillage, watershed management, forestry, and water policy, will enable adaptation to current vulnerabilities, but their potential for dealing with the expected climate change is yet to be recognized fully.

Instigation of climate change into the planning process is still a distant goal. There is growing engagement of the government with bilateral and multilateral approach on climate change issue for the above development programs, but this is of recent origin. Involvement of relevant ministries, and more importantly the states, in this agenda will be vital. A small beginning has been made with the agriculture ministry setting up a separate unit to ascertain impact of climate change on the agriculture sector. World Bank is also engaged with some state information about the need to adapt and make the best use of the existing developmental programs in mitigating climate change.

National Action Plan for Climate Change (NAPCC) was adopted by central government in 30 June 2008. It advocates a strategy that promotes, firstly, for the adoption of climate change policy. Secondly, it further enhances ecological sustainability with development of the country. The action plan also identifies measures that promote sustainable development with reducing effect of climate change. Some states such as Himachal Pradesh, Kerala, Delhi, and Gujarat have established the state-level action plan on climate change.

Conclusion

The uniform climate change scenario in India indicates a significant loss in wheat production and possibly moderate gains in rice production. The loss of wheat yields accounts for 11 %, primarily in Southern and Central India, with more moderate losses in Northern region. Further, the most important outcomes have been that these impacts are seen to be most negative in the most productive districts, indicating losses in wheat production. These results indicate significant losses in wheat production and possibly moderate gains in rice production. However, given the substantial loss in output of rice, these estimates are likely to translate into potentially substantial decreases in rice production, at least at the local scale. So it is clear that in the absence of significant changes in agricultural practices and technology, climate change is likely to lead to an increase in food insecurity for India's poor, because of decreased yield at both the regional and national level.

Based on the above model estimated results, there are significant negative impacts for wheat yield that were 6 % at 0.5 °C increase in temperature and up to 10 % losses for 1 °C increase in winter temperature. However, result indicated that loss of rice productivity in minimum at all levels of quantiles but in the case of highest quantiles and even mildly are considered positive impact has been realized in many regions, primarily the Eastern region, a result which is also consistent with the Guiteras (2008), and with a recent study of Welch et al. (2010). These results are robust to flexible functional forms (quadratic, orthogonal polynomials and fully nonparametric) for growing degree days and to different plausible thresholds for GDD. For rice, however, impacts of uniform warming of up to 2 °C were shown to be mildly positive, with increases in yield (up to 17 % at the lowest quantile) concentrated in the lower quantiles, or in the least productive areas, while impacts at the upper quantile are seen to be minimal or even mildly negative. This is likely to translate into a moderate increase in production of rice. The results here are consistent with many estimates of changes in rice yields obtained while using various crop models under a variety of climate change scenarios as well as in a recent study of rice cultivation in Asia under irrigated conditions by Welch et al. (2010).

References

Aggarwal PK (2009) Vulnerability of Indian agriculture to climate change: current state of knowledge. Presented at the National Workshop, New Delhi

Attri S, Rathore L (2003) Simulation of impact of projected climate change on wheat in India. Int J Climatol 23:693–705, 5.5.2

Buchinsky M (1995) Estimating the asymptotic covariance matrix for quantile regression models a Monte Carlo study. J Econ 68(2):303–338, 3.4

Deschenes BO, Greenstone M (2007) The economic impacts of climate change: evidence from agricultural output and random fluctuations in weather. Am Econ Rev 97(1):354–385, 1, 5.4

Fitzenberger B (1998) The moving blocks bootstrap and robust inference for linear least squares and quantile regressions. J Econ 82(2):235–287, 3.4

Fitzenberger B, Koenker R, Machado JAF (2002) Economic applications of quantile regression. Physica Verlag, New York, 3.2, 10, 3.4, 5

Foley A et al (2011) Solutions for a cultivated planet. Nature 478:337–342

Guiteras R (2008) The impact of climate change on Indian agriculture', Unpublished manuscript, 1, 2.1, 2.2, 4.2.1, 5.1, 5.4

Harding M, Lamarche C (2009) A quantile regression approach for estimating panel data models using instrumental variables. Econ Lett 104(3):133–135. 3.2, 5.6 India. Glob Environ Chang 18(4): 655–665

Kapetanios G (2008) A bootstrap procedure for panel data sets with many cross-sectional units. Econ J 11 (2):377–395, 3.4, B

Koenker R (2004) Quantile regression for longitudinal data. J Multivar Anal 91(1):74–89, 3.2, and 3.4

Koenker R (2005) Quantile regression. Working paper, Cambridge University Press, 2, 3.5

Koenker R (2010) Additive Models for Quantile Regression: An Analysis of Risk Factors for Malnutrition in India. Advances in Social Science Research Using R, pages 23–33, .3.5

Koto K, Galvao A Jr, Montes-Rojas G (2010) Asymptotics and bootstrap inference for panel quantile regression models with fixed effects, Working paper no. TR-10. Hiroshima University, Hiroshima

Krishnamurti TN, Kishtawal CM, Zhang Z, Larow T, Bachiochi D, Williford CE, Gadgil S, Surendran S (2000) Multimodel ensemble forecasts for weather and seasonal climate. J Climate 13:4196–4216

Krishnamurthy KCB (2010) The distributional impacts of climate change on Indian agriculture: a quantile regression approach 'Job Market Paper. www.sipa.columbia.edu/~cb2322/job_market/JMP.pdf. Accessed 21 Nov 2011

Krishnamurthy CKB (2012) The distributional impacts of climate change on Indian agriculture: a quantile regression approach, Working paper, 69/2012. MSE, India

Kumar KS (2010) Climate change studies in Indian agriculture. EPW, 45 (26–27)

Kumar KS, Parikh J (2001) Indian agriculture and climate sensitivity. Glob Environ Chang 11(2):147–154

Mall RK, Aggarwal PK (2002) Climate change and rice yields in diverse agro-environments of India I. Evaluation of impact assessment models. Clim Chang 52(3):315–330, 5.5.2

Mall RK, Singh R et al (2006a) Impact of climate change on Indian agriculture. Clim Chang 78:445–748, a. 5.4, 5.5.2

Mall RK, Singh R et al (2006b) Impact of climate change on Indian agriculture: a review. Clim Chang 78 (2):445–478, b. 5.4, 5.5.2

Mendelsohn R, Nordhaus WD, Shaw D (1994) The impact of global warming on agriculture: a Ricardian analysis. Am Econ Rev 84(4):753–771, 1

Powell D (2010) Unconditional quantile regression panel data with exogenous and endogenous repressors. Working paper no WR 701–1. RAND Population Research Center

Rosen A (2009) Set identification via quantile regression in short panel, Working paper, no. CWP, 26/09. Cemmap, London

Rosenzweig MR, Binswanger HP (1993) Wealth, weather risk and the composition and profitability of agricultural investments. Econ J 103(416):56–78

Sanghi A, Medelshohn R, Dinar A (1998) Measuring the impact of climate change on Indian agriculture, Technical report-402. World Bank, Washington, DC

Schlenker W, Roberts MJ (2006) Nonlinear effects of weather on corn yields. Appl Econ Perspect Policy 28(3):391, 1, 4.2.1

Schlenker W, Hanemann W, Fisher A (2009) The impact on global warming on US agriculture: an econometric analysis of optimal growing conditions. Rev Econ Stat 88(1):113–125

Stern N (2007) The economics of climate change: the stern review. Cambridge University Press, Cambridge, pp 265–280

Welch J, Vincent J, Auffhammer M, Moya P, Dobermann A, Dawe D (2010) Rice yield in topical/sub-topical Asia exhibit large but opposing sensitivities to minimum and maximum temperature. Proc Natl Acad Sci U S A 107(33):145–162

Climate Change and Hill Agriculture in Northwest Himalaya

J.C. Bhatt, R. Arunkumar, and J. Stanley

Abstract

Northwest Himalayan agriculture is strongly influenced by climate change/variability, and the agriculture is often affected due to several climatic/weather components. This chapter reviews the magnitude of climate change in few important places in terms of change in temperatures, rainfall and glaciers retreat. Due to the fragile nature of mountain ecosystem, the impact of climate change/variability is higher, and this severity can be seen through the different biotic, abiotic stresses. Because of changes in climatic conditions, the microclimate of crop ecosystem is expected to change, which influences the pest and disease spectrum and also its dynamics. The changing climate may favour some pests (sucking pest) and diseases (wheat yellow rust, rice blast, etc.) and suppress some others. New pest (rice brown plant hopper) and disease (maize *Phyllosticta* and zonate leaf spot) infestations are already been reported in Himalayan hills. The shift in crop season may influence the availability and abundance of pollinators which is a major concern for pollination in cross-pollinated crops. Impact assessment of climatic components with rice-wheat cropping system showed that the higher mean, maximum and minimum temperatures during winter season resulted in poor wheat grain yield, whilst lower mean, maximum and minimum temperatures resulted in poor rice grain yield. Due to weather variability, the drought, cold and terminal heat stresses and extreme weather events often resulted in severe yield losses in rice, garden pea, wheat, ragi, lentil, bhindi, rajmash, French bean, buckwheat and horse gram. Perhaps, most seriously, due to the observed and predicted climate change, there is high uncertainty in crop production in the near future, and adequate attention has to be given to sustain the NW Himalayan agriculture.

J.C. Bhatt (✉) • R. Arunkumar • J. Stanley
Vivekananda Parvatiya Krishi Anusandhan Sansthan
(ICAR), Almora 263 601, India
e-mail: jagbhatt@yahoo.com

A.K. Singh et al. (eds.), *Climate Change Modelling, Planning and Policy for Agriculture*,
DOI 10.1007/978-81-322-2157-9_17,

Keywords
Agriculture • Abiotic stress • Biotic stress • Climate change • Northwest Himalayas

Introduction

Agriculture contributes nearly 13.7 % of gross domestic product (GDP) of India. Paroda and Kumar (2000) reported that by 2020, food grain requirement will be almost 30–50 % more than the demand during 2000 AD. Besides increased competition for land, water and labour, the Indian agriculture is facing one of the most important challenges due to increasing climate change/variability. The Intergovernmental Panel on Climate Change (IPCC) of the United Nations has reported that the earth temperature has increased by 0.74 °C between 1906 and 2005 due to increase in anthropogenic emission of greenhouse gases. By the end of this century, temperature increase is likely to be 1.8–4.0 °C. This would lead to more frequent hot extremes, floods, droughts, cyclones and gradual recession of glaciers, which in turn would result in greater instability in food production. For Indian regions (south Asia), the IPCC has projected 0.5–1.2 °C rise in temperature by 2020, 0.88–3.16 °C by 2050 and 1.56–5.44 °C by 2080 depending on the scenario of future development (IPCC 2007). It is estimated that crop production losses in India by 2100 AD could be 10–40 % despite the beneficial effects of higher CO_2 on crop growth. It is projected that by the end of the twenty-first century, rainfall over India will increase by 15–40 % and the mean annual temperature by 3–6 °C (NATCOM 2004). Several Indian studies on this context confirm a decreasing trend of agricultural crop production with climate change (Aggarwal and Sinha 1993; Rao and Sinha 1994; Saseendran et al. 2000; Mall and Aggarwal 2002; Aggarwal 2003). Aggarwal (2008) reported that with every rise of 1 °C temperature throughout the growing period with current land use, there will be projected loss of 4–5 million tonnes in wheat production. Severe drought during 2002 resulted in a loss of more than 10 % in food production (Samra and Singh 2002). The projected increase in these events could result in greater instability in food production and threaten livelihood security of farmers. Whilst the above review indicates gross effects of climate change on crops in Indian plain areas, adequate attention has not been given to several specific ecosystems like Western Ghats, coastal areas, North-Eastern region and Himalayan ranges.

Among these aforesaid, four regions particularly the northwest Himalayas, part of Himalayas consisting of three states, viz., Uttarakhand, Himachal Pradesh and Jammu and Kashmir, contribute very important ecology and economy to Indian agriculture, but the effects of climate change/variability on agricultural impacts in this region have not received much adequate attention. Singh et al. (2010) reported that the mountains are the most fragile environments on earth and they are rich repository of biodiversity and water and providers of ecosystem goods and services on which downstream communities, both regional and global, rely. They also reported that the effect of climate change on the Himalayan region is of high magnitude and the hill and mountain farmers are expected to be more vulnerable to any shift in climate due to their dependence on natural resources and poor risk-bearing ability with lack of credits.

Northwest Himalayan Hill Agriculture

The total geographical area, the per cent forest area and the per cent net area sown of NW Himalayas are 33,139 ha, 53.66 and 15.48, respectively. The annual rainfall of NW Himalayas is 350–3,000 mm per year along with dry and cold weather (Gupta 2006). Kamal

(2003) has broadly defined the agriculture in the northwest Himalayas covering all land-based activities such as cropping, animal husbandry, horticulture, forestry and their linkages and support system and is a prime source of sustenance for most mountain communities. Indian Network for Climate Change Assessment (INCCA 2010) recently reported that there are five major farming systems prevalent in northwest Himalayas, namely, (i) cereal-based production system (rice, wheat, millets, maize); (ii) horticulture or agrihorti-based production system; (iii) vegetables, floriculture and mushroom-based production system; (iv) livestock-based production system; and (v) agrihorti-silvi-pastoral-based production system. The per cent net area sown, forests, permanent pastures and others of NW Himalayas are 15, 54, 14 and 17 %, respectively. In the western Himalayan region, wheat is the main crop and rice, maize, millets, barley and buckwheat, pulses and oil seeds were also widely grown (Partap 2011).

Observed Climate Change in Northwest Himalayas

The average temperature of the last 53 years (1955–2007) of a Lesser Himalayan Hill station, viz., Almora, located at 29°35′ N and 79°35′ E at an elevation of 1,640 m from the m.s.l., showed an increasing trend. The average temperature of Almora, i.e. 17.55 °C (1955–2007), has increased up to 0.46 °C during the last 53 years. This preliminary observation suggests that the average temperature is rising in the state. Rainfall records of three places, i.e. Almora, Manora Peak and Hawalbagh, indicated a decreasing trend in the annual rainfall (UCCC 2011).

Due to climate change, there has been increasing trend of occurrence of drought incidences in the recent past. The Hawalbagh rainfall records postulate this fact which reveals that during 1964–2000, the total incidences of drought were 16, out of which 5 were severe, but during 2001–2009 within 8 years, 7 droughts occurred and out of which 3 were severe (Panday et al., 2009). Maximum and minimum temperatures were reported to have increasing and decreasing trends, respectively (Panday et al. 2003).

Also in Hawalbagh farm, the following weather changes were observed, i.e. (1) decrease in monsoonal rainfall and increase in pre-monsoon and post-monsoon rainfall, (2) extreme events like severe drought for a period of more than 28 days during *kharif* 2009 and followed by flooding rainfall (rainfall > 900 mm during the post-monsoon), (3) severe flooding (during 2010) due to heavy rainfall > 1,300 mm, (4) more than 30 days of frosting during December 2010 to February 2011, (5) decreased sunshine hours followed by high-intensity solar radiation and (6) severe soil erosion and landsliding due to extreme events during 2009, 2010 and 2011.

The Gangotri glaciers of Uttarakhand showed a retreating trend of 26.5 m/year, 17.1 m/year and 12.5 m/year during 1935–1971, 1971–2004 and 2004–2005, respectively (Kumar et al. 2008; Bali et al. 2009). They also reported a similar retreating trend in Pindari and Dokriani glaciers of Uttarakhand. Upreti (2010) reported that the increased anthropogenic pressure along the slopes in the mountain areas has changed the existing land use pattern resulting to different hazardous events like landslides leading to significant loss of lives, resources and property. Based on three decades of data, he also reported that due to global warming, the epiphytic and green algae containing lichens were found increasing in comparison to cyanolichens. Joshi et al. (2008) also reported that there was an increase in *Trentepohlia* lichen species in thinned-out forest areas of temperate areas of temperate Himalayas.

Predicted Climate Change in Himalayas

Indian Network for Climate Change Assessment (INCCA 2010) recently reported the following predictions of climate change in Himalayas in 2030s: The mean annual temperature is projected to increase from 0.9 ± 0.6 °C to 2.6 ± 0.7 °C in the 2030s. The net increase in temperature ranges from 1.7 to 2.2 °C with respect to the 1970s.

The minimum temperatures are projected to rise by 1–4.5 °C, and the maximum temperatures may rise by 0.5–2.5 °C. The annual rainfall in the Himalayan region is likely to vary between 1,268 ± 225.2 and 1,604 ± 175.2 mm in 2030s. The projected precipitation is likely to increase by 5–13 % in 2030s with respect to 1970s. The number of rainy days in the Himalayan region may increase by 5–10 days on an average in the 2030s, and an increase of more than 15 days is projected for the eastern part of the Jammu and Kashmir. The intensity of rainfall is also likely to increase by 1–2 mm/day.

With the reduction in rainfall, the rain-fed agriculture will suffer the most. Horticultural crops like apple are also showing decline in production and area coverage particularly due to decline in snowfall (Partap and Partap 2002). A remarkable increase in the area under off-season vegetable cultivation has been noticed in the present cropping pattern in Bajaura Valley (1,500–2,200 m). The area under food crops and fruits has shifted to off-season vegetable cultivation. Decline in area under apple and other fruit was comparatively higher in marginal and small farms. The productivity of apple has recorded a decrease of about 2–3 % over the past years, and the maximum decrease of about 4 % has been noticed in marginal farms. The total farm income on an average has shown a marginal reduction of about 4 % in the present period. Off-season vegetables have shared more than 84 % of the area under field crops in Theog Region (above 2,000 m). The area under cereals has declined to the extent of 80 %. The productivity of cash crops like potato and peas has declined by more than 11–15 % over a decade period.

The water yield in the Himalayan region, mainly covered by river Indus, is likely to increase by 5–20 % in most of the areas, with some areas of Jammu and Kashmir and Uttarakhand showing an increase of up to 50 % with respect to the 1970s. The impact of increase in precipitation in this region has been reflected in an almost similar pattern of increase in the ET. There is an increase in the drought development for those areas of various regions that have either projected decrease in precipitation or have enhanced level of evapotranspiration in the 2030s. Similarly, the weeks belonging to moderate soil moisture stress show an increase in severity of drought from baseline to the mid-century scenario, which is self-evident.

It is evident from the depiction that the moderate to extreme drought severity has been pronounced for the Himalayan region, where the increase is more than 20 % in many areas despite the overall increase in precipitation. Possible floods have been projected using the daily outflow discharge in each subbasin as generated by the SWAT model, ascertaining the change in magnitude of flood peaks above 99th percentile flow in 1970s and in 2030s. Change in peak discharge equal to or exceeding at 1 % frequency in the 1970s and 2030s for various regions indicates that the flooding varies from 10 % to over 30 % of the existing magnitudes in most of the regions. This has a very severe implication for existing infrastructure such as dams, bridges and roads in the areas and will require appropriate adaptation measures to be taken up.

Impact of Climate Change in NW Himalayan Hill Agriculture

Shifting Cropping Patterns

Trend analysis of last few decades showed that fruit trees, vegetables and agricultural crops were most affected at the lower altitudes in mountain regions, whilst farmers have shifted from apple to vegetable crops like cauliflower, cabbage, peas, carrot and other fruit crops like pomegranate, kiwi and pear cultivation in mid elevations. At the highest altitudes, only slight reduction is noticed in the productivity of some crops like potato and pulses at some locations, and farmers benefit from this the most due to increased temperature and consequently lengthened growing period. Here, the farmers have shifted traditional agricultural crops such as buckwheat, barley, finger millet, grain amaranth and chenopod to apple, potato, hops, garden pea and other off-season vegetables and medicinal plants which fetch them high prices (Sharma and Rana 2005). However, there apprehension that these crops will perform well until the irrigation water received through glaciers melt is available. But, if the glaciers' receding continues, the water course may change, and some areas may become water scare in the cold arid regions (Rana 2010).

Rice-Wheat Cropping System

Based on the relationship obtained between the yield data and climatic components of zero tillage experiment of rice – wheat cropping system – Panday et al. (2011) reported that the higher mean, maximum and minimum temperatures and sunshine hours during *kharif* season for a period of 6 years resulted in better rice grain yield, whilst lower mean, maximum and minimum temperatures and sunshine hours during *kharif* season for a period of 3 years resulted in poor rice grain yield compared to that of average rice grain yield obtained for the period of 2001–2009. They also reported that the higher mean, maximum and minimum temperatures and sunshine during winter season for a period of 4 years resulted in poor wheat grain yield, whilst lower mean, maximum and minimum temperatures and sunshine during the same season for a period of 5 years resulted in poor wheat grain yield compared to that of average wheat grain yield obtained for the period of 2001–2009. A decreasing trend in minimum and maximum temperatures in *kharif* season and increasing trend in minimum and maximum temperatures in winter season and decreasing trend in rainfall during annual and *kharif* season are being observed.

Rice

Tripathi and Chintamanie (2011) reported that the crop yield is directly influenced by climate variability as exhibited in the time series data of Uttarakhand rice yields. Further, they also reported that the rain-fed rice productivity in the Uttarakhand (formerly part of Uttar Pradesh) (1979–2007) is affected by rainfall variability, i.e. the interannual rainfall variability and the late-onset and early termination and high- and low-intensity rainfall negatively affected the rice yield. They also reported that the period 1979–1985 were drier than the period 1985–2005, and therefore, during the period 1979–1985, crops would have required more irrigation water than usual and may have resulted in the reduced yield; hence, the cropping sequence and schedule can be planned to take advantage of available soil moisture.

Climate Change on Seed Production of Cabbage

The analysis of the impact of climatic changes during 1981–2004 on the seed yield of cabbage var. Golden Acre in the upper Kullu Valley in Hindu Kush Himalayas indicated that the average maximum temperature of May month increased by 1.58 °C and the minimum temperatures for the months of April and August increased by 2.03 and 2.16 °C, respectively (Kumar et al. 2009). They also recorded 40 % reduction in seed production due to erratic rainfall and increased temperature. The rainfall during August has decreased, and during September, it has increased resulting in late onset of autumn, thereby suggesting that the planting of cabbage should also be delayed at least by a fortnight to avoid incidence of soft rot and to get increased seed yield.

Invasive Weedy Species

Rana (2010) reported that the hill regions had not witnessed much biological invasions till 1970s, but in the last 30 years, the three most invasive species, viz., *Lantana camara*, *Parthenium hysterophorus* and *Ageratum conyzoides*, have invaded and altered community structure and population dynamics of native flora and fauna of the Shivalik hills. Other weeds like *Ageratina adenophora*, *Bidens pilosa*, *Polygonum polystachyum*, *Solanum chacoense* and

Cyclanthera brachystachya are at an early stage of invasion even at higher elevations, attributed mainly to rising temperatures. Species like Lantana and Parthenium have not only outnumbered the native vegetation but have shifted upwards. The vegetation analysis in Shivalik hills showed that Lantana constituted 28.32 % of the total shrub species, whilst Ageratum (21.42 %) and Parthenium (20.51 %) together accounted for 41.93 % of the total herb species. The populations of economic plants like *Carissa spinarum*, *Adhatoda vasica*, *Dodonea viscose*, *Cassia tora*, many grasses, medicinal herbs and wild flowers have reduced drastically (Rana 2010).

Biotic Stress

Due to climate change/climate variability, a change in the biotic (pest and diseases) and abiotic stress have been recorded and expected particularly in major crops grown in NW Himalayas.

Pests of Major Crops

Climate Change on Insect Pests of Rice

Insect pests extremely depend on temperature and expected to respond rapidly to climate change by shifting their geographic distributions to take advantage of new niches that become available. Stem borer, leaf folder and white grubs are the major insect pests of rice in NW Indian Himalayas.

Stem Borer Stem borer is reported to cause up to 52 % yield loss in rice in Uttarakhand. The incidence of stem borer is positively correlated with *kharif* rainfall and negatively correlated to first hr RH. Thus incidence of stem borer is expected to be less.

Leaf Folder Rainfall during the crop season (August) is reported to have a positive impact on leaf folder. Since the August rainfall is decreasing significantly with less number of rainy days, leaf folder incidence may also reduce.

White Grubs White grubs are important and devastating pest of upland rice in hills. An average crop loss of 30 % and up to complete crop failures are reported due to white grubs in rice. Sushil et al. (2004) reported that the higher temperature during May–June and low temperature during December, January and February favoured the white grub beetle emergence, whilst rains during April, October and November disfavoured beetle emergence. Stanley et al. (2009) reported temperature variation during December to February and rainfall during April and October and November are highly detrimental for survival of white grubs since it breaks diapause (resting) during April and enters into diapause during October–November. The expectation is the change in dominant species, wherein white grubs include 78 species reported only in NW Indian Himalayas.

Brown Planthopper Increase in maximum temperature and cessation of rainfall attribute to incidence of BPH (brown planthopper) in rice. Migration of BPH from hotter parts to cooler areas due to global warming was reported in Japan (Heong et al. 1995). Incidence of BPH in low hills of Uttarakhand during *kharif* 2009 and heavy incidence for the first time in mid hills during *kharif* 2010 can be attributed to the increase in maximum temperature, cessation of rainfall and delay in sowing time (Stanley et al. 2009).

Diseases of Major Crops

Climate Change on Wheat Diseases Resurgence of Wheat Yellow Rust in Northern Hills and Tarai Region

Sudden outbreak of yellow rust in the northern part of the country has been recorded during *Rabi* 2010–2011 that ranged from 5 to 80S. The disease on majority of areas started in mid February and the severity reached up to 80S in the last week of March. In Uttarakhand hills, the severity reached up to 80S with high prevalence in local wheat varieties. However, the recommended varieties like VL Gehun 804, VL Gehun 829 and VL Gehun 907 showed resistance. In Himachal Pradesh, yellow rust was observed ranging from 10 to 80S in parts of Kangra, Hamirpur, Bilaspur and Mandi districts. Some of the varieties like HPW 184, HPW 42, HPW 155 and VL 907 were resistance and free from disease. In Jammu and Kashmir, prevalence of yellow rust was observed ranging from 5 to 80S in Jammu, Samba, Kathua and Akhnoor, Chatha, RS Pura, Ghaghwal, Mathura Chak, Gial band, Merheen and Udaywala with severity up to 60S. Varieties like PBW 343, PBW 550, PBW 502 and DBW 17 were most affected with 30–60 % prevalence. Raj 3,077 and Raj 3,765 were found resistant and less affected by yellow rust (10S) (Kant and Jain 2011).

Climate Change on Rice Diseases

High humidity is reported to cause fungal epidemics. Increase in UV B radiation makes rice more susceptible to blast especially in cooler regions where high humidity and low temperature prevails. Increased winter temperature increases blast and sheath rot incidence in rice.

Blast Weather conditions having minimum temperature and average temperature between 15–20 °C and 22–25 °C, respectively, along with relative humidity ≥90 % and higher rainfall with more number of rainy days are reported congenial for blast development in rice at Uttar Pradesh hills (Bhatt 1992). Due to climate change, we will have more number of days (4.5 days) with both the conducive weather parameters, which increases the incidence of blast.

Brown Spot Brown spot has positive correlation with September rainfall and August maximum temperature. A temperature of 25–30 °C and a RH of 86–100 % are favourable to brown spot incidence in rice. Due to changed climatic situations, crop will be susceptible or climate will be favouring brown spot pathogen for another 10 days during rice growing period.

Climate Change on Barley Diseases

Cold arid trans-Himalayan Ladakh region is highly vulnerable to climate change, and serious environmental threats often result in reduction in crop productivity due to agricultural pest and plant pathogens. Vaish et al. (2011) reported that the yellow rust, powdery mildew, leaf spot blotch/blight, covered smut, loose smut, foot/root rot and cereal cyst nematode causing molya disease were observed during their field survey and the yellow rust, molya and foot/root rot were found as the most destructive diseases in barley. They also reported that the enhanced temperature would lower the occurrence of yellow rust, whilst the occurrence of leaf spot blotch/blight might become severe.

New Diseases on Crops

Occurrences of zonate leaf spot (Hooda et al. 2009) and sclerotium rot of maize caused by Gloeocercospora sorghii and Sclerotium rolfsii were also recorded first time in Hawalbagh farm located at Almora in Uttarakhand.

Climate Change on Pollinators

Insects are the first victims of the climate change because of their narrow range of temperature tolerance (Connor 2008). Low temperature and high humidity have double effect on reducing the bee activity and slowing the release of pollen (Joshi and Joshi 2010). Temperature, relative humidity and wind affect the quantity and concentration of nectar and as a result the flower attractiveness to bees (Somerville 1999).

When the changing climate observed in NW Himalayas is correlated with the bioecology of pollinators, decrease in rainfall and increase in maximum temperature in winter months will not affect the insect pollinator per se but on the crop growth and flower production. Decrease in minimum temperatures during December and January (peak winter months) will have a high negative impact on pollinators even on the over-wintering wild bees, wasps, butterflies and flies (Stanley et al. 2010). Honeybees have to spend much of its energy in heating the comb and the brood (Kleinhenz et al. 2003), thus less efficient in pollination. There will have a greater impact on the pollinator abundance. Increased first-hour relative humidity will have a negative effect on flower formation, blooming and pollen release, since RH of 70 % or less is required for anther dehiscence and pollen release. A climate-driven mismatch between the times when the flowers open and when pollinators emerge from hibernation is an important factor which can irrevocably alter the intimate relationship between plants and pollinators that have coevolved over the past thousands of years.

Effect of Weather Variability on Major Hill Crops

Vivekananda Parvatiya Krishi Anusandhan Sansthan (VPKAS, Almora, Uttarakhand, India) known for its contribution in developing the quality protein maize (QPM 9) through marker-assisted selection (Gupta et al. 2009) has also conducted few studies regarding the effect of weather variability on field crops through different dates of sowing (DOS) (Table 1). Different DOS is considered to be the simplest technique to predict the probable crop responses to the climate change/variability. These studies will be more helpful to make contingency planning to fit proper variety to extreme weather conditions like raising and decreasing temperatures, varying rainfall and other weather components.

Responses of Lentil Varieties to Normal and Late Sown Condition

Five lentil genotypes, viz., VL 125, VL 126, VL 133, VL 507 and VL 514, sown under normal and late sown (1 month late) condition showed that late sown condition resulted in significant reduction in seed yield of 23 % compared to that of normal sown. The grain yield reduction in late sown lentil genotypes could be plausibly due to initial cold stress and terminal heat stress compared to that of normal sown condition. Significant reduction of 10 % in leaf relative water content (RWC), leaf area 45 % (60 DAS) and flowers 45 % (120 DAS) were obtained under late sown condition. Under late sown, there was general significant increase in anthocyanin content, and total carotenoid content in all the varieties, and the VL 514 showed the highest values in all the observed physiological parameters, and this was reflected in better yield. Six per cent reduction in grain protein content was observed under late sown condition compared to normal sown, and the low yielded VL 133 possessed higher protein content under late sown condition.

Responses of Rice Genotypes to Drought and Cold Stresses

Drought Stress Seventy-three diverse rice genotypes grown under irrigated and drought condition (grain filling stage for short period) showed that under drought condition, there was a significant decrease in all the physiological parameters, viz., chlorophyll content, chlorophyll ‘b’ content, total chlorophyll content, total carotenoid content, photosystem II efficiency and photosynthetic efficiency.

Cold Stress Experimental results obtained from 73 diverse rice genotypes under normal sown (cold-free environment) and late sown (cold stress environment during grain filling stage) showed that under cold stress condition, there was a significant decrease in all the physiological parameters, viz., chlorophyll ‘a’ content, chlorophyll ‘b’ content, total chlorophyll content, total carotenoid content and grain yield. Significant decline in per cent sterility of grain was also recorded under cold compared to ambient condition. IRCTN-91-94, IR 3941–23, Sinsatsu, RCPL 1-2C and few IRCTN genotypes were recorded with better fertility compared to other genotypes, and this could be due to its reproductive tolerance to cold stress.

Table 1 Reduction in grain yield of different crops due to delayed sowing

Crop	Yield (Q/ha) Normal sown	Late sown	Percent reduction	Reasons for yield reduction
Wheat (var. VL 804)	49.8	40.0	19.7	Terminal heat stress
Soybean (var. VLS 73)	13	6.7	48.5	Drought
Lentil (var. VL 507)	19.2	13.7	28.6	Initial cold and terminal heat stress
Garden pea (var. VL 7)	25.9	14.2	45.2	Initial cold and terminal heat stress
Field pea (var. VL 42)	31.9	20.2	36.7	Initial cold and terminal heat stress

Responses of Garden Pea to Normal and Late Sown Condition

Three garden pea varieties, viz., VL7, VL10 and VL11, under five different dates of sowings (Oct 15, Nov 1, Nov 15, Dec 1 and Dec 15, 2010) revealed that due to frost injury and *Alternaria* blight disease, Oct 15 sown crop recorded very low yield. However, Nov 15 sown crop recorded maximum yield due to low frost and low *Alternaria* damage.

Responses of Finger Millet to Normal and Late Sown Condition

Different dates of sowings (July 5, July 15, July 25 and August 5) in 5 varieties of finger millet, viz., VL mandua 146, VL mandua 149, VL mandua 315, VL mandua 324 and PRM 2, showed that July 5 recorded significantly higher grain yield (33.82 q ha^{-1}) as compared to other dates. Among the different varieties tested, VL 146 gave significantly higher grain yield as compared to other varieties. VL 315, VL 146, VL 324 and VL 146 were found most suitable varieties for sowing on July 5, July 15, July 15 and August 5, respectively.

Responses of Wheat Varieties to Late Sown Condition

Eight wheat varieties, viz., VL 616, VL 738, VL 802, VL 804, VL 829, VL 832, VL 892 and VL 907, grown under four dates of sowings (Jan 10, Feb 4, March 1 and March 26, 2010–2011) showed that due to terminal heat stress, significant differences in all physiological parameters were recorded in all wheat genotypes. Wheat varieties VL 907 and VL 802 performed well when sown on Jan 10, whereas VL 802 and VL 804 performed better on March 26 in terms of grain yield. Temperature stress resulted in decreasing the number of effective tillers in all wheat genotypes, and VL 616 recorded with more noneffective tillers resulted in poor yield. However, VL 802 recorded with more effective tillers along with higher yield under late sown conditions.

Extreme Weather Events

Crop failures due to extreme weather changes such as flood, drought, prolonged dry spell and water logging are important consequence of climate change (Bhatt 2010). Due to cyclonic storms/heavy rains/flash flood/landslides during 2010–2011, nearly 0.26, 0.14 and 5.02 lakh hectares of cropped area was affected in Himachal Pradesh, Jammu Kashmir and Uttarakhand, respectively (Fig. 1). Obasi (2001) stated that in recent years, the global loss of US $ 50–100 billion annually is caused due to these natural hazards together with the loss of life of about 2,50,000. However, these increased losses may be either due to a real increase in the frequency of the extreme weather events or due to increased vulnerability of cities, towns and the associated infrastructure and installations which have grown rapidly to meet the needs of a growing population. Philip and Daniel (1976) reported that Uttarakhand recorded the highest frequency of hailstorm next to that of Assam Valley.

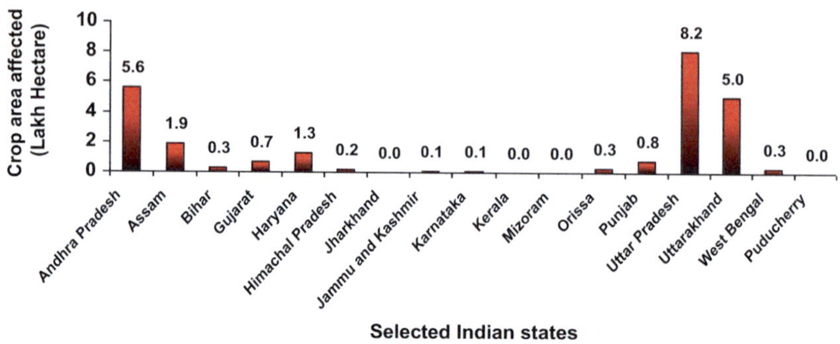

Fig. 1 Selected state-wise cropped areas affected due to cyclonic storms/heavy/rains/flash floods/landslides/cloudburst in India during 2010 (www.indiastat.com)

Effect of Extreme Weather Events on Few Hill Crops

Due to severe drought stress, soil moisture less than 8 % was recorded in Hawalbagh, VPKAS, Almora, during the 2009 *kharif* season which led to severe decline in rice production, i.e. rice grain yield of about 250 kg to 1,200 kg per hectare. Soybean grain yield was also significantly low during 2009. Due to heavy and unevenly distributed rainfall of 1369.5 mm recorded in Hawalbagh (Almora), a drastic grain yield loss was recorded in French bean, bhindi, rajmash, horse gram and buckwheat.

Winter x Spring Wheat Derivatives for the Changing Climate

For breaking yield stagnation, exploitation of gene pools of winter and spring wheat has been considered as one of the novel approaches. Winter wheat possesses enormous variability for tillering, leaf size, spike length, grain size, grain number in addition to tolerance to drought as well as heat and better N and P efficiency even under low input conditions. More than 2,000 winter wheat accessions are being maintained in VPKAS, Almora. They can be of special significance under the low input, rain-fed and drought-prone areas as they are known for their drought tolerance. Winter wheat gene pool assumes great importance for enhancing wheat production by insulating future varieties with the biotic and abiotic stresses along with these attributes. Some of the advance winter x spring wheat derivatives were tested in the changed climatic conditions in northern hill zone, and encouraging results were obtained (Kant et al. 2011).

Fodder Availability and Climate Change

INCCA (2010) reported that due to warming of temperature, the upward shifts in agriculture may result in loss of permanent pastures and grassland (*Bugyals*) to arable cultivation, which is already very low. This may cause lower availability of fodder and can adversely affect livestock sector and agriculture. Under these circumstances, the dual-purpose wheat cultivars VL *Gehun* 829 and VL 840 which produced at par green fodder, grain and straw yields after cutting at 70 and 85 DAS for green fodder could be one way to ensure fodder and food security (Bisht et al. 2008).

In a nut shell, it is clear that the changing climate is affecting the various components of agriculture, and hence, it is essential to study the magnitude of climate change/variability in northwest Himalayas along with scientific interventions with the policy support to combat the yield losses which could feed the burgeoning population.

References

Aggarwal PK (2003) Impact of climate change on Indian agriculture. J Plant Biol 30:189–198

Aggarwal PK (2008) Global climate change and Indian agriculture: impacts, adaptation and mitigation. Indian J Agric Sci 78:911–919

Aggarwal PK, Sinha SK (1993) Effect of probable increase in carbon dioxide and temperature on productivity of wheat in India. J Agric Meteorol 48:811–814

Bali R, Agarwal KK, Ali SN, Srivastava P (2009) Is the recessional pattern of Himalayan glaciers suggestive of anthropogenically induced global warming. Arab J Geosci. doi:10.1007/s12517-010- 0155–9

Bhatt JC (1992) Epidemiology of foliar blast of rice in Uttar Pradesh hills. Oryza 29:350–354

Bhatt JC (2010) Impact of climate change on hill farming. In: International workshop on mountain biodiversity and impacts of climate change with special reference to Himalayan Biodiversity Hotspot. GB Pant Institute of Himalayan Environment & Development, Kosi-Katarmal, Almora, Uttarakhand, pp 5–8

Bisht JK, Kant L, Srivastva AK (2008) Cutting management of dual purpose wheat cultivars: a new approach for increasing fodder availability. Cereal Res Commun 36:177–187

Connor S (2008) Insects' will be climate change's first victims. The Independent, (Tuesday, 6 May 2008). http://www.independent.co.uk/news/science/insects. Assessed 16 June 2010

Gupta HS (2006) Hill agriculture: status of research and production potential. In: Gupta HS, Srivastava AK, Bhatt JC (eds) Sustainable production from agricultural watersheds in North West Himalayas. Venus Printers and Publishers, New Delhi, pp 1–21

Gupta HS, Agrawal PK, Mahajan V, Bisht GS, Kumar A, Verma P, Srivastava A, Saha S, Babu R, Pant MC, Mani VP (2009) Quality protein maize for nutritional security: rapid development of short duration hybrids through molecular marker assisted breeding. Curr Sci 96:230–237

Heong KL, Song YH, Pimamarn S, Zhang R, Bae SD (1995) Global warming and rice arthropod communities. In: Peng S (ed) Climate change and rice. Springer-Verlag, New York, 326–335 p

Hooda KS, Pant SK, Mahajan V, Khati P, Bhatt JC (2009) A first record of zonate leaf spot in maize from Uttarakhand, India. J Mycol Pl Pathol 39:541

INCCA (2010) Indian network for climate change and assessment. Report: climate change and India: a 4 × 4 assessment in India. http://chimalaya.org/2010/11/17/report-climate-change-and-india-a-4×4-assessment-in-india/. Assessed 20 Nov 2012

IPCC (2007) Summary for policymakers. In: Parry ML, Canziani OF, Palutikof JP, van der Linden PJ, Hanson CE (eds) Climate change 2007: impacts, adaptation and vulnerability. Contribution of working group II to the fourth assessment report of the intergovernmental panel on climate change. Cambridge University Press, Cambridge, pp 7–22

Joshi NC, Joshi PC (2010) Foraging behaviour of *Apis* spp on apple flowers in subtropical environment. N Y Sci J 3:71–76

Joshi S, Upreti K, Punetha N (2008) Change in Lichen flora of Pindari Glacier during the last three decades. Ann For 16:168–169

Kamal MY (2003) Mountain agriculture in transition. In: Sharma R, Pande RK, Shah D, Kumar A (eds) The challenges in sustainable mountain agriculture. pp 37–46

Kant L, Jain SK (2011) Resurgence of wheat yellow rust in northern hill and tarai regions. VPKAS Newsl 15: 1–2

Kant L, Jain SK, Mahanta D, Agrawal PK, Bhatt JC (2011) Do winter x spring wheat derivatives have the capability to tolerate the climate change conditions? In: 3rd international group meeting on wheat productivity enhancement under changing climate, February 9–12, 283 p

Kleinhenz M, Bujok B, Fuchs S, Tautz J (2003) Hot bees in empty broodnest cells: heating from within. J Exp Biol 206:4217–4231

Kumar K, Dumka RK, Miral MS, Satyal GS, Pant M (2008) Estimation of retreat rate of Gangotri glacier using rapid static and kinematic GPS survey. Curr Sci 94:258–262

Kumar PR, Yadav SK, Sharma SR, Lal SK, Jha DN (2009) Impact of climate change on seed production of cabbage in North Western Himalayas. World J Agric Sci 5:18–26

Mall RK, Aggarwal PK (2002) Climate change and rice yields in diverse agro-environments of India: evaluation of impact assessment models. Climate Change 52:315–330

NATCOM (2004) India's Initial National Communication to UNFCCC, Ministry of Environment & Forests, Government of India

Obasi GOP (2001) WMOs contribution to the development of Meteorology, Keynote lecture on the occasion of Sir Gilbert Walker Memorial Lecture and Award Ceremony, New Delhi

Panday SC, Singh RD, Srivastava AK (2003) Change in climate and its consequence on crop management and production. In: 6th agricultural science congress. Multi-Enterprise Systems for Viable Agriculture. IISS, Bhopal, 203 p

Panday SC, Bisht KKS, Srivastava AK, Gupta HS (2009) Climate change in mid Himalaya. In: 4th world congress on conservation agriculture, innovations for improving efficiency, equity and environment, New Delhi, 417 p

Panday SC, Srivastava AK, Mahanta D, Mina BL, Agrawal PK, Bhattacharya R, Singh KP, Kumar A, Bhatt JC (2011) Impact assessment of climate change on rice and wheat grain yield for the Himalaya region of Uttarakhand. In: International conference on preparing agriculture for climate change, Ludhiana, 43 p

Paroda RS, Kumar P (2000) Food production and demand in South Asia. Agric Econ Res Rev 13:1–24
Partap T (2011) Sustainability factors of hill agriculture and opportunities. J Hill Agric 2:1–13
Partap U, Partap T (2002) Warning signals from the apple valleys of the Hindu Kush- Himalayas: productivity concerns and pollination problems. ICIMOD Publication, Kathmandu, 104 p
Philip NM, Daniel CEF (1976) Hailstorms over India, IMD Meteorological Monograph, Climatology No. 10
Rana JC (2010) Climate change impacting floral diversity and cropping patterns – a study in the western Himalayan region. In: International workshop on mountain biodiversity and impacts of climate change with special reference to Himalayan Biodiversity Hotspot. GB Pant Institute of Himalayan Environment & Development, Kosi-Katarmal, pp 90–95
Rao GD, Sinha SK (1994) Impact of climatic change on simulated wheat production in India. In: Rosenzweig C, Iglesias I (eds) Implications of climate change for international agriculture: crop modelling study. EPA, Washington, DC, pp 1–10
Samra JS, Singh G (2002) Drought management strategies. Indian Council of Agricultural Research, 68 p
Saseendran SA, Singh KK, Rathore LS, Singh SV, Sinha SK (2000) Effects of climate change on rice production in the tropical humid climate of Kerala, India. Clim Chang 44:495–514
Sharma BD, Rana JC (2005) Plant genetic resources of western Himalaya- status prospects. Bishen Singh Mahendra Pal Singh. Dehradun, India, 467 p
Singh AK, Ngachan SV, Meghalaya U, Rawal R, Singh KK, Sathapathy S (2010) Food security and climate change in Himalayan region: status and strategies for livelihood improvement. www.bhutanclimatesummit.org.bt. Assessed 14 Oct 2012
Somerville D (1999) Honeybees in cherry and plum pollination. New South Wales Agriculture. Agnote. http://www.dpi.nsw.gov.au/_data/assets/pdf_file/0019/117109/bee-cherry-plum-pollination.pdf. Assessed 16 Nov 2010
Stanley J, Aditya JP, Bhatt JC, Pandey SC, Agrawal PK (2009) Change in pest spectrum of rice under changing climatic situations in NW Himalayan hills. In: National conference on pest biodiversity in rice and their management under changed climate, CRRI, Cuttack, India
Stanley J, Khusboo, Pandey SC, Gupta JP, Subbanna ARNS, Jain SK, Bhatt JC (2010) Effect of climate change on insect pollinators of onion and mustard in North Western Indian Himalayas. In: International workshop on mountain biodiversity and impacts of climate change with special reference to Himalayan Biodiversity Hotspot. GBPIHED, Kosi-Katarmal, pp 141–145
Sushil SN, Pant SK, Bhatt JC (2004) Light trap catches of white grub and its relation with climatic factors. Ann Plant Prot Sci 12:254–256
Tripathi SK, Chintamanie B (2011) Agronomic impacts of climate variability on rice production with special emphasis on precipitation in South Western Plains of Uttarakhand. In: Attri SD (ed) Challenges and opportunities in agrometeorology. Springer, Berlin/Heidelberg. doi:10.1007/978-3-642-19360-6_41
UCCC (2011) Uttarakhand centre for climate change. Interim Report. Kumaon University, Nainital. www.ucccku.com. Assessed 14 Nov 2011
Upreti DK (2010) Himalayan lichens as an indicator of climate change. In: International workshop on mountain biodiversity and impacts of climate change with special reference to Himalayan biodiversity Hotspot. G.B.PIHED, Kosi-Katarmal, Uttarakhand, pp 90–95
Vaish SS, Bilal AS, Prakash K (2011) First documentation on status of barley diseases from the high altitude cold arid Trans-Himalayan Ladakh region of India. Crop Prot 30:1129–1137
www.indiastat.com

Climate Change and Mitigation Options for Sustainable Agriculture in Rajasthan

R.P. Jangir and Surendra Singh

Abstract

Climate change is the greatest challenge before the global society impacting the ecology, economy, and society in several ways. Rajasthan has reason to be concerned about the impact of climate change as its large population depends upon climate-sensitive sectors like agriculture and forestry for livelihood. Rajasthan shows a significant warming of 0.5 °C which is comparable to the global mean trend of 0.3 °C and all India mean of 0.4 °C per 100 years. The climate of Rajasthan has exhibited a continuing trend towards desiccation, particularly after the severe drought of 1987. For example, in February and March, 2006 and 2008, there was high-temperature stress, and there was heavy frost in January, 2007 and 2010. In December and January, 2008–2009, there was unusual rise in temperature, and the yields of wheat and mustard were adversely affected. Recently, in kharif, 2009, there was severe drought in most of the areas causing large-scale crop failure in western Rajasthan. Impact of climate change on agriculture is visualized in terms of increased problem of water stress in major kharif crops; heat stress in wheat, barley, mustard, and gram; occurrence of frost in mustard, gram, and pea; virus- and root-related diseases; sucking pests, mites, leaf minor, and gram pod borer in selected crops; emergence of new weeds; etc. These are creating suboptimal and stressful environment for agricultural crop production. Green

R.P. Jangir (✉)
S. K. Rajasthan Agricultural University, Bikaner, Rajasthan 334006, India
e-mail: jangirrp52@gmail.com

S. Singh
A.R.S, Durgapura, Jaipur, India

A.K. Singh et al. (eds.), *Climate Change Modelling, Planning and Policy for Agriculture*,
DOI 10.1007/978-81-322-2157-9_18, © Springer India 2015

agriculture, water-saving agriculture including water harvesting and land treatment for in situ moisture conservation, adoption of integrated farming system, use of crop simulation models, emergency response system, and crop and weather insurance are some of the mitigation options available for ameliorating adverse impact of climate change for sustainable agriculture.

Keywords
Climate change • Agriculture • Mitigation options • Heat stress • Frost

Introduction

Climate change is one of the greatest challenges of our time. Rajasthan is the largest state in India with two third of its area as semiarid. The semi-arid agriculture of Rajasthan is extremely vulnerable to weather and climate. In recent past, there was substantial loss of crop in the region due to extreme and unusual weather conditions. Semi-arid ecosystem has reason to be concerned about the impact of climate change as its large population depends upon climate-sensitive sectors like agriculture and forestry for livelihood. Any adverse impacts of water availability would threaten food security and rural households (Dash and Hunt 2007).

Thus, climate change is a complex alteration of climate, subtle and continuous yet important through its consequences on vegetation of various types that thrived under constant or relatively unchanged climate. Some of the main effects of climate change with specific reference to agriculture and food production especially during the last decade are increased occurrence of drought, frost, flood, and heat stress; increased frequency of pest and diseases; etc. (Venkateshwaralu and Shanker 2009) (Fig. 1).

Climate Change in Rajasthan

In recent years, the incidence of climatic extreme events like drought, frost, heat stress, and even sometimes flood has become frequent in the state. Climatic changes have already set in, resulting in changes in rainfall and temperature. The analysis of temperature data of the last 100 years indicates a rise of 0.6 °C in Rajasthan's temperature (Anonymous 2007). The analysis further reveals that climate of Rajasthan has exhibited a continuing trend towards desiccation, particularly after the severe drought of 1987. For example, in February and March, 2006 and 2008, there was high-temperature stress, and there was heavy frost in January, 2007 and 2010. In December and January, 2008–2009, there was unusual rise in temperature, and the yields of wheat and mustard were adversely affected. Recently, in *kharif*, 2009, there was severe drought in most of the areas causing large-scale crop failure in western Rajasthan (Anonymous 2011). Some specific examples of climate change events in state are as under:

Impact on Ground Water Status

Lowering of ground water has become concern for the agricultural community in the state. In 1984, out of 236 blocks, 203 were safe (white category) and 32 blocks were dark. In 2008, out of 236 blocks, 31 blocks remained safe and 195 blocks come under dark zone. Thus, due to variability in rainfall and overexploitation of groundwater, water becomes the most scarce commodity for the agriculture (Anonymous 2010).

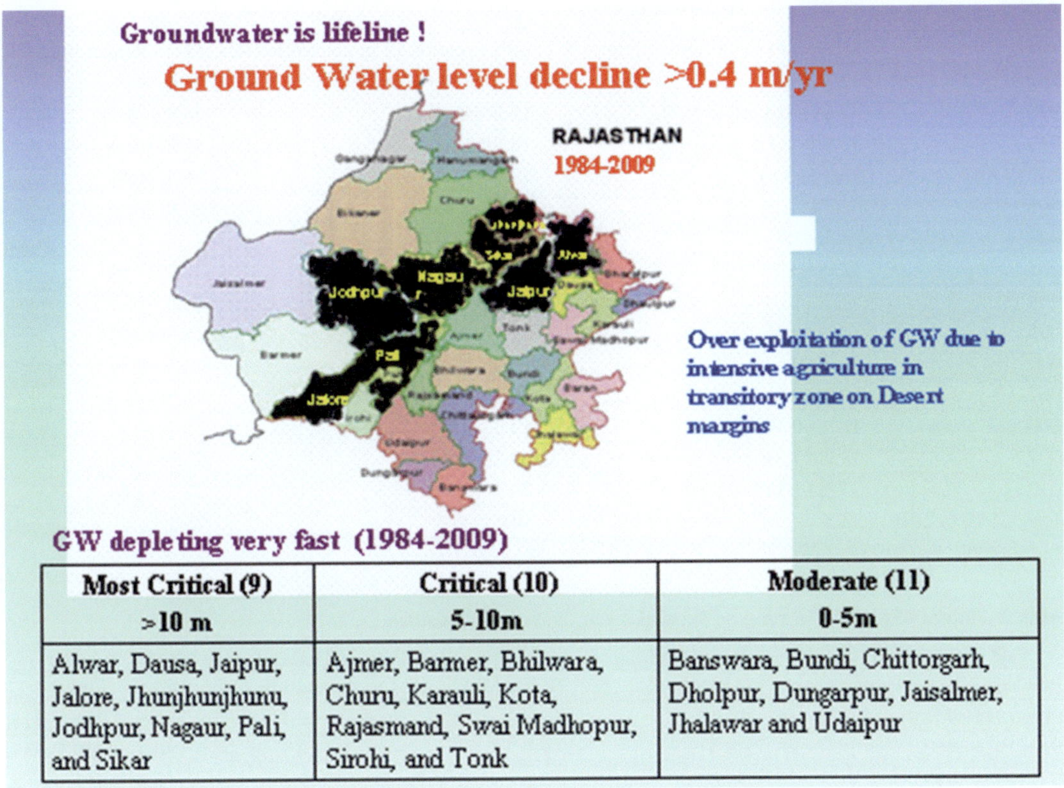

Most Critical (9) >10 m	Critical (10) 5-10m	Moderate (11) 0-5m
Alwar, Dausa, Jaipur, Jalore, Jhunjhunjhunu, Jodhpur, Nagaur, Pali, and Sikar	Ajmer, Barmer, Bhilwara, Churu, Karauli, Kota, Rajasmand, Swai Madhopur, Sirohi, and Tonk	Banswara, Bundi, Chittorgarh, Dholpur, Dungarpur, Jaisalmer, Jhalawar and Udaipur

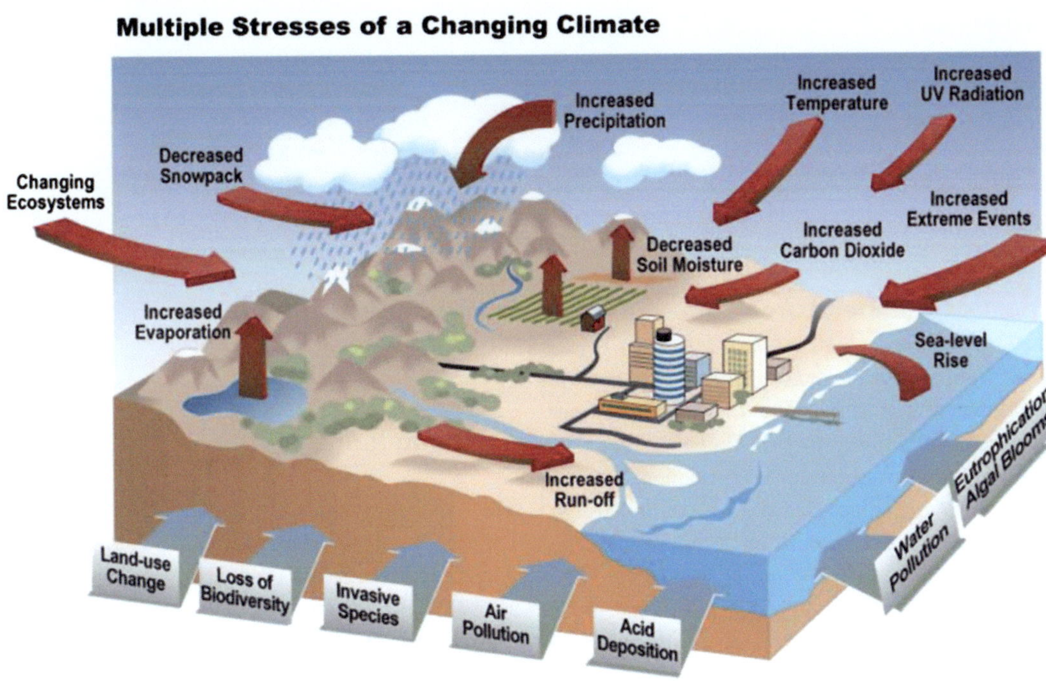

Fig. 1 Various kinds of stresses arise due to the impact of climate change

Table 1 Temperature trend in the state 1998 onwards

Year	Mean maximum temperature (°C)	Mean minimum temperature (°C)	Mean temperature (°C)
Normal	32.7	18.4	25.6
1998	32.6 (−0.1)	19.3(+0.9)	26.0 (+0.4)
1999	32.8 (+0.1)	18.8(+0.4)	25.8(+0.2)
2000	33.3 (+0.5)	19.5(+1.1)	26.4(+0.8)
2001	32.9 (+0.3)	18.8(+0.4)	25.9(+0.3)
2002	34.4(+1.8)	20.4(+2.0)	27.4(+1.8)
2003	32.9(+0.2)	19.4(+1.0)	26.2(+0.6)
2004	33.7(+1.0)	19.9(+1.5)	26.8(+1.3)
2005	32.8(+0.1)	18.9(+0.5)	25.9(+0.3)
2006	33.3(+0.6)	19.8(+1.4)	26.5(+0.9)
2007	33.2(+0.5)	19.8(+1.4)	26.5(+0.9)
2008	32.8(+0.1)	19.3(+0.9)	26.1(+0.5)
2009	33.9(+1.2)	20.2(+1.8)	27.0(+1.5)
2010	33.8(+1.1)	20.5(+2.1)	27.2(+1.6)
Mean (1979 to 2010)	33.1	19.1	26.1

Table 2 Impact of global warming on decadal mean rainfall of Rajasthan

Decade	Monsoon rainfall (June–Sept.)	Winter rainfall (Oct.–Jan.)	Pre-monsoon rainfall (Feb.,–May)	Decadal mean rainfall	C.V (%)	Range
1901–1911	451.4	14.1	26.6	492.1	35.5	261.4–814.4
1911–1921	467.1	33.4	34.3	534.7	46.4	239.1–811.8
1921–1931	499.0	30.0	42.1	549.6	16.8	412.1–688.9
1931–1941	494.5	22.0	25.9	542.4	21.3	371.0–681.8
1941–1951	529.2	22.1	22.8	632.7	19.2	458.0–777.9
1951–1961	468.2	28.0	17.9	511.2	27.0	312.6–703.7
1961–1971	477.2	13.9	24.9	516.0	19.4	363.0–689.7
1971–1981	578.1	32.0	22.2	632.4	25.9	411.5–834.3
1981–1991	459.1	29.0	33.0	521.1	21.1	314.5–727.3
1991–1901	531.3	34.1	26.4	591.8	18.6	403.5–757.2
2001–1911	487.8	8.2	35.8	532.4	24.6	267.1–718.7
Mean	494.8	24.3	28.4	550.6	25.1	261.4–834.3

Impact on Temperature

From 1998 onwards, mean temperature of state has shown rising trend. The mean temperature has increased from 0.2 to 1.6 °C (Table 1). Likewise, the mean maximum and mean minimum temperature has also shown increasing trend. Similarly, the mean maximum temperature during the month of February and March of the zone III a has also risen from 0.6 °C (2003) to 7.9 °C (2009). The temperature of February and March is most critical for the ripening of wheat crop in the state (Anonymous 2011).

Monsoon Rains

The monsoon rains in the state had shown wide variability in the last decades (Table 2). The decadal variability of monsoon rains ranged from 19.2 (1941–1951) to 46.4 (1911–1921). The mean rainfall in 100 years ranged from 261.4 to

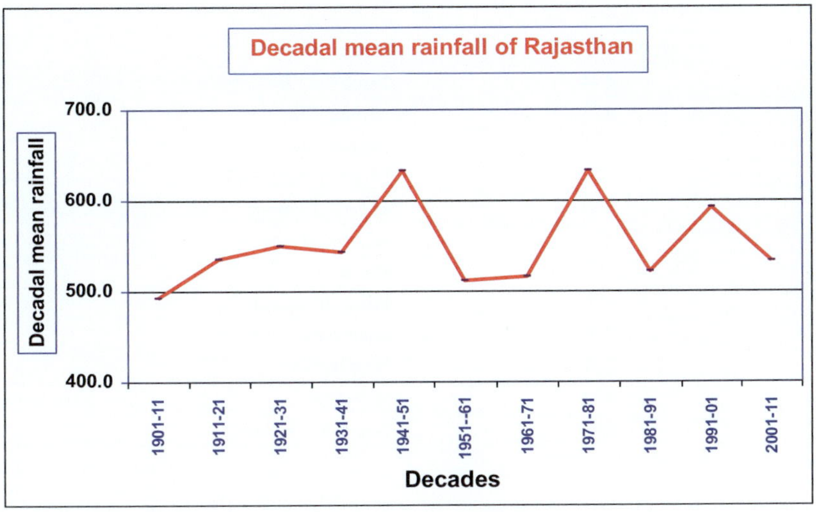

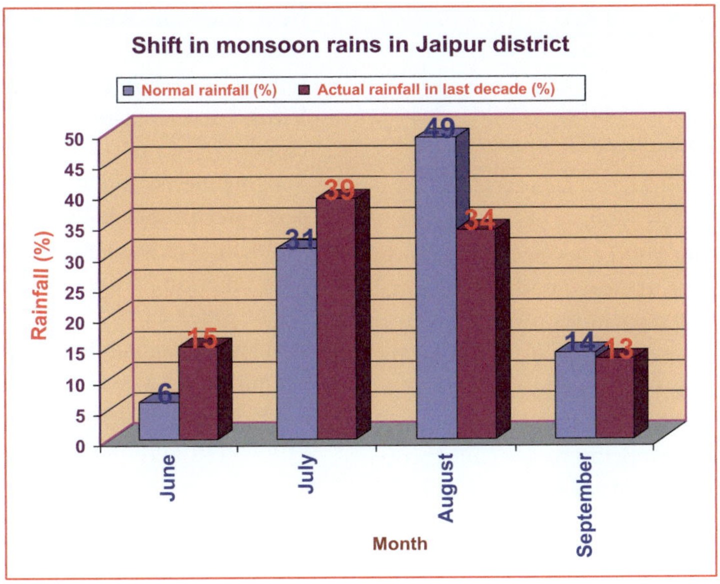

834.3 mm. The mean decadal pre-monsoon rainfall has shown increasing trend, while winter rains in overall has shown decreasing trend.

The last decade rainfall data (2001–2010) shows that the peak month of monsoon rains in Jaipur district, i.e., August, is shifting towards July. Extreme events such as low temperature, maximum temperature, rainy days, and 1 day rains (>100 mm) are more pronounced in the last decade. Arid districts, i.e., Barmer, Jaisalmer, Bikaner, and Jodhpur, are witnessing flood and more rains in the last years (Tables 3 and 4).

Impact of Climate Change on Agriculture

A recent simulation study has shown that a rise in temperature by 1 °C can lead to a decline in

Table 3 Maximum rains recorded in 1 day

Year	Rainfall (mm)	Place	Date
1995	494	Rupawas, Bharatpur	8 August 1995
1996	475	Mount Abu	29 July 1996
1997	305	Mount Abu	28 July 1997
1998	327	Mansi Dam, Tonk	14 July 1998
1999	300	Bilara, Jodhpur	31 August 1999
2000	385	Bijolia, Bhilwara	21 July 2000
2001	400	Baran	2 July 2001
2002	196	Hindoli, Bundi	9 August 2002
2003	270	Banswara	28 July 2003
2004	391	Hamirgarh	12 August 2004
2005	336	Bharatpur	12 July 2005
2006	451	Kalibore, Sirohi	20 August 2006

Table 4 Increasing rainfall in western districts of Rajasthan in the last 5 years

Year	Barmer	Jaisalmer	Jodhpur	Jalore	Bikaner
2006	750.0	399.5	150.7	321.1	216.40
2007	245.2	356.8	248.2	422.5	270.10
2008	229.2	400.2	301.5	309.5	169.70
2009	201.0	356.8	497.6	562.1	263.60
2010	521.2	409.5	489.0	489.0	296.5
Normal	171.1	164.5	324.9	438.1	247.6

wheat production by 250 kg/ha in Rajasthan. In Indian mustard, the decline was 100 kg/ha/degree rise in temperature in Rajasthan, whereas in chickpea (gram) the decline was 200 kg/ha. There are concomitant declines in biomass yield also (Kalra et al. 2008). In winter dry land crops, CO_2 enrichment and temperature rise could probably result in depletion of soil water at a rapid during vegetative phase of growth leading to moisture stress during grain filling and consequently poor harvest index and low productivity. In north India, warming would reduce some losses in yield by early pod set in winter grain legume like gram and lentil and by arresting frost damage in rape seed and mustard (Sharma 2009). Dadheech et al. (2009) had revealed that a 1 °C increase in temperature would reduce the duration of wheat crop by 1 week, which lead to loss in yield by 400–500 kg/ha.

The following are the important examples of impact of climate change on agriculture visualized in the state and need attention for sustaining agricultural production:

- Problem of water stress in major kharif crops
- Heat stress in wheat, barley, mustard, and gram
- Impact of frost in mustard, gram, and pea
- Virus- and root-related diseases in major crops
- Sucking pests, mites, leaf minor, and gram pod borer in selected crops
- Problem of new emerging weeds in field crops
- New emerging areas of salinity due to lowering of water table
- Problem of nematode in pulse crops
- Terminal heat stress in pearl millet
- Increasing intensity of YMV in pulses
- Problem of gummosis in cumin

Strategies for Mitigating the Impact of Climate Change for Sustainable Agricultural Production

Adaptation can complement mitigation as a cost-effective strategy to reduce the vulnerability of the natural and socioeconomic systems to projected climate change risks (Sathaye et al. 2006). Early actions on adaptation are always justifiable to avoid possible eco-disaster from the view point of precautionary principle and ecological prudence.

Key strategies for sustaining agricultural production in light of climate changes are as under:

- Green agriculture for resilience to climate change
- Imparting redox homeostasis for crop tolerance to environmental stress
- Agroforestry for soil restoration and climate change mitigation
- Adoption of water-saving agriculture including water harvesting and land treatment for in situ moisture conservation
- Adoption of integrated farming system approach for maximizing farm income
- Developing stress-tolerant varieties and genotypes
- Use of crop simulation models for decision support system
- Emergency response system (ERS) based on advanced information and information communication technology (ICT) for effective utilization of weather-based agro-advisory using modern tools like GIS and remote sensing
- Crop weather insurance
- Drought contingency planning

Technologies Generated for Mitigating the Adverse Impact of Climate Change

Heat Stress in Wheat

On the basis of experiments conducted during 2007–2008 to 2010–2011, the following recommendations are generated for the farmers of the state for mitigating the adverse impact of heat stress in wheat:

1. Spray of thioglycolic acid (100 ppm) and salicylic acid (100 ppm) at jointing and ear emergence stage are recommended for mitigating the adverse impact of heat stress in wheat.
2. Water stress at tillering stage in wheat is useful in mitigating the impact of heat stress in wheat.
3. Raj-4037, Raj-4083. and Raj-4161 are newly released varieties of wheat tolerant to heat stress.

Frost Management in Mustard and Gram

1. Spray of 0.1 % H_2SO_4 and thiourea (500 ppm) is recommended at pre-flowering and pod formation stage for mitigating the adverse impact of frost in mustard and gram.
2. RGN-13 and RGN-45 varieties of mustard are tolerant to frost.

In conclusion, it can be said that in the facet of climate change, looming large sustainable agriculture in the modern era will have to be based on appropriate use of eco-technology, biotechnology, and information technology. Blending of traditional farming with modern technologies can help to reduce the ill effects of climate change through efficient utilization of natural resources for sustainable agricultural production.

References

Anonymous (2007) Climate: variability & change background material towards preparation of Adaptation strategy for the state of Rajasthan. Department of environment, GOR, Rajasthan Pollution control board

Anonymous (2010) Vital agriculture statistics of Rajasthan. Commissionaire of Agriculture. GoR, Pant Krishi Bhavan, Jaipur

Anonymous (2011) Consolidated progress report on "Adaptation Strategies for mitigating the adverse impact of climate change in late sown wheat

Dadheech RC, Solanki NS, Tiwari RC (2009) Climate change: a threat and challenge to agriculture. Paper presented in conference on effect of environment change on agriculture held at RCA, Udaipur

Dash SK, Hunt JC (2007) Variability of climate change in India. Curr Sci 93(6):783–788

Kalra N, Chakraborty D, Sharma A, Rai HK, Jolly M, Chander S, Ramesh Kumar P, Bhadraray S, Barman D, Miltal RB, Lal M, Sehgal M (2008) Effect of increasing temperature on yield winter crops in north west India. Curr Sci 94:82–88

Sathaye J, Shukka PR, Ravindanath NH (2006) Curr Sci 90(3):314–325

Sharma DD (2009) Impact of atmospheric carbon dioxide shift in agriculture. Paper presented in conference on effect of environment change on agriculture held at RCA, Udaipur

Venkateshwaralu B, Shanker AK (2009) Climate change and agriculture: adaptation and mitigation strategies. Indian J Agron 54(2):226–230

Trends in Climatic Change in the Last 50 Years at Seven Agro-climatic Regions of Tamil Nadu

T. Sivakumar, P.T. Suraj, and P.C. Jayashree

Abstract

The climate change issue is part of the larger challenge of sustainable development and one of the most important global environmental challenges facing humanity which go far beyond its effect on the environment. In order to understand the climatic change happening in the different agro-climatic regions of Tamil Nadu and to develop strategies to mitigate climatic stress and to optimise productivity monthly, Maximum and Minimum Temperature and Relative Humidity Data during the period 1955–2005 were obtained from the Indian Meteorological Department, Pune. From the basic temperature data, mean maximum, mean minimum and Temperature Humidity Index was computed for each month for the seven agro-climatic regions of Tamil Nadu. Mean maximum temperature was observed at the month of May, and Cauvery Delta zone showed maximum temperature of 38.22 ± 1.33 °C with significant difference of $P < 0.01$ between the regions. Mean minimum temperature was observed at the month of January, and hilly zone showed minimum temperature of 5.59 ± 1.31 °C with a significant difference of $P < 0.01$ between the regions. The long-term mean and annual compounded growth rates of T_{max}, T_{min} and THI were worked out on the basis of the representative areas selected for the study. The annual growth rate of maximum and minimum temperatures showed different patterns for different agro-climatic zones in the study area. The T_{max} showed an increase in all the agro-climatic zones except in north-western zone and southern zone. T_{min} also showed a positive growth pattern in all agro-climatic zones except in north-western zone and southern zone. The annual growth rate of THI showed different patterns for different agro-climatic zones. Five agro-climatic zones, viz., north-eastern zone, western zone and hilly zones, Cauvery Delta zone and high rainfall zone, were showing a positive

T. Sivakumar (✉) • P.T. Suraj • P.C. Jayashree
Department of Livestock Production and Management, Madras Veterinary College, Chennai 600007, Tamil Nadu, India
e-mail: tsk63@rediffmail.com

A.K. Singh et al. (eds.), *Climate Change Modelling, Planning and Policy for Agriculture*, DOI 10.1007/978-81-322-2157-9_19, annual compounded growth rate for both morning and evening THI. In north-western zone the growth in THI was limited to morning and the evening THI showed a negative growth. Southern zone showed a negative annual compounded growth rate for both morning and evening THI.

Keywords
Annual compounded growth rate • Climate change • T maximum • T minimum • Temperature Humidity Index • Tamil Nadu

Introduction

Livestock and agriculture sectors are primarily dependent on the timely availability of water and favourable climate. Total rainfall received in a particular area is a key factor in determining the amount of water available to meet various demands in agricultural and livestock divisions. Climate changes may influence rainfall patterns, thus increasing the demand for availability of water with the danger of rising incidents of droughts. The southwest monsoon (SW) brings about 80 % of rainfall over the country and is the major source for the availability of freshwater for drinking and irrigation. Climate change over the Indian boundaries, especially the SW monsoon, would have a significant impact on water resources management in agricultural production and livestock industry. In view of the above, a number of studies have attempted to investigate the trend of climatic variables for the country. These studies have looked at the trends on the country scale, regional scales and at the individual stations (Jain and Kumar 2012).

In irregular distribution of rainfall and the difference between water availability and demand, large storage reservoirs are required to redistribute the natural flow in accordance with the requirements of specific regions (Mehrotra 1999). Changes in rainfall due to global warming will influence the hydrological cycle and the pattern of stream-flows and demands. Changes in run-off and its distribution will depend on likely future climate scenarios (IPCC 2007). Climate change disrupts temperatures, as well as rainfall patterns in the densely populated regions that would have enormous significance for livelihood and well-being of the people of the region. Climate change will have environmental and social impacts that will likely increase uncertainty in water supplies and agricultural production for people across India Geethalakshmi et al. (2011). The cascading effects of rising temperatures are already affecting water availability, biodiversity, ecosystem boundaries and global feedbacks (Amin et al. 2004).

The trend analysis of rainfall, temperature and other climatic variables on different spatial scales will help in the construction of future climate scenarios. A combination of adaptive and preventive measures is urgently required in conjunction with climatological data to assess climatic impact and provide a rational basis for decisions on long-range management and housing strategies for livestock producers. It is the poorest who has the least resources and capacity to adapt who will be hit hardest by the changing climate. They rely largely on climate-sensitive activities largely in the agriculture sector. Supporting these people to adapt to the effects of climate change is critical in order to achieve sustainable development. India's immense geographic diversity adds to the difficulty of developing an adaptation strategy. Approaches will need to be tailored to meet local vulnerabilities and conditions. This is why it is important to conduct in-depth vulnerability assessment to identify those areas of India which will be most affected by climate change. Similarly, a combination of long-term climate changes and expected production changes provide a rational basis for prediction of the impact of potential global climate changes on livestock production (Tadross et al. 2005). The status on weather events lies scattered in the scientific and technical papers and in the research work of many authors, and if put together, they will help the research community for further analysis. The aim of this study is thus to provide figures as well as composite (average) curves for the main meteorological parameters to identify and develop strategies to mitigate climatic stress and to optimise productivity monthly Maximum and Minimum Temperature and Relative Humidity Data during the period 1955–2005.

Materials and Methods

Past data on temperature (maximum and minimum) and relative humidity for the past 50 years (1955–2005) were collected from the National Data Centre, Indian Meteorological Department, Shivajinagar, Pune 411 005, Maharashtra, for understanding the climate change occurring in the seven different agro-climate zones of Tamil Nadu. Temperature Humidity Index (THI) was calculated for morning and evening from mean dry bulb and wet bulb temperature by using the formula THI = 0.72 (dry bulb temp + wet bulb temp). From the basic temperature data, mean maximum (T_{max}) and mean minimum (T_{min}) along with their standard deviation (SD) and significance were calculated by using SPSS 17. The long-term mean and annual compounded growth rates of T. maximum, T. minimum and THI (morning and evening) were also worked by using the formula (LOGEST(DATA − 1)*100). Cluster analysis was performed with the primer software version 6. It is also essential to mention that there were some missing data in some months. Data were considered to be missing when the data were not recorded. To maintain the continuity, the gaps were filled up by the time mean values of the existing years. After completion collecting data were compiled, tabulated and analysed according to the objectives of the study.

Results and Discussion

Studying climate change has been one of the most challenging problems around the world because of both its practical value in meteorology and for scientific research. Every sign points to the facts that there is a recognised need for accurate estimates of the temperature on a variety of temporal and spatial scales (Tadross et al. 2005). The THI morning in the seven agro-climatic regions of Tamil Nadu is presented in Table 1. All the seven agro-climatic regions showed significant difference ($P < 0.01$) in THI morning. Similarly, THI evening also showed significant difference ($P < 0.01$) between different agro-climatic regions that are presented in Table 2. The T_{max} in the seven agro-climatic regions of Tamil Nadu are presented in Table 3. Mean maximum temperature was observed at the month of May, and Cauvery Delta zone showed maximum temperature of 38.22 ± 1.33 °C and showed significant difference ($P > 0.01$) between the regions. The T_{min} in the seven agro-climatic regions of Tamil Nadu are presented in Table 4. Mean minimum temperature was observed at the month of January, and hilly zone showed minimum temperature of 5.59 ± 1.31 °C and showed significant difference ($P > 0.01$). The annual growth rate of maximum and minimum temperatures showed different patterns for different agro-climatic zones in the study area. The annual compounded growth rate of T_{max} and T_{min} is shown in Figs. 1, 2, 3, 4, 5, 6 and 7. The T_{max} showed an increase in all the agro-climatic zones except in north-western zone, southern zone and Cauvery Delta zone. T_{min} also showed a positive growth pattern in all agro-climatic zones except in north-western zone and Cauvery Delta zone. The annual growth rate of THI showed different patterns for different agro-climatic zones that are presented in Figs. 8, 9, 10, 11, 12, 13 and 14. Five agro-climatic zones, viz., north-eastern zone, western zone, hilly zones, southern zone and high rainfall zone, were showing a positive annual compounded growth rate for both morning and evening THI. In the north-western zone, the growth in THI was limited to morning and the evening THI showed a negative growth and Cauvery Delta zone showed a negative growth in THI morning.

Temperatures over a particular area may vary seasonally or annually depending upon latitude,

Table 1 Mean ± SD THI morning in the seven agro-climatic regions of Tamil Nadu

Months	North-eastern zone	North-western zone	Western zone	Hilly zone	Cauvery Delta zone	Southern zone	High rainfall zone	Overall
Jan	69.72 ± 3.30	69.67 ± 1.87	71.14 ± 2.76	53.59 ± 1.29	72.29 ± 1.09	73.74 ± 1.31	74.59 ± 0.94	69.60 ± 6.20
Feb	71.36 ± 3.69	71.18 ± 3.71	72.46 ± 2.28	55.08 ± 1.17	73.89 ± 1.34	74.84 ± 1.26	75.84 ± 1.06	71.03 ± 6.31
Mar	74.50 ± 3.84	74.41 ± 4.00	75.07 ± 1.87	57.53 ± 2.62	76.88 ± 1.09	77.36 ± 1.03	78.29 ± 0.71	73.87 ± 6.44
Apr	78.16 ± 4.32	77.63 ± 4.28	77.53 ± 1.22	60.61 ± 3.02	79.95 ± 0.72	79.88 ± 0.83	80.09 ± 0.69	76.82 ± 6.42
May	78.94 ± 4.88	77.87 ± 4.30	77.70 ± 1.39	60.77 ± 5.17	80.21 ± 0.74	80.69 ± 0.88	79.97 ± 0.84	77.21 ± 6.72
Jun	77.83 ± 4.19	76.36 ± 4.13	76.34 ± 1.76	60.29 ± 0.74	79.13 ± 0.59	79.95 ± 1.02	77.99 ± 0.91	76.06 ± 6.24
Jul	76.94 ± 3.98	75.44 ± 4.00	75.35 ± 1.95	58.64 ± 3.74	78.23 ± 0.68	79.01 ± 0.87	77.10 ± 0.72	75.07 ± 6.51
Aug	76.53 ± 3.85	75.12 ± 3.96	75.25 ± 1.80	59.35 ± 0.57	77.82 ± 0.63	78.98 ± 1.01	77.05 ± 0.74	74.92 ± 6.12
Sep	76.34 ± 3.84	75.25 ± 3.99	75.53 ± 1.71	59.53 ± 0.74	77.51 ± 0.61	78.80 ± 1.02	77.59 ± 0.72	74.92 ± 6.04
Oct	75.52 ± 3.70	74.81 ± 3.94	75.26 ± 1.43	59.47 ± 0.81	77.07 ± 0.61	77.44 ± 4.41	77.70 ± 0.64	74.37 ± 6.07
Nov	73.25 ± 3.50	72.78 ± 3.80	73.95 ± 1.84	57.91 ± 1.22	75.34 ± 0.82	76.23 ± 0.97	76.78 ± 0.64	72.70 ± 5.85
Dec	70.74 ± 3.38	70.29 ± 3.61	71.93 ± 2.52	55.53 ± 1.75	73.29 ± 1.02	74.53 ± 1.35	75.28 ± 0.90	70.53 ± 6.03

Table 2 Mean ± SD THI evening in the seven agro-climatic regions of Tamil Nadu

Months	North-eastern zone	North-western zone	Western zone	Hilly zone	Cauvery Delta zone	Southern zone	High rainfall zone	Overall
Jan	75.75 ± 1.30	75.78 ± 1.71	76.22 ± 1.57	60.51 ± .87	77.09 ± 0.97	77.60 ± 1.45	77.34 ± .77	74.81 ± 5.20
Feb	78.28 ± 1.87	78.35 ± 1.73	78.53 ± 1.38	62.04 ± 1.10	79.43 ± 1.25	79.43 ± 1.25	78.32 ± 0.90	76.94 ± 5.41
Mar	80.95 ± 2.49	80.83 ± 1.57	80.86 ± 1.36	63.64 ± 1.02	82.27 ± 1.16	81.68 ± 1.02	80.20 ± 0.74	79.30 ± 5.70
Apr	82.86 ± 2.70	82.26 ± 1.31	82.12 ± 1.29	64.41 ± .96	84.34 ± 1.07	82.99 ± 1.29	81.39 ± 0.78	80.79 ± 5.98
May	82.95 ± 3.34	82.42 ± 1.39	81.10 ± 1.66	64.58 ± 1.13	85.05 ± 1.30	83.40 ± 1.47	80.85 ± 0.88	80.87 ± 6.12
Jun	81.78 ± 2.29	80.55 ± 1.37	78.59 ± 1.87	61.60 ± 1.00	83.40 ± 1.16	82.04 ± 1.29	78.76 ± 0.82	79.03 ± 6.44
Jul	80.91 ± 2.07	79.73 ± 1.27	77.74 ± 1.85	60.63 ± .83	82.64 ± 1.16	81.89 ± 1.32	77.98 ± 0.75	78.18 ± 6.46
Aug	80.71 ± 1.95	79.51 ± 1.15	77.86 ± 1.68	60.96 ± .66	82.47 ± 1.07	81.12 ± 1.16	77.82 ± 0.76	78.10 ± 6.30
Sep	80.19 ± 1.74	79.45 ± 1.23	78.35 ± 1.51	61.73 ± .84	82.05 ± 1.18	81.07 ± 1.27	78.33 ± 0.72	78.14 ± 5.98
Oct	78.41 ± 1.28	77.91 ± 1.13	77.72 ± 1.07	61.51 ± .75	80.18 ± 1.00	79.24 ± 4.39	78.52 ± 0.73	76.82 ± 5.77
Nov	76.33 ± 1.07	76.10 ± 1.26	76.71 ± 1.22	60.63 ± .93	77.85 ± 0.94	78.08 ± 0.99	78.11 ± 0.54	75.31 ± 5.28
Dec	75.11 ± 1.07	74.87 ± 1.48	75.60 ± 1.53	60.20 ± .97	76.63 ± 0.86	77.16 ± 1.40	77.63 ± 0.86	74.29 ± 5.16

Table 3 Mean ± SD T_{max} in the seven agro-climatic regions of Tamil Nadu

Months	North-eastern zone	North-western zone	Western zone	Hilly zone	Cauvery Delta zone	Southern zone	High rainfall zone	Overall
Jan	29.89 ± 1.13	30.89 ± 1.44	30.40 ± .87	21.10 ± 1.19	30.42 ± 0.71	30.84 ± .86	30.86 ± .58	29.52 ± 3.09
Feb	32.73 ± 1.32	33.57 ± 1.34	33.06 ± 1.09	21.66 ± 1.19	32.89 ± .99	33.19 ± 1.12	31.63 ± .56	31.73 ± 3.68
Mar	35.69 ± 1.39	36.2 ± 1.15	35.62 ± 1.08	22.68 ± 1.19	35.57 ± 0.93	35.82 ± 1.18	32.18 ± .53	34.06 ± 4.23
Apr	37.44 ± 1.42	37.35 ± 1.05	36.39 ± 1.09	23.00 ± 1.18	37.53 ± 1.12	36.92 ± 1.46	32.72 ± .56	35.25 ± 4.59
May	38.13 ± 1.65	36.84 ± 1.29	34.79 ± 1.31	22.27 ± 1.15	38.22 ± 1.33	37.66 ± 1.56	32.43 ± .76	35.25 ± 5.00
Jun	35.79 ± 1.33	34.66 ± 1.02	32.18 ± 1.03	18.57 ± 1.06	37.14 ± 0.90	36.26 ± 1.24	30.59 ± .82	33.15 ± 5.50
Jul	34.33 ± 1.26	33.62 ± .89	31.08 ± .95	17.20 ± .94	36.25 ± 0.94	35.30 ± 1.09	30.28 ± .74	32.07 ± 5.55
Aug	33.76 ± 1.21	33.06 ± .90	31.51 ± .77	17.59 ± .76	35.83 ± .81	35.39 ± 1.03	30.39 ± .61	31.93 ± 5.29
Sep	33.54 ± 1.22	32.89 ± 1.19	32.35 ± 0.95	18.90 ± .82	34.89 ± 1.06	35.62 ± 1.22	30.53 ± .64	32.00 ± 4.85
Oct	31.87 ± .97	31.53 ± 0.97	31.48 ± 0.90	19.09 ± .76	32.63 ± 0.91	33.62 ± 0.90	30.45 ± .48	30.68 ± 4.19
Nov	29.72 ± .97	30.15 ± 1.25	29.81 ± 0.85	19.27 ± .82	30.36 ± 0.82	30.99 ± 1.14	30.19 ± .52	29.05 ± 3.52
Dec	28.81 ± 1.05	29.59 ± 1.55	29.18 ± .95	20.33 ± 1.26	29.42 ± 0.81	29.42 ± 0.81	30.27 ± .70	28.53 ± 3.08

Table 4 Mean ± SD T_{min} in the seven agro-climatic regions of Tamil Nadu

Months	North-eastern zone	North-western zone	Western zone	Hilly zone	Cauvery Delta zone	Southern zone	High rainfall zone	Overall
Jan	17.42 ± 1.48	18.55 ± 1.29	18.34 ± .92	5.59 ± 1.31	20.46 ± .74	21.19 ± 1.44	23.44 ± .53	18.12 ± 4.82
Feb	18.82 ± 1.54	19.66 ± 1.35	19.44 ± 1.06	6.59 ± .90	21.36 ± .98	21.88 ± 1.46	23.86 ± .66	19.12 ± 4.71
Mar	21.17 ± 1.33	21.79 ± 1.41	21.49 ± .80	8.83 ± .85	25.35 ± .92	23.63 ± 0.92	25.14 ± .59	21.12 ± 4.54
Apr	24.08 ± 1.21	24.55 ± 1.02	23.43 ± .59	10.73 ± .96	26.04 ± .60	25.66 ± .94	26.17 ± .48	23.44 ± 4.54
May	25.15 ± 1.36	24.75 ± .88	23.43 ± .58	11.58 ± 1.00	26.80 ± .65	26.46 ± 1.06	26.16 ± .62	24.01 ± 4.50
Jun	24.73 ± 1.32	24.03 ± .68	22.41 ± .54	11.03 ± .67	26.63 ± .55	26.18 ± 1.11	24.58 ± .65	23.41 ± 4.52
Jul	24.17 ± 1.25	23.44 ± .65	21.81 ± .41	10.89 ± .58	26.11 ± .51	25.80 ± .97	24.04 ± .78	22.93 ± 4.42
Aug	23.75 ± 1.08	23.07 ± .62	21.83 ± .41	10.86 ± .72	25.64 ± .44	25.68 ± .98	24.00 ± .58	22.69 ± 4.31
Sep	23.24 ± .94	22.80 ± .72	21.93 ± .37	10.39 ± .73	24.86 ± .44	25.18 ± .98	24.32 ± .41	22.34 ± 4.29
Oct	22.26 ± .99	22.22 ± .74	21.81 ± .50	10.08 ± .93	24.15 ± .45	24.29 ± .96	24.31 ± .42	21.73 ± 4.19
Nov	20.38 ± 1.30	20.78 ± .91	20.73 ± .72	8.62 ± 1.30	21.30 ± .67	23.07 ± 1.45	23.77 ± .49	20.40 ± 4.35
Dec	18.25 ± 1.46	19.15 ± 1.17	18.96 ± .92	6.85 ± 1.51	24.02 ± 0.38	21.78 ± 1.45	24.02 ± .38	18.83 ± 4.62

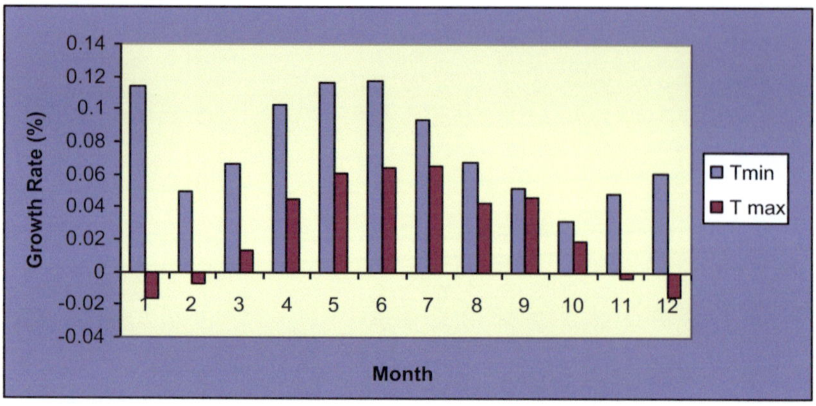

Fig. 1 Growth in T_{max} and T_{min} in north-eastern zone

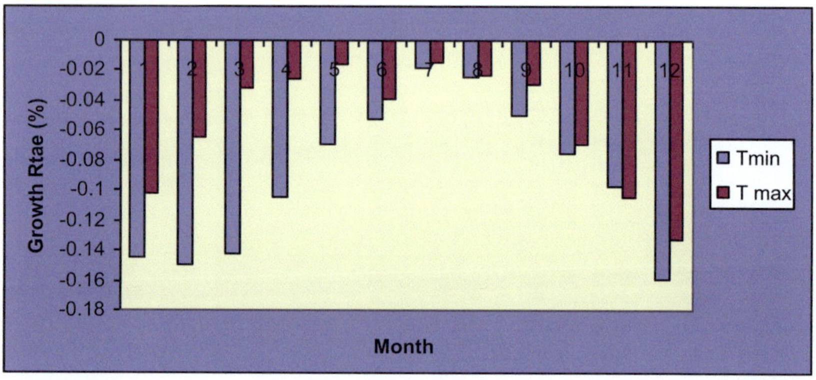

Fig. 2 Growth in T_{max} and T_{min} in north-western zone

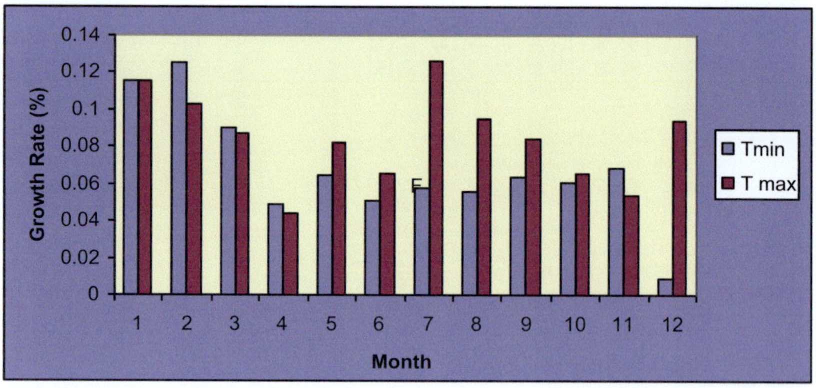

Fig. 3 Growth in T_{max} and T_{min} in western zone

altitude and location in relation with geographical features such as a river, lake, sea and mountains. Significant increase in the global mean air temperature during the past century, perhaps one of the important quoted texts in the aspects of climate change and identification of the temperature trends and their projection, has been the subject matter of interests in large number of studies (Jain and Kumar 2012). In many temperature trend studies particularly in India, main emphasis is directed on the analysis of annual and seasonal temperature data for a single

Fig. 4 Growth in T_{max} and T_{min} in hilly zone

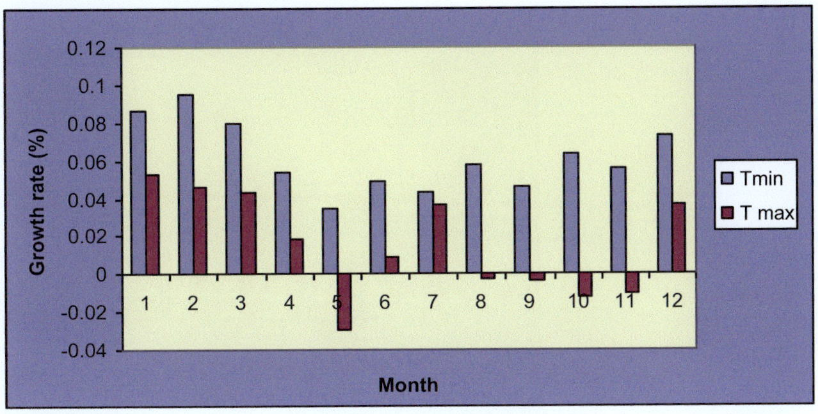

Fig. 5 Growth in T_{max} and T_{min} in southern zone

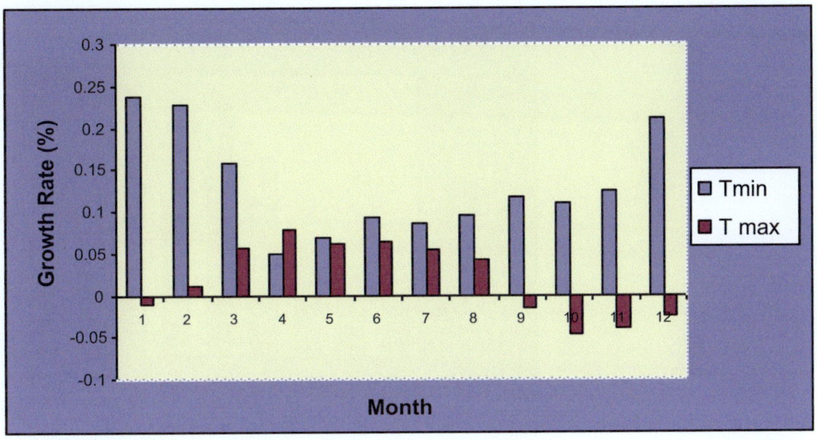

Fig. 6 Growth in T_{max} and T_{min} in Cauvery Delta zone

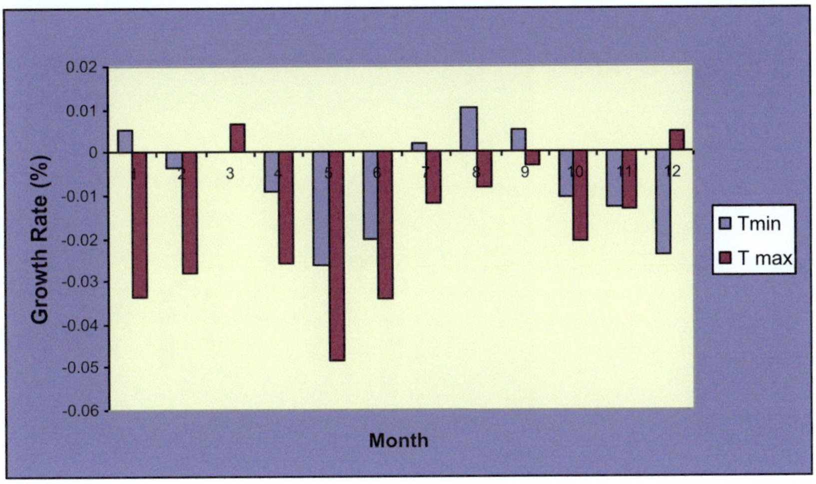

Fig. 7 Growth in T_{max} and T_{min} in hilly zone

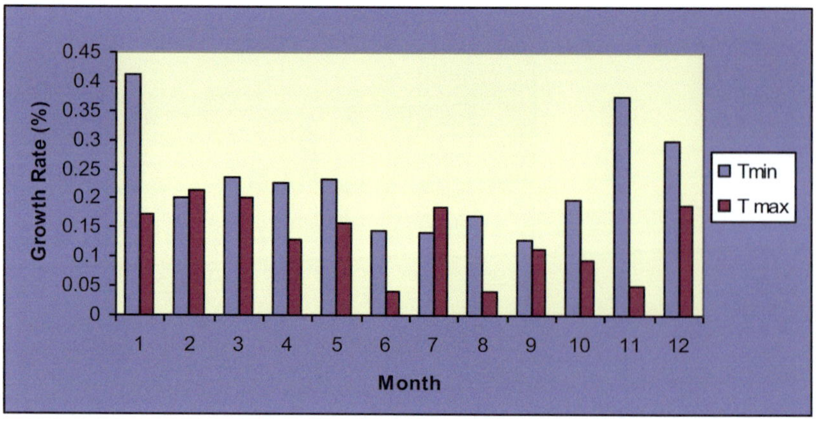

Fig. 8 Growth in THI in north-eastern zone

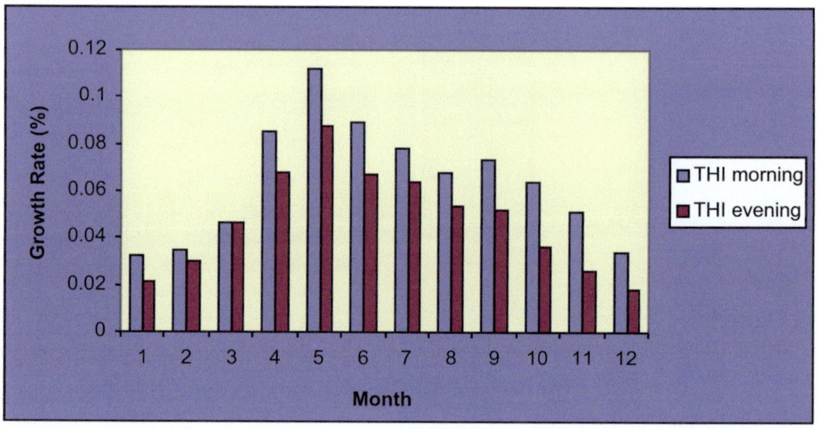

Fig. 9 Growth in THI in north-western zone

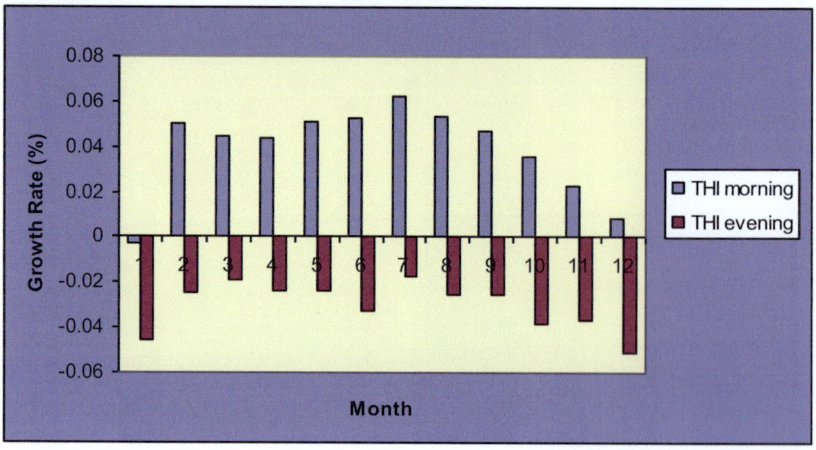

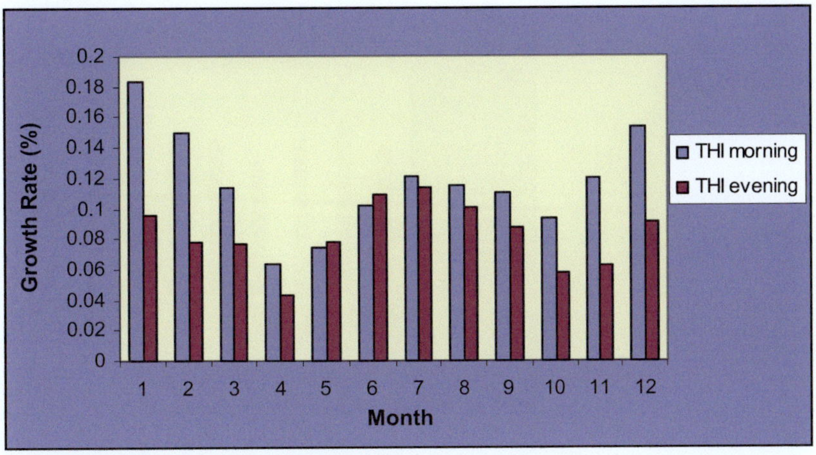

Fig. 10 Growth in THI in western zone

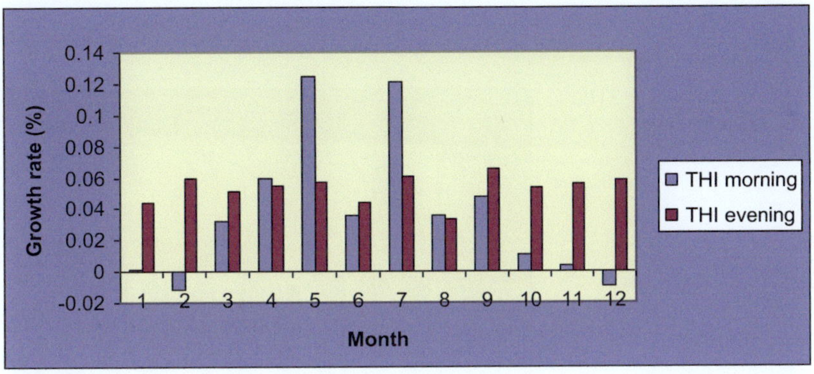

Fig. 11 Growth in THI in hilly zone

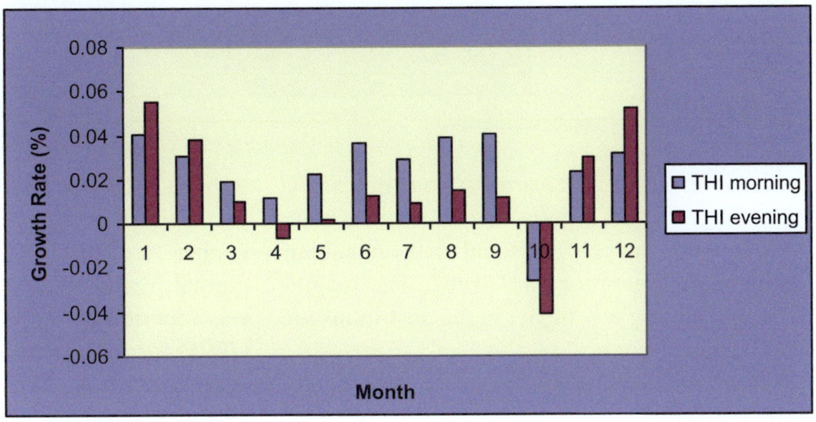

Fig. 12 Growth in THI in southern zone

station or a group of stations, as in the case of the present study, such studies date back to at least 50 years. The first of this kind of studies that was carried out by Pramanik and Jagannathan (1954) did not find any general tendency for an increase or decrease in temperatures in relation with trends in the annual mean, maximum and minimum temperatures over the whole country. The study conducted by Srivastava et al. (1992) delivered the diurnal asymmetry of temperature trends over India which is quite different from that over many other parts of the globe were the first

Fig. 13 Growth in THI in Cauvery Delta zone

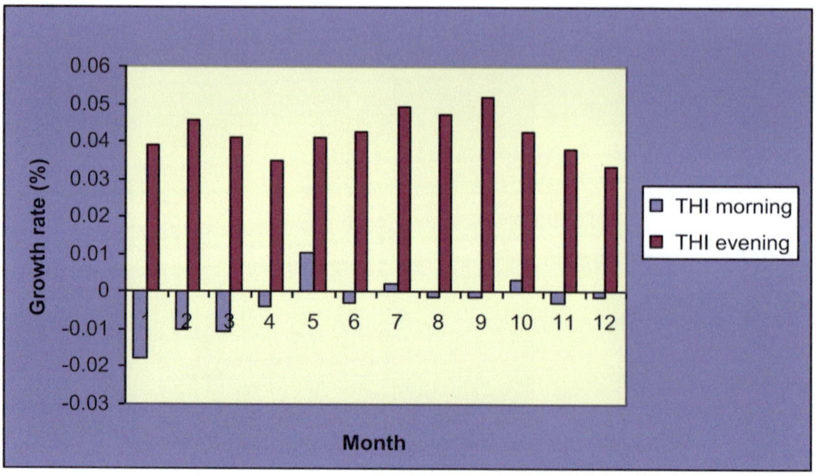

Fig. 14 Growth in THI in high rainfall zone

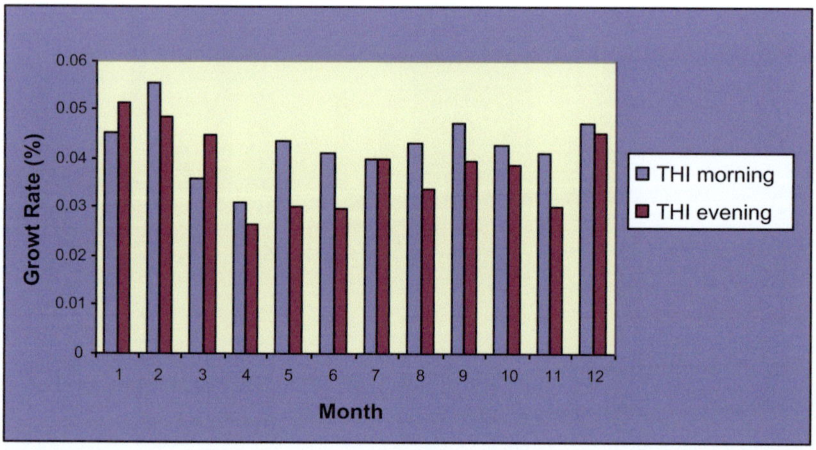

indications. Significant warming trend of 0.57 °C per 100 years was reported by Pant and Kumar (1997) who analysed seasonal and annual air temperature series for 1881–1997. The magnitude of warming was higher in the post-monsoon and winter seasons.

Decreased trend in mean surface air temperature for the year 1901–1982 over the northwest Indian meteorological subdivision regions of Punjab, Haryana, west Rajasthan, east Rajasthan and west Madhya Pradesh was reported by Pant and Hingane (1988). Highly significant warming trend in the mean maximum, mean minimum and average mean temperatures of the Mahanadi river basin was reported by Rao (1993) in seven stations between 1901 and 1980. Increased maximum temperature and trendless minimum temperature, resulting in the rise in mean and diurnal range of temperature trend analyses, was reported by Rupakumar and Krishankumar (1994) for maximum and minimum temperature data derived at 121 stations in India during 1901–1987. Pal and Al-Tabbaa (2010) conducted research in long-term trends and variations of the monthly maximum and minimum temperatures in various climatological regions in India and reported an increased monthly maximum temperature unevenly over the last century.

Better understanding in spatial and temporal distribution with changing patterns in temperature is a basic and important requirement for the planning and management in the broad aspects of climate change. Temperature data in the present study, however, show nearly repetitive rising trends even though the behaviour of the maximum and the minimum temperature at different locations was different. In trend analysis of the present studies, the results significantly depend upon the period of data and stations used. An observation in trend detection pertains that most of the stations were located in rural areas. Thus, the study of trends using this data may be the acceptable depiction of the reality, and this aspect needs to be addressed globally. These concerns, in fact, highlight the importance of identifying a network of baseline stations for change detection studies. More studies to detect teleconnections between Indian rainfall and temperatures are needed. The Ministry of Environment and Forests, GOI (MOEF 2010), carried out an assessment for health in four climate-sensitive regions of India, namely, the Himalayan region, the Western Ghats, the coastal areas and the north-eastern region, and the impact of climate change in the 2030s on key sectors of the Indian economy, namely, agriculture, livestock, water, natural ecosystems and biodiversity, was the main focus.

Hence, there is a need of an incorporated countrywide agenda for detecting change, impact assessment and its adaptation and mitigation in the agriculture and livestock sectors. While interpreting the results of trend analysis from the present study, the observations of Cohn and Lins (2005) are worth repeating: 'that reported trends are real yet insignificant indicates a worrisome possibility: natural climatic excursions may be much larger than we imagine. So large, perhaps, that they render insignificant the changes, human-induced or otherwise, observed during the past century' (Jain and Kumar 2012). To sustain crop productivity and livestock farming during climate change, farmers need different adaptation and mitigation strategies as different options. The mitigation strategies for agriculture and livestock sectors should include present farmer's practices with a long-term strategy to cope with extreme weather conditions. Practical measures for adaptation to climate change can significantly trim down many of the adverse impacts and thus contribute to the upliftment of livelihood security zones of the vulnerable rural population. Regional models can be developed for downscaling the scenarios in climate change of rural areas for the future, and the sensitivity analysis for all possible scenarios of climate change using weather model should be taken up for assessing the impact of future climate and for developing adaptation strategies.

Conclusion

The findings imply that there is definite climatic change in the different agro-climatic regions of Tamil Nadu and also significant ($P > 0.01$) difference in change between the regions. Droughts, hot extremes and heat waves are known to impact agricultural production and the livelihood of the local livestock farmers seriously. As climate change projections form the basis for assessing the impact on crop production and developing adaptation strategies, reliable future changes with reduced level of uncertainty are increasingly important. Careful intervention needs to be adopted for sustainable livestock production and to improve productivity based on the specific needs of the region.

Acknowledgement The authors acknowledge their sincere thanks to the National Data Centre, Indian Meteorological Department, Shivajinagar, Pune - 411 005, Maharashtra for providing the data.

References

Amin MGM, Ali MH, Islam AKMR (2004) Agro-climatic analysis for crop planning in Bangladesh. Bangladesh J Agric Eng 15(1 & 2):1–40

Cohn TA, Lins HF (2005) Nature's style: naturally trendy. Geophys Res Lett 32(23):402. doi:10.1029/2005GL024476

Geethalakshmi V, Lakshmanan A, Rajalakshmi D, Jagannathan R, GummidiSridhar, Ramaraj AP, Bhuvaneswari K, Gurusamy L, Anbhazhagan R (2011) Climate change impact assessment and adaptation strategies to sustain rice production in Cauvery basin of Tamil Nadu. Cur Sci 101(3):342–347.

Retrieved from: http://www.currentscience.ac.in/Volumes/101/03/0342.pdf
IPCC (2007) Summary for policymakers. In: Solomon SD et al (eds) Climate change 2007: the physical science basis. Cambridge University Press, Cambridge. Retrieved from: http://www.ipcc.ch/pdf/assessment-report/ar4/wg1/ar4-wg1-spm.pdf
Jain SK, Kumar V (2012) Trend analysis of rainfall and temperature data for India. Cur Sci 102(1):38–49. Retrieved from: http://www.currentscience.ac.in/Volumes/102/01/0037.pdf
Mehrotra R (1999) Sensitivity of runoff, soil moisture and reservoir design to climate change in central Indian river basins. Clim Chang 42:725–757. doi:10.1023/A:1005496932343
MOEF (2010) Climate change and India: a '4 × 4' assessment. Ministry of Environment and Forests, New Delhi. Retrieved from: http://envfor.nic.in/sites/default/files/fin-rpt-incca_0.pdf
Pal I, Al-Tabbaa A (2010) Long-term changes and variability of monthly extreme temperatures in India. Theor Appl Climatol 100:45–56. doi:10.1007/s00704-009-0167-0
Pant GB, Hingane LS (1988) Climatic changes in and around the Rajasthan desert during the 20th century. Int J Climatol 8:391–401. doi:10.1002/joc.3370080406
Pant GB, Kumar KR (1997) Climates of South Asia. Wiley, Chichester. doi: 10.1002/(SICI)1097-0088(199804)18:5<581::AID-JOC267>3.0.CO;2
Pramanik SK, Jagannathan P (1954) Climatic changes in India rainfall. Indian J Meteorol Geophys 4:291–309
Rao PG (1993) Climatic changes and trends over a major river basin in India. Clim Res 2:215–223. Retrieved from: http://www.int-res.com/articles/cr/2/c002p215.pdf. doi:10.3354/cr002215
Rupakumar K, Krishankumar PGB (1994) Diurnal asymmetry of surface temperature trends over India. Geophys Res Lett 21:677–680. doi:10.1029/94GL00007
Srivastava HN, Dewan BN, Dikshit SK, Rao PGS, Singh SS, Rao KR (1992) Decadal trends in climate over India. Mausam 43:7–20
Tadross MA, Hewitson BC, Usman MT (2005) The interannual variability of the onset of the maize growing season over South Africa and Zimbabwe. J Clim 18:3356–3372. doi:http://dx.doi.org/10.1175/JCLI3423.1

The Effect of Changing Climate and Land Use/Land Cover on Water Resources in Hard Rock Region of Maharashtra State

S.D. Gorantiwar, Y.V.N. Krishnamoorthy, D.S. Pandit, A.K. Joshi, and S. Das

Abstract

The climate, land use and land cover show the changing behaviour over a period of time. These factors influence the water resources. However, it is necessary to know their impact on the quantification of surface water and groundwater resources to enable to plan and manage these resources appropriately. Therefore, it is necessary to build the water resource models for the catchment area. Water resource modelling of an entire catchment is a complex phenomenon but of a great importance for obtaining a better quantitative understanding of water uses and arriving at important water management decisions. Therefore, the use of water resource models such as MIKE SHE in conjunction with remote sensing and GIS techniques is helpful in planning and management of land and water resources and managing water resources.

S.D. Gorantiwar (✉)
Department of Irrigation and Drainage Engineering, Mahatma Phule Krishi Vidyapeeth, Rahuri 413712, Maharashtra, India
e-mail: sdgorantiwar@gmail.com

Y.V.N. Krishnamoorthy
Indian Institute of Remote Sensing, Indian Space Research Organisation, Dept. of Space, Govt. of India, 4-Kalidas Road, Post Box No. 135, Dehradun - 248 001, Uttarakhand, India
e-mail: director@iirs.gov.in

D.S. Pandit
Water Resources, DHI (India) Water & Environment Pvt Ltd, A1, Abhimanshree Housing Society Pashan Road, Pune, India
e-mail: djp@dhigroup.com

A.K. Joshi
Regional Remote Sensing Centre, Central, Indian Space Research Organisation, Dept. of Space, Govt. of India, Amravati Road, Nagpur, India

S. Das
Remote Sensing, Maharashtra Remote Sensing Application Centre (MRSAC), VNIT Campus, South Ambazari Road, Nagpur – 440 011 Maharashtra, India

A.K. Singh et al. (eds.), *Climate Change Modelling, Planning and Policy for Agriculture*,
DOI 10.1007/978-81-322-2157-9_20, © Springer India 2015

In this study, it is proposed to use the surface water-groundwater resource model MIKE SHE for the estimation of groundwater resources, and the methodology was developed for this purpose.

The MIKE SHE model was calibrated for a small catchment using remote sensing data and GIS. Mapping of the study area was carried out with the help of Geocoded (1:50000) IRS-1C/1D LISS-III and PAN images. The input data that was used for preprocessing and set-up preparation of the MIKE SHE model were catchment boundary map, topography, land use/land cover map, soil distribution map, climatological data and crop characteristic (LAI, root depth) and hydraulic properties of soils in the catchment. Roughness coefficient of overland and channel flow, initial water depth, maximum profile water balance error and storing time steps were considered as calibration parameters. The roughness coefficient was varied in the range of 5 to 15, initial water depth in 0 to 0.1 mm, maximum profile water balance error in 0.0001 to 0.02 and storing time steps in 120 to 720 h, respectively. The groundwater fluctuations were simulated for 31 well locations for each combination of calibration parameters. The RMSE between the simulated and actual groundwater fluctuations was calculated, and the set of calibration parameters that yielded the lowest values of RMSE (3.83) was selected for the calibrated MIKE SHE model. These were 10, 0.01 mm, 0.0001 and 360 h for roughness coefficient, initial water depth, maximum profile water balance error and storing time steps, respectively.

Dhor-nani catchment which is a small part of the upper Godavari river basin in Ahmednagar district of Maharashtra state (upstream of Jaikwadi dam) was used to evaluate the effect of changes in land use/land cover and climate on surface water and groundwater resources. Two model set-ups of calibrated MIKE SHE were prepared for the years 1998 and 2005, respectively. The remote sensing images of IRS 1-D (Wifs) for the year 1998 (January) and IRS 1-D (A-Wifs) for the year 2005 (November), climate and other data on various aspects of the catchment – namely, topography, soil distribution and cropping pattern of the respective years, i.e. 1998 and 2005 – were used.

The results of Dhor-nani catchment for the years 1998 and 2005 for influence of change in land use/land cover and climate on water resources indicate that groundwater levels and overland water depths were influenced by change in land use/land cover and climate. The groundwater levels were shallower in the year 1998 compared to 2005, and the depth of overland water increased in 2005 as compared in the year 1998. This may be due to increased values of rainfall (climate) and the changes in land use/land cover between 2 years.

Keywords
Water resource model • MIKE SHE • Remote sensing • GIS

Introduction

Water, being the critical input for plant growth, plays a key role in enhancing agricultural production. Water will continue to exert overriding impact on crop production, and therefore, massive efforts are being made to augment water resources and use them efficiently in many countries. Global climate change will have a major long-term impact on average water availability and the occurrence of floods and droughts, and trends are predictable for each region of the world. Continuous failure of monsoon, increasing demand and over-exploitation lead to depletion of groundwater level, which in turn tends to increase both the investment and the operational costs. New developments of grid-based water resource modelling have been spurred by increasing access to radar, remote sensing (RS) and geographical information system (GIS) techniques. These new developments allow water resource modellers to build and use more comprehensive distributed models. Various software tools have been developed to study the different aspects of water resource planning and allocation, operation and management. In line with the requirements of water management agencies, attention is turning to the development of integrated tools with broad range of capabilities that include water resource planning under different climatic and allocation policies (Golian 2000). In a densely populated country like India, groundwater resource is in high demand.

The MIKE SHE software in conjunction with RS and GIS is one of the important tools in estimating the groundwater resources in a systematic way. A MIKE SHE model was built incorporating a root zone component, a comprehensive three-dimensional groundwater component and a river component. The surface water-groundwater interactions within the catchments could then be deciphered. MIKE SHE is distributed, physically based hydrologic modelling system for the simulation of major processes occurring in the land phase of hydrological cycle, including interception, evaporation, overland, channel flow, snowmelt, unsaturated and saturated zone flow and surface water-groundwater interactions. Several researchers in the past used MIKE SHE model for water resource management (Refsgaard et al. 1992; Demetriou and Punthakey 1999; Singh and Subramanian 1999). In this study, the methodology was developed to use the MIKE SHE model for the estimation of water resources and use the model for evaluating the effect of changing climate and land use/land cover on water resources.

Methodology

The MIKE SHE model (Abbott et al. 1986) was used for analysing the effect of changing climate and land use/land cover on water resources in hard rock region of Maharashtra state. The model was calibrated for the small catchment for which the requisite data were available and then was used for estimating the surface water and groundwater resources in response to different climatological conditions for the catchment under study. This section describes the MIKE SHE model.

MIKE SHE is a comprehensive deterministic, distributed and physically based modelling system for the simulation of all major hydrological processes occurring in the land phase of the hydrological cycle (Abbott et al. 1986). The basic MIKE SHE module is the MIKE SHE WM that describes the water movement in the area under study. The overall model structure is illustrated in Fig. 1. MIKE SHE WM has been designed with a modular programme structure comprising six process-oriented components each describing the major physical process in individual parts of the hydrological cycle and in combination describing the entire hydrological cycle. The components are interception/evapotranspiration (ET), overland zone (OC), unsaturated zone (UZ), saturated zone (SZ), snowmelt (SM) and exchange between aquifer and rivers (EX).

Interception

The interception process is modelled by introducing an interception storage expressed as a function of leaf area index (Jensen 1983):

$$I_{\max} = C_{\text{int}} * \text{LAI} \tag{1}$$

where

C_{int} = interception parameter, mm

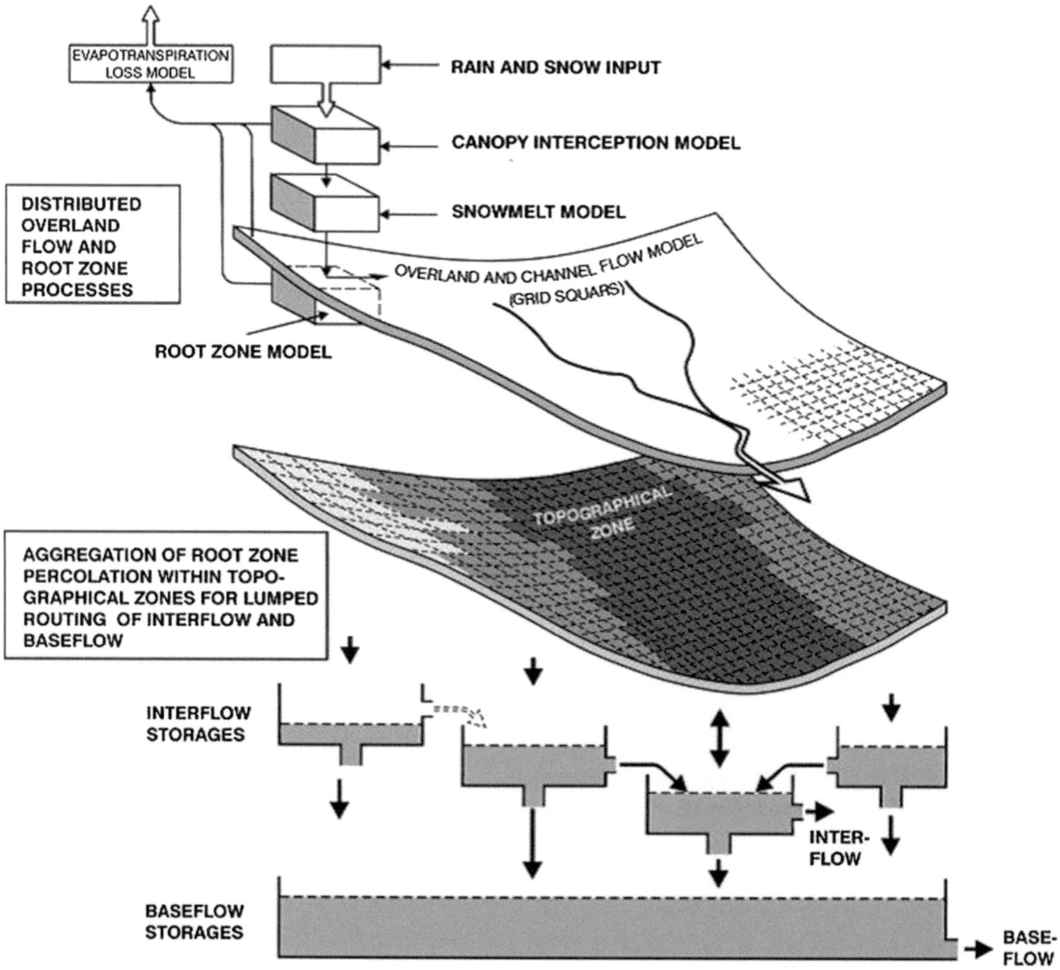

Fig. 1 Schematic representation of the components of the MIKE-SHE (Butts et al. 2005)

LAI = leaf area index
I_{max} = intercepted water storage capacity, mm

Evapotranspiration

The actual evapotranspiration is calculated based on the potential evapotranspiration using the Kristensen and Jensen model (Kristensen and Jensen 1975). It is given by Eq. (2):

$$E_{at} = f_1(\text{LAI})\ f_2(\theta)\ \text{RDF}\ E_p \quad (2)$$

where

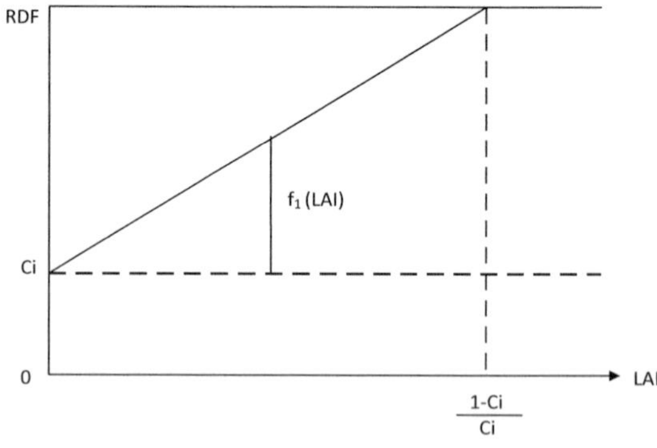

Fig. 2 Leaf area function, f_1, (LAI) as a function of leaf area index (DHI 2005)

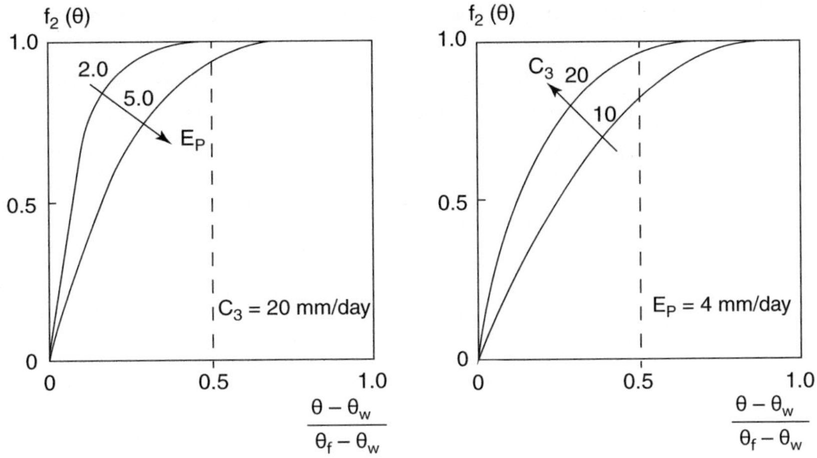

Fig. 3 The soil moisture function $f_2(\theta)$ for constant C_3 (20 mm/day) and varying E_p (4 mm/day), (DHI 2005)

E_{at} = actual transpiration
RDF = root distribution function
f_1 (LAI) = function shown in Fig. 2
f_2 (θ) = function shown in Fig. 3 and given as in Eq. (3)
E_p = potential evaporation

$$f_2(\theta) = 1 - \left(\frac{\theta_F - \theta}{\theta_F - \theta_W}\right)^{\frac{C_3}{E_P}} \quad (3)$$

where
θ_F = volumetric moisture content at field capacity
θ_W = volumetric moisture content at wilting point
θ = volumetric moisture content
C_3 = empirical parameter, mm/day

Equation (2) is applied to all nodes in the root zone. Figure 4 illustrates the variation in soil evaporation, i.e. E_S/E_P, with the soil moisture content in the upper layer, and it is expressed in equation form as follows:

$$E_s = E_p f_3(\theta) + \left(E_p - E_{at} - E_p f_3(\theta)\right) f_4(\theta) \left(1 - f_1(\text{LAI})\right) \quad (4)$$

where E_s = soil evaporation

Functions f_3 (θ), f_4 (θ) are expressed as follows:

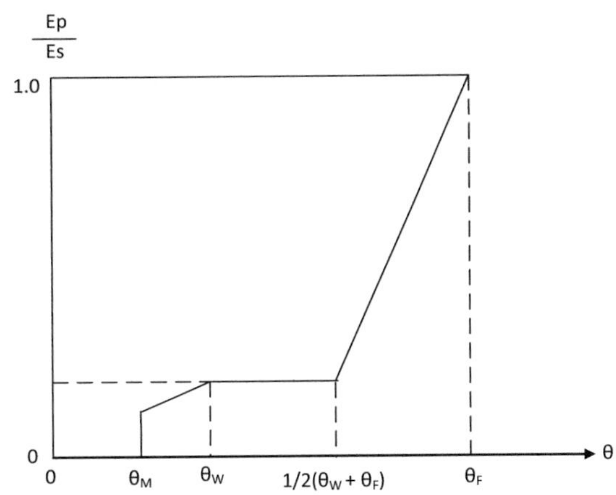

Fig. 4 Soil evaporation E_s in relation to E_p as a function of θ in the top layer when f(LAI) = 0, (DHI 2005)

$$f_3(\theta) = \begin{cases} C_2 & \text{for} \quad \theta \geq \theta_W \\ C_2 \dfrac{\theta}{\theta_w} & \text{for} \quad \theta_r \leq \theta \leq \theta_W \\ 0 & \text{for} \quad \theta \leq \theta_r \end{cases} \quad (5)$$

$$f_4(\theta) = \begin{cases} \dfrac{\theta - 1/2(\theta_W + \theta_F)}{\theta_F - 1/2(\theta_W + \theta_F)} & \text{for}\, \theta \geq 1/2(\theta_W + \theta_F) \\ 0 & \text{else} \end{cases} \quad (6)$$

where
θ_r = residual moisture content
C_2 = empirical parameter

Overland and Channel Flow Component

The overland flow and the channel flow are modelled by approximations of the Saint-Venant equations of continuity and momentum (DHI 2005):

$$q = \frac{\partial h}{\partial t} + \frac{\partial (uh)}{\partial x_i} + \frac{\partial (vh)}{\partial x_j} \quad (7)$$

$$\frac{\partial h}{\partial x_i} = S_o x_i - S_f x_i \quad x_i \text{direction} \quad (8)$$

$$\frac{\partial h}{\partial x_j} = S_o x_j - S_f x_j \quad x_j \text{direction} \quad (9)$$

The Strickler/Manning-type law for each friction slope is used, giving the following relations between velocities and flow depths:

$$uh = k_{x_i} I_{x_i}^{1/2} h^{5/3} \quad (10)$$

$$vh = k_{x_j} I_{x_j}^{1/2} h^{5/3} \quad (11)$$

where
h = local water depth exceeding any detention storage, m
x_i, x_j = space coordinates, m
t = time, s
u, $v(x_i, x_j, t)$ = flow velocities, m/s
$q(x_i, x_j, t)$ = source/sink per unit horizontal area, $m^3/s/m^2$
S_{oi}, S_{oj} (x_i, x_j) = bed slope in x and y directions
K_i, K_j (x_i , x_j) = Strickler roughness coefficients in x and y directions, $m^{1/3}/s$
I_i, I_j (x_i, x_j) = gradients of water surface levels in x and y directions
The channel flow is calculated by equivalent set of equations, but in one dimension only:

$$Q = \frac{\partial A}{\partial t} \frac{\partial (Au)}{\partial x} \quad (12)$$

$$\frac{\partial h}{\partial x} = S_{ox} - S_{fx} \quad (13)$$

where
A = cross-sectional area of the channel, m^2
Q = source/sink (lateral inflow and outflow from overland flow, drainage flows and exchange with the aquifer), m/s

Unsaturated Zone Component

The unsaturated zone (UZ) component plays an important role in the MIKE SHE WM model as all other components depend on boundary data from UZ components. It links the two- and three-dimensional surface and subsurface flow processes together. In most comprehensive mode, MIKE SHE solves Richards' (DHI 2005) equation for one-dimensional vertical flow, which includes the effects of gravity, soil suction and soil evaporation and is given by Eq. (14):

$$C\frac{\partial \psi}{\partial t} = \frac{\partial}{\partial z}\left(K\frac{\partial \psi}{\partial z}\right) + \left(\frac{\partial K}{\partial z}\right) - S \quad (14)$$

where
$\Psi(z, t)$ = pressure head, m
t = time, s
z = vertical space coordinate, m
C = soil water capacity, m^{-1}
$K(\theta, z)$ = hydraulic conductivity, m/s
θ = soil moisture content
$S(z, t)$ = source/sink term (e.g. root extraction), s^{-1}

A coupling procedure between the unsaturated and the saturated solutions is included to compute the correct soil moisture and the water table dynamics in the lower part of the soil profile. The procedure consists of bookkeeping of an accumulated mass balance error, E_{cum}, for the entire grid square column. If $|E_{cum}|$ exceeds certain specified limits E_{max}, action is taken for adjustment of the water table and a redistribution of moisture content in the lower part of the UZ until

$$|E_{cum}| < E_{max}.$$

Saturated Zone Component

The saturated zone (SZ) component of MIKE SHE WM calculates the saturated subsurface flow in the catchment. MIKE SHE allows for a fully three-dimensional flow in a heterogeneous aquifer with shifting conditions between unconfined and confined conditions. The partial differential equation of groundwater flow in three dimensions is given by Eq. (15):

$$\frac{\partial}{\partial x_i}\left(K_{ij}\frac{\partial h}{\partial xj}\right) = S_S\frac{\partial h}{\partial t} + R \quad (15)$$

where
$h(x_i)$ = hydraulic head, m
x_i = space coordinate, m
$K(X_{ij})$ = hydraulic conductivity, m/s
$S_S(X_{ij})$ = specific storage
$R(X_{ij})$ = volumetric flow rate via sources/sinks per unit volume, $m^3/s/m$

The equation is solved numerically by a finite difference method using a modified Gauss-Seidel iterative scheme, providing a value for hydraulic head in time at each computational node.

For solving Richards' equation, two important hydraulic functions need to be determined experimentally for the soil types which characterise the individual soil profiles within the area. They are soil moisture retention curve $Pf(\theta)$ and hydraulic conductivity function $K(\theta)$. The Ψ-θ relationships were used from the available literature for the four soil types of the catchments. The model derives the hydraulic conductivity from the respective saturated hydraulic conductivities. The parameters that need to be specified are soil moisture at saturation (θ), soil moisture at effective saturation (θ_{eff}), capillary pressure at field capacity (pF_{fc}), capillary pressure at wilting point (pF_W), residual soil moisture content (θ_r), threshold value of capillary pressure (pF_{-thr}), exponent in hydraulic conductivity function (n) and saturated hydraulic conductivity (K_s).

Study Area

Pimpalgaon Ujjaini catchment was used for the calibration, and the effect of changing climate and land use/land cover on water resources in hard rock region of Maharashtra state was analysed for Dhor-nani catchment.

Pimpalgaon Ujjaini Catchment

Pimpalgaon Ujjaini catchment which is located at Pimpalgaon Ujjaini village in Ahmednagar district of Maharashtra lies between latitude 18.15° to 18.20°N and longitude 74.05° to 74.10°E, with an area of around 30.73 km^2. The average annual rainfall within the catchment is 642 mm, and the mean maximum and minimum temperatures are 35 °C and 10 °C recorded in summer and winter, respectively. The average ground surface elevation of Pimpalgaon Ujjaini catchment ranged from 680 to 940 m above mean sea level. The maximum area of the catchment has a slope in the range of 1–3 %. The soils of the study area have been divided into five land capability classes (Patil 2008). The different vegetation types simulate different water uptake patterns within the catchment, which influenced the vertical distribution of moisture content in the unsaturated zone on the different land parcels. The land use pattern and vegetation characteristics are necessary to simulate the actual evapotranspiration over the whole area. Crop or plant growth is not explicitly modelled in MIKE SHE model, since this model is essentially a catchment model, and specific details about crop development are excluded. However, time series of root growth and leaf area index are needed for each vegetation type occurring in the study area for the calculation of the actual evapotranspiration.

Dhor-nani Catchment

This is a small catchment selected from the upper Godavari river basin in Ahmednagar district of Maharashtra state (upstream of Jaikwadi dam). The catchment lies between 75053′00″ to 75054′55″ E longitudes and 19032′20″ to 19033′54″ N latitudes, with an altitude ranging from 458 to 490 m above MSL covering an area of about 920 km^2. The recorded values of average maximum and minimum temperatures in °C, evaporation in mm/day, maximum and minimum relative humidity in percent, mean annual rainfall in mm, wind speed in km/h and bright sunshine hours of Dhor-nani catchment are 32.0, 15.9, 6.0, 88.3, 43.5, 550, 5.9 and 7.8, respectively. The elevation of Dhor-nani catchment varies from 450 to 800 m above mean sea level. The soils of the study area have been divided into five land capability classes. These are described as follows:

IIe – These are nearly levelled, very gently sloping pediment surfaces with moderate erosion hazards. Most of the area in this unit is under agro-horticulture or double cropped. The soils of this unit are deep to very deep and clay textured and have good water holding capacity.

IIIe – These are very gently sloping pediment surfaces with severe erosion. The soils of this unit are moderately deep, clay textured with lesser water holding capacity as compared to type IIe.

IVes – The unit comprises of shallow, gravelly clay loam to gravelly sandy loam soils developed on gentle slopes of mesa and pediments with severe erosion hazards. This soil shows low water holding capacity and high infiltration rate.

VIes – This unit comprises of shallow, gravelly sandy loam soils developed on moderate strong slopes of escarpments with very severe erosion hazards and low water holding capacity.

VIIes – These are moderate strong, steep to very steep slopes of escarpments with very severe erosion hazards. The soils of this unit are shallow, gravelly sandy loam soil with less water holding capacity.

Crops like sorghum, sugar cane and pulses are dominating in Dhor-nani catchment. Fields are used for single *kharif* cropping or single *rabi* cropping or double cropping. Most of the area comes under scrub and fallow land in this catchment.

Input Data Handling and Preprocessing and Processing for MIKE SHE

The technical guidelines provided by the user manual of the MIKE SHE Integrated Hydrological Modeling System, DHI Water and Environment (2005) have been adopted for set-up preparation and simulating all the processes in land phase hydrological cycle of the MIKE SHE

Table 1 The details of satellite data used for Pimpalgaon Ujjaini catchment

S. no.	Satellite	Sensor	Date of pass
1.	IRS-1D	LISS-III	7 January 1998
2.	IRS-1D	PAN	7 January 1998

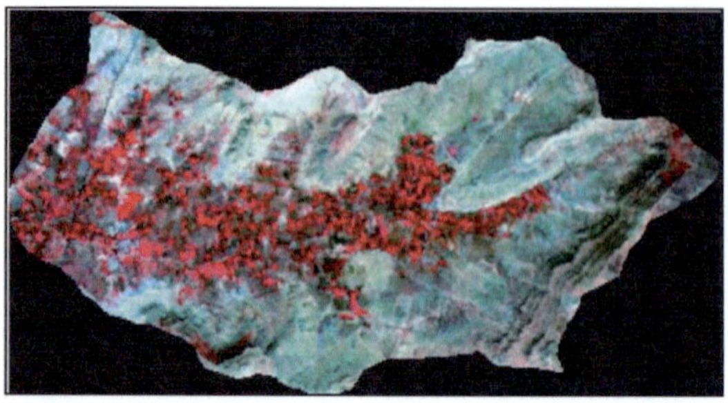

Fig. 5 FCC of January 1998 LISS-III camera of Pimpalgaon Ujjaini catchment

water resource model. Land use/land cover classification and mapping of the study area were carried out with the help of Geocoded (1:50,000) IRS-1C/1D LISS-III and PAN images. In order to establish an integrated picture of the saturated zone (presence of the geological layer and lenses) in the model for estimation of groundwater resources, the geology/lithology map has been prepared. All thematic maps were digitised (in continuous mode in the vector format and the digitised values were then edited).

Prior to simulation by MIKE SHE, it is required to prepare a data set that is required for running MIKE SHE. Preparing a data set ready for simulation run with the MIKE SHE requires a great deal of preprocessing. MIKE SHE operates with three types of data files: type dfs0, type dfs2 and digitised data. They are all formatted in ASCII files, which can be edited by a normal text editor. Type dfs0 data files (time series) consist of series of individual values or scalars. Each time step in the series can be different. Type dfs2 data (matrix data) consist of a two-dimensional array, e.g. surface topography in each grid square, soil-type grid code, land use grid code and meteorological station grid code.

The data acquired through modern techniques such as remote sensing is main source of input for preparation of the data files required for

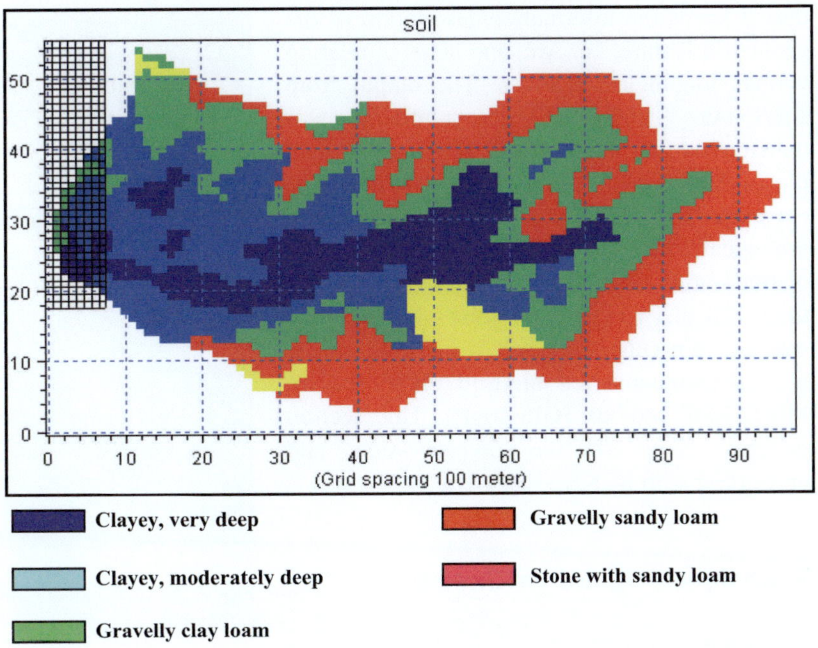

Fig. 6 Soil map of Pimpalgaon Ujjaini catchment

Table 2 Area under different classes of land use/land cover for Pimpalgaon Ujjaini catchment

S. no.	Land use/land covers	Area (ha.)	Percentage (%)
1.	*Kharif*	59.77	1.92
2.	*Rabi*	360.84	11.60
3.	Double crop	345.86	11.15
4.	Fallow	263.55	8.47
5.	Scrub	1,849.74	59.48
6.	Barren	102.29	3.28
7.	Plantations	91.75	2.95
8.	Water bodies	15.25	0.50
9.	Built-up land	20.30	0.65

Table 3 Details of satellite data of Dhor-nani catchment

S. No.	Satellite	Sensor	Date of pass
1.	IRS-1D	Wifs	January 1998
2.	IRS-1D	A-Wif	November 2005

MIKE SHE processing. The digitisation programmes are one of the tools that have not been included in the menu system of MIKE SHE, since it is usually done with the help of other GIS software. In the process of digitisation, the maps of catchment boundary, topography, soil-type distributions, land use, location of the open wells, groundwater depth and Thiessen polygon related to the precipitation network of area were first scanned and then digitised using Arc/INFO, a GIS software. Here Arc stands for spatial data, i.e. it helps in creating different geometric features such as line, point and polygon, and integrating them. INFO stands for information;

thus it keeps the information/description of the geometric feature, i.e. Arc, and helps in relating various topological features. The output from Arc/INFO digitising programme was transformed into the format of the digitised data file of MIKE SHE using the text editor. The coverages of soil, land use/land cover map, meteorological station maps, topography (DEM), etc. of study area were further processed to prepare the ASCII file. The polygrid/point grid command was used for generating ASCII file. The ASCII file was then converted into grid .dfs2 format in MIKE SHE toolbox (GIS tools). This section presents some of the preprocessing and processing with MIKE SHE.

Fig. 7 FCC of January 1998 Wifs camera of Dhor-nani catchment

Pimpalgaon Ujjaini Catchment

The details of satellite data used for Pimpalgaon Ujjaini catchment are presented in Table 1. The FCC of January 1998 LISS-III camera of Pimpalgaon Ujjaini catchment is shown Fig. 5, and the soil map of Pimpalgaon Ujjaini catchment is shown in Fig. 6. Area under different classes of land use/land cover for Pimpalgaon Ujjaini catchment obtained from the analysis of remote sensing images is presented in Table 2.

Fig. 8 FCC of January 2005 A-Wifs camera of Dhor-nani catchment

Dhor-nani Catchment

The details of satellite data used for Dhor-nani catchment is presented in Table 3. The satellite data was used for generation of thematic and derived maps. Figures 7 and 8 present the false colour composite (FCC) of January 1998 Wifs camera and false colour composite (FCC) of November 2005 A-Wifs camera of Dhor-nani catchment, respectively.

Figure 9 shows the land use/land cover map of Dhor-nani catchment for the years 1998 and 2005.

Figures 10, 11 and 12 show the boundary map, topography map and soil map of Dhor-nani catchment, respectively.

There are three rain gage stations in the catchment/adjoining catchment. The spatial distribution of rainfall as obtained from the model is shown in Fig. 13.

The spatially varying land use cover (vegetation) of the Dhor-nani catchment was prepared and is divided into seven classes. The spatial distribution of land use for the years 1998 and 2005 is shown in Figs. 14 and 15, respectively.

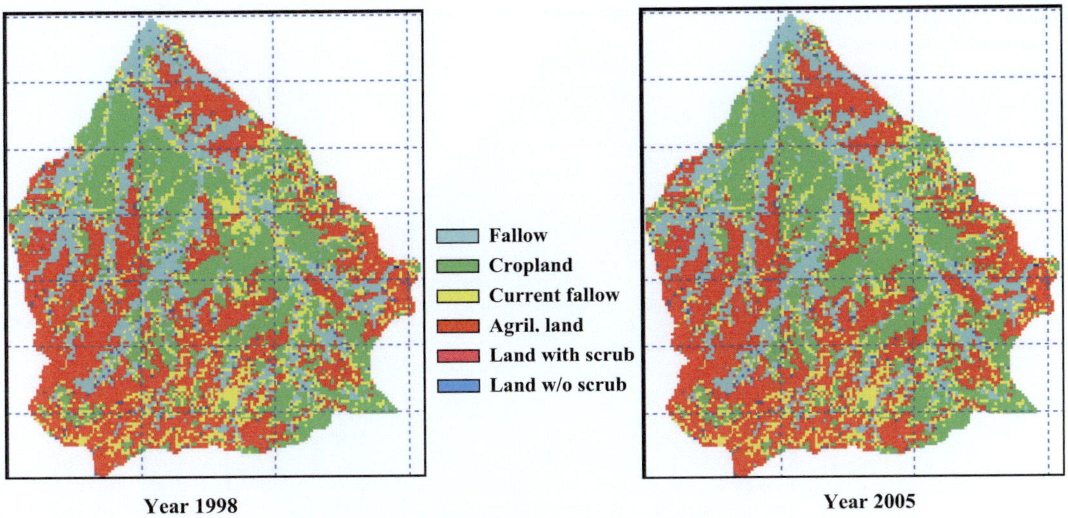

Fig. 9 Land use/land cover maps of Dhor-nani catchment

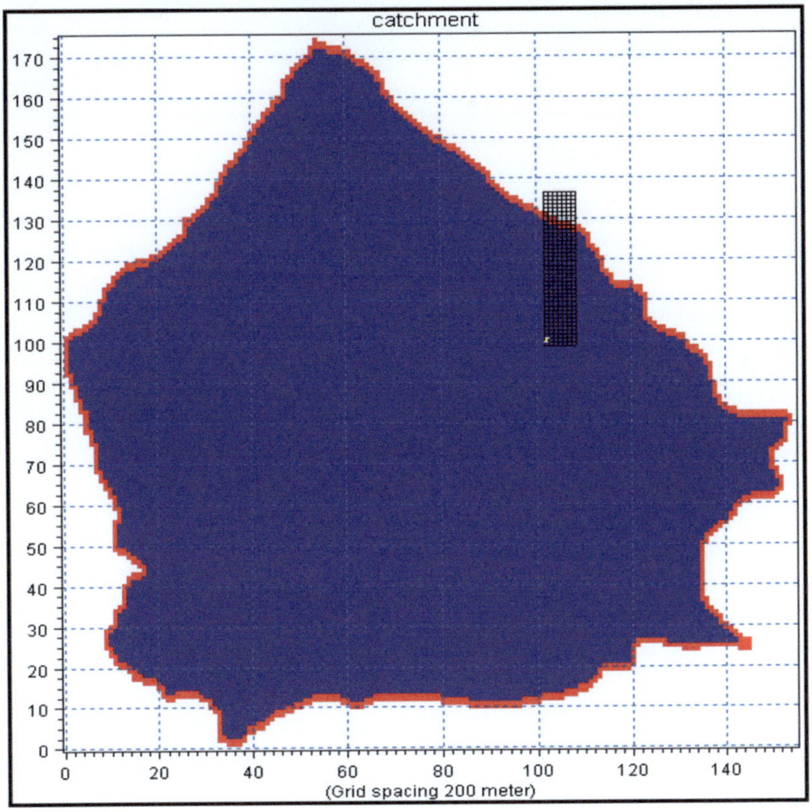

Fig. 10 Boundary map of Dhor-nani catchment (*.dfs2 file*)

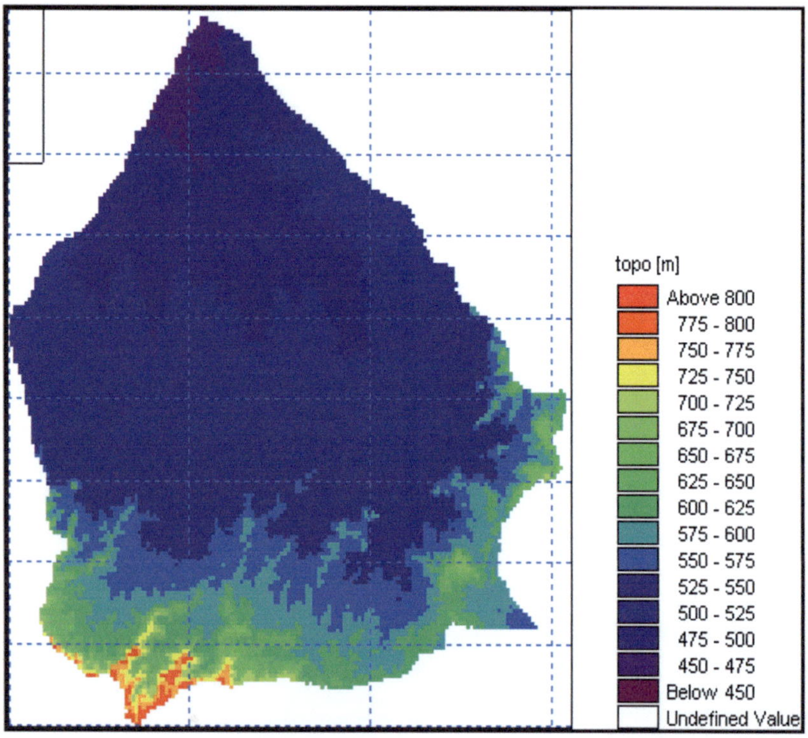

Fig. 11 Topography map of Dhor-nani catchment (*.dfs2 file*)

Model Calibration

The MIKE SHE model was calibrated for Pimpalgaon Ujjaini catchment. The field data collected for the years 2002 and 2003 were used for this purpose. Calibration was performed from 15 May 2002 to 15 May 2003. The model was calibrated for Pimpalgaon Ujjaini catchment by varying following parameters:

1. Roughness coefficient for the overland and channel flows
2. Initial water depth and detention storage
3. Storing time steps and maximum profile water balance error

A quantitative comparison of the temporal and spatial variations between the simulated and observed data of groundwater levels was used in the calibration process. During the calibration runs, the selected parameters were varied manually until model performed satisfactorily in terms of selected performance criteria. The model performance was evaluated by comparing the observed and simulated groundwater levels. In this study, calibration for unsaturated zone was performed considering the information of the area available from the soil survey report. As the information of saturated zone is limited, parameters were varied to a wide range.

Performance Criteria

To evaluate the performance of the model, the simulated data were compared with the observed ones. The performance can be visually interpreted by plotting the simulated and observed data simultaneously on a single plot. However, statistical test can give the quantitative

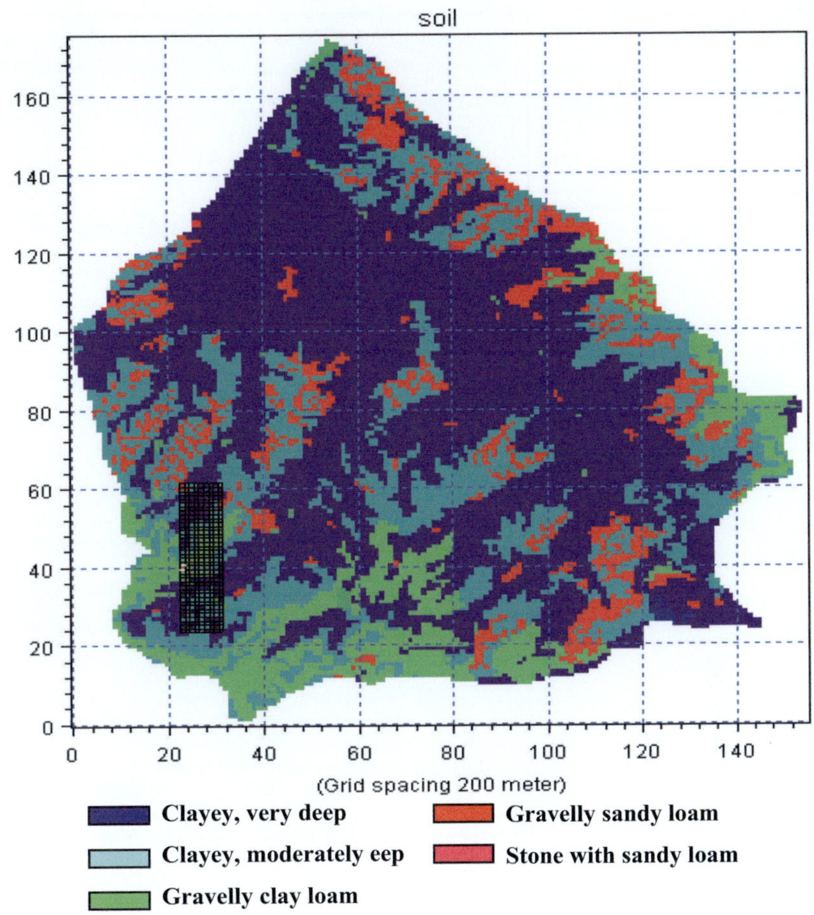

Fig. 12 Soil map of Dhornani catchment

performance of the model. Here the root mean square error (RMSE) criteria were used.

Root Mean Square Error (RMSE)

The RMSE value indicates the extent to which the simulations are overestimating or underestimating observed values, expressed as a percentage of the average value of the observations. RMSE is given by Eq. (16):

$$\mathrm{RMSE} = \sum_{i=1}^{n} \left[\frac{(\mathrm{S}_i - \mathrm{O}_i)^2}{n} \right]^{1/2} \quad (100/\overline{\mathrm{O}}) \qquad (16)$$

where

O_i = the observed groundwater level for ith observation, m

S_i = the simulated water ground level for ith observation, m

$\bar{\mathrm{O}}$ = the average of the observed groundwater level value

n = the total number of observations

Results and Discussion

Model Calibration

In the process of calibration, the roughness coefficient for overland flow was varied in between the range of 5 and 15. Initial water depth values were adjusted from 0 to 0.1 mm. The storing time steps varied from 120 to 720 h, and the maximum profile water balance error was varied in between 0.02 and 0.0001. The calibration of the model

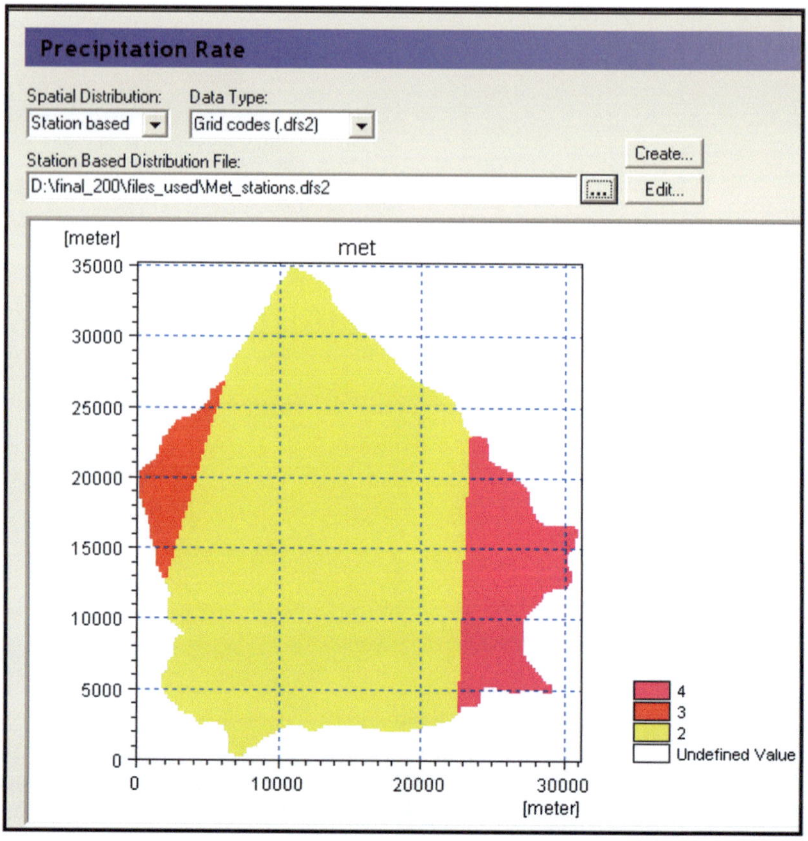

Fig. 13 Screen showing loaded .dfs2 file of precipitation for Dhor-nani catchment

was performed on the pre- and post-monsoon water table fluctuations at 31 different locations in the catchment area. For this purpose, the model estimated the groundwater fluctuation during the calibration period, i.e. from May 2002 to April 2003, at these 31 locations (wells). These model-estimated groundwater fluctuations were compared with the actual groundwater fluctuations during this period by the parameter root mean square error (RMSE). The RMSE values were estimated for each set of calibration parameters for the period of calibration for all the wells. Then average value of RMSE for all wells was computed. The lowest value of average RMSE (3.83) was obtained for the following set of calibration parameters:

1. Roughness coefficient for the overland and channel flows (10)
2. Initial water depth (0.01 mm)
3. Storing time steps (360 h)
4. Maximum profile water balance error (0.0001)

The visual observation shows that the observed and simulated values match well for 24 stations. However, for seven there is significant difference between the observed and simulated values; the RMSE is greater than 5.0 for these wells. This may be due to the presence of local lenses which are not considered in the study. RMSE values show no significant difference between the actual and simulated groundwater levels. Thus, the MIKE SHE is considered as calibrated. However, a detailed calibration may further improve the simulation accuracy. The example result of the MIKE SHE model for observed and estimated groundwater level at well ID 1W1 is presented in Fig. 16. The continuous line shows the model-estimated

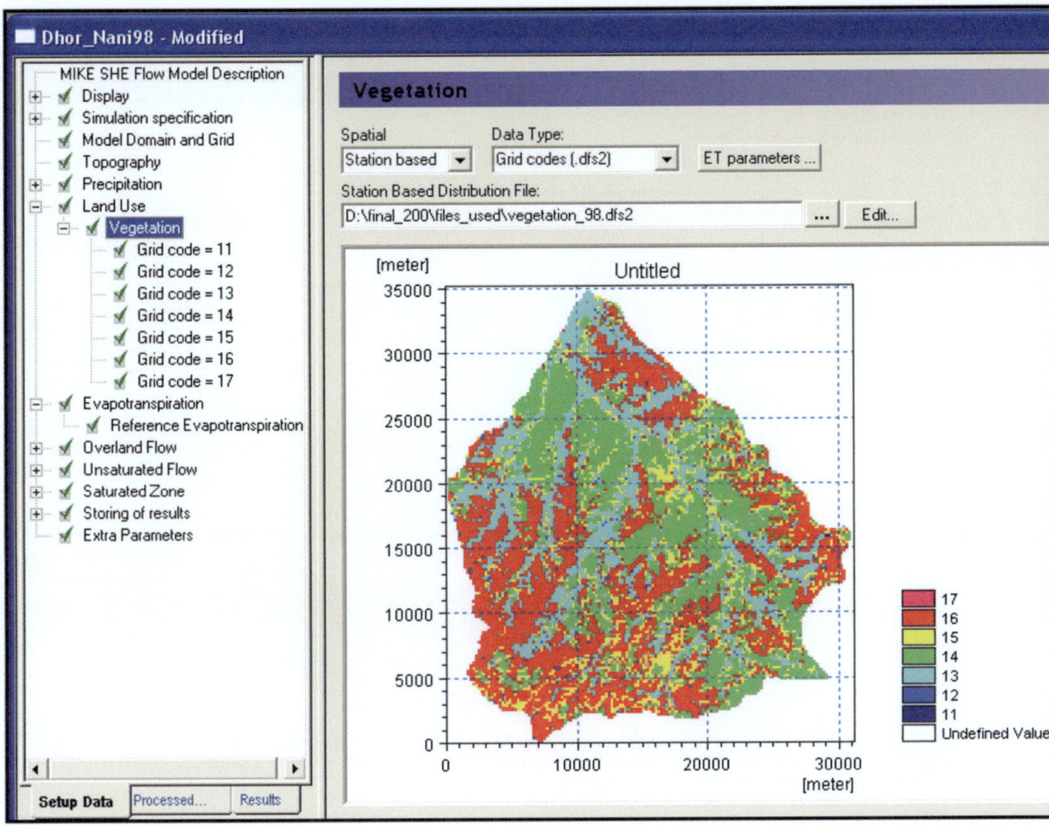

Fig. 14 Screen showing loaded .dfs2 file of land use grid codes for Dhor-nani catchment (1998) in MIKE SHE model

fluctuations, whereas the points show the observed values.

Results of MIKE SHE Model for Change in Land Use/Land Cover and Climate on Water Resources of Dhor-nani Catchment

The simulation was run for the Dhor-nani catchment using the input data of the years 1998 and 2005. Groundwater table and depth of overland water at different locations in the catchment were simulated with the help of calibrated MIKE SHE model. The estimated depths to groundwater table for the years 1998 and 2005 are shown in Figs. 17 and 18 for one representative location. The groundwater levels for both years are compared in Fig. 19. Figure 19 shows that groundwater table is shallower for the year 1998 compared to 2005 except for few days. Overall for year 1998, groundwater is shallower for 75 % of the days in year. The trend was observed for all the locations for which the depths to groundwater tables were simulated. This may be due to the fact that there is more rainfall in the year 1998 compared to 2005. The values of rainfall for both years are shown in Figs. 20 and 21.

The overland flows were also estimated with the help of calibrated MIKE SHE model for the years 1998 and 2005, which are shown in Figs. 22 and 23, respectively.

It is seen from the figures that the depth of overland water increased in 2005 as compared in year 1998. This may be due to variation in rainfall. The changes in groundwater level and overland water between 1998 and 2005 indicate that there is effect of change in land use/land cover and climate on water resources.

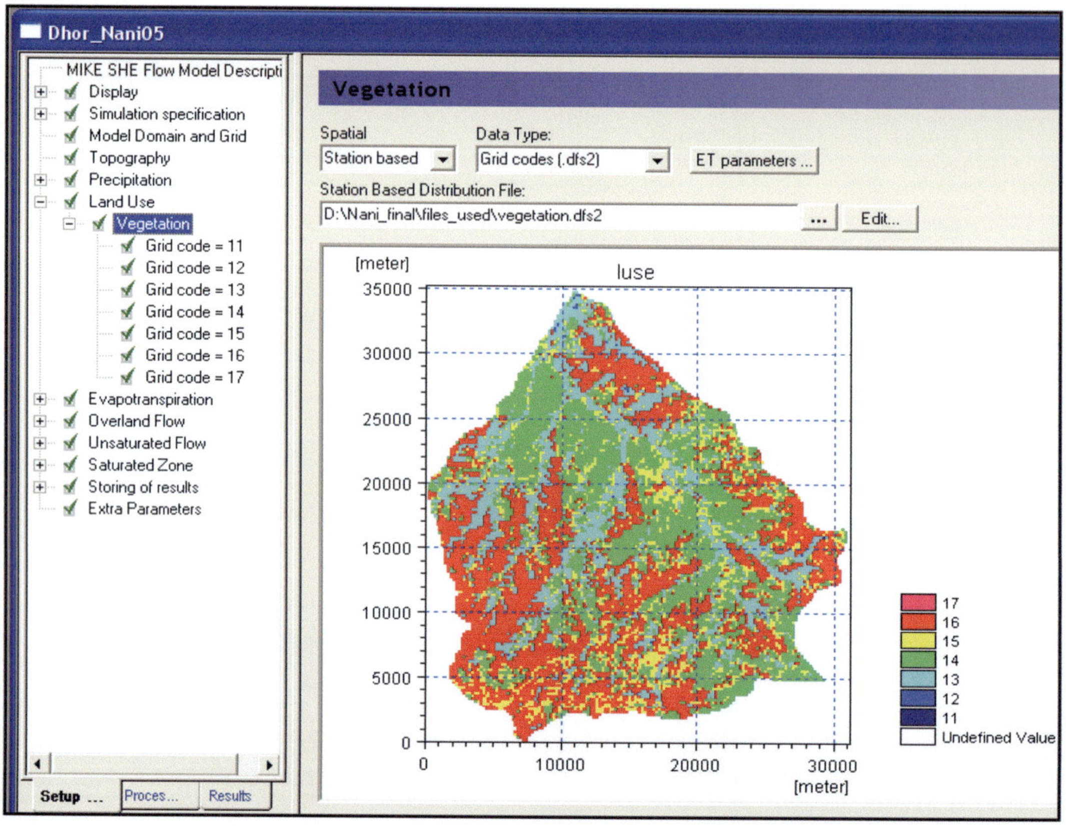

Fig. 15 Screen showing loaded .dfs2 file of land use grid codes for Dhor-nani catchment (2005) in MIKE SHE model

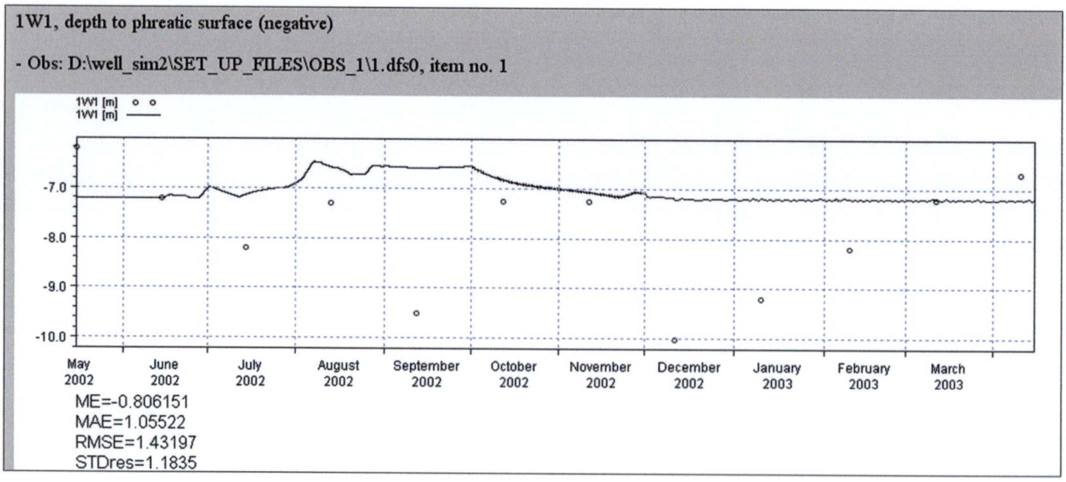

Fig. 16 The example result of the MIKE SHE model for observed and estimated groundwater level at well – 1W1

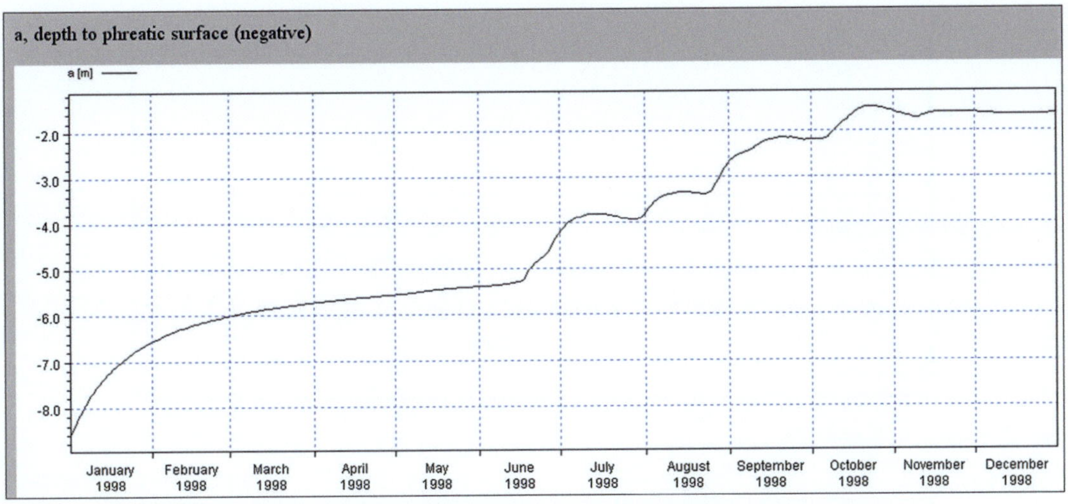

Fig. 17 MIKE SHE model-estimated groundwater table for Dhor-nani catchment for the year 1998

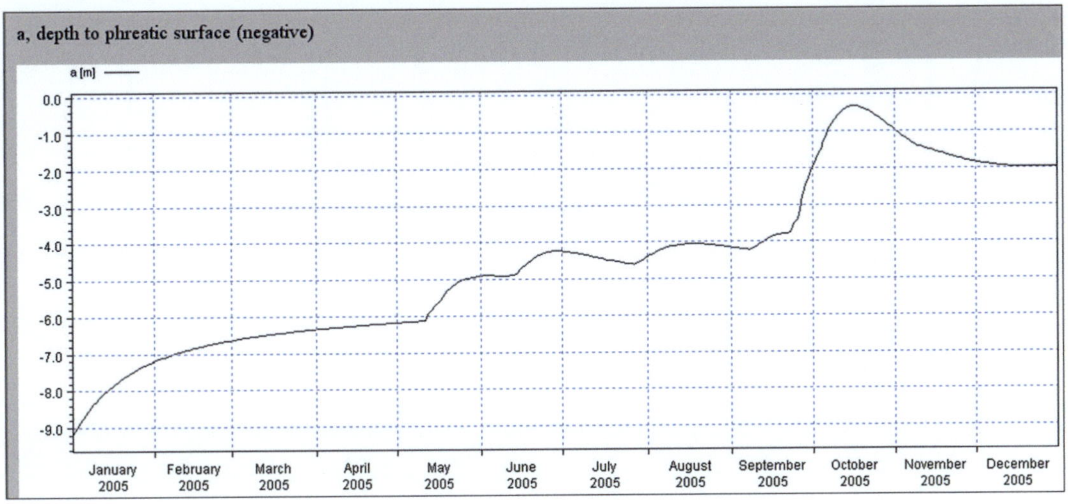

Fig. 18 MIKE SHE model-estimated groundwater table for Dhor-nani catchment for the year 2005

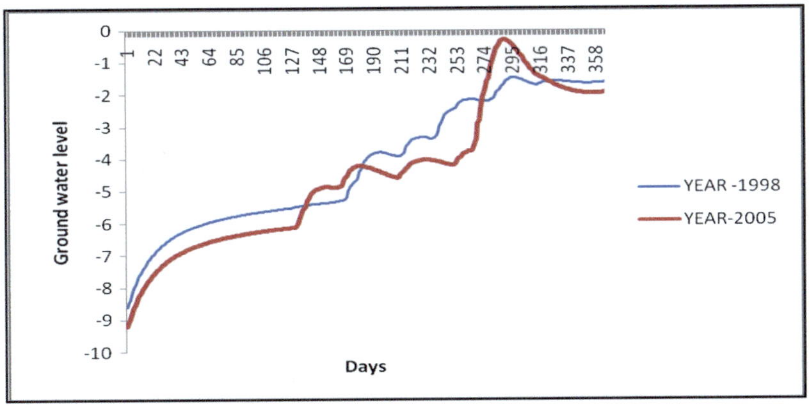

Fig. 19 Comparison of estimated groundwater tables for Dhor-nani catchment for the years 1998 and 2005

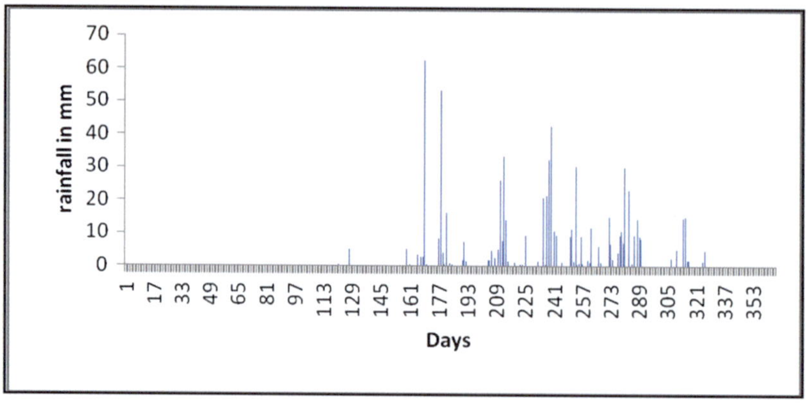

Fig. 20 Rainfall over year 1998

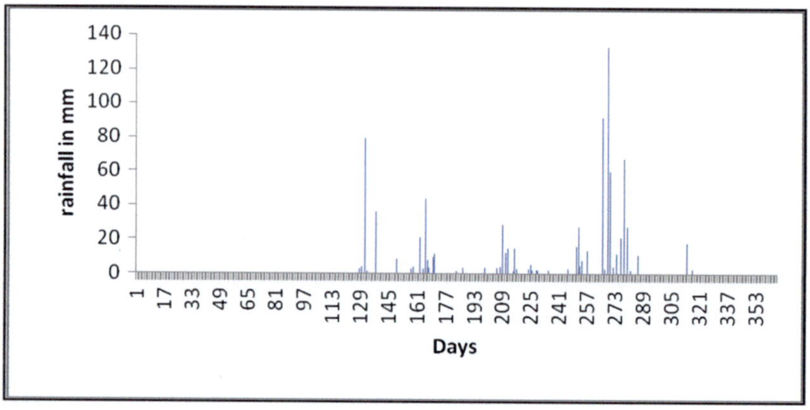

Fig. 21 Rainfall over year 2005

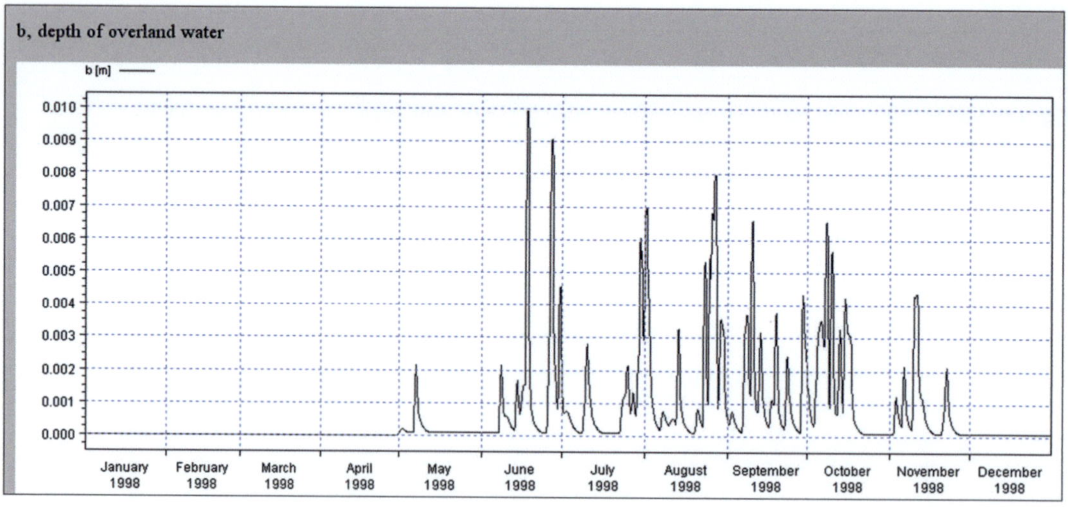

Fig. 22 MIKE SHE model-estimated depth of overland water for Dhor-nani catchment for the year 1998

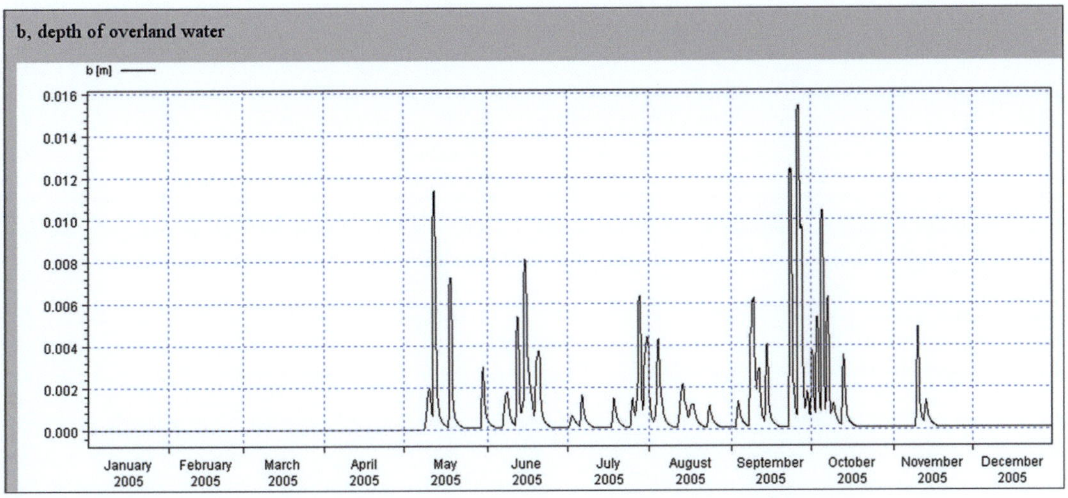

Fig. 23 MIKE SHE model-estimated depth of overland water for Dhor-nani catchment for the year 2005

Conclusions

The physically based MIKE SHE model can be used for the estimation of water resources in hard rock region of catchment as it can simulate the hydrological processes as accurately as possible upon its calibration. The MIKE SHE model was calibrated for Pimpalgaon Ujjaini catchment. The calibrated MIKE SHE model was used to evaluate the effect of changing land use/land cover and climate on water resources of Dhor-nani catchment. The results indicated that the integrated modelling tool MIKE SHE in conjunction with RS and GIS technique gives realistic output than the traditional modelling tools and found to be more reliable and rigorous decision-making tool for assessing water availability. The model has been responsive to evaluate the effect of changing climate and land use/land cover on water resources.

References

Abbott MB, Barthurst JC, Cunge JA, O'Connell PEO, Rasmussen J (1986) An introduction to European Hydrological System – Systeme Hydrologique Europeen "SHE" 1: history and philosophy of a physically based distributed modelling system. J Hydrol 87:45–59

Butts MB, Overgaard J, Viaene P, Dubicki A, Strońska K, Szalinska W (2005) Flexible process-based hydrological modelling framework for flood forecasting–MIKE SHE. In: Proceedings of the international conference "Innovation, advances and implementation of flood forecasting technology" Tromsø, Norway, pp 17–19

Demetriou C, Punthakey JF (1999) Evaluating sustainable groundwater management options using the MIKE SHE integrated hydro geological modelling package. Environ Model Softw 14:129–140

DHI (2005) MIKE-SHE: an integrated hydrological modelling system. DHI Water and Environment, Horsholm, Denmark

Golian LM (2000) Utilizing Internet resources by educational professionals in the new millennium. Inf Technol Libr 19(3):136

Jensen KH (1983) Simulation of water flow in the unsaturated zone including the root zone. Technical series paper no. 33. Institute of Hydrodynamics and Hydraulic Engineering, Technical University of Denmark, Copenhagen, Denmark

Kristensen KJ, Jensen SE (1975) A model of estimating actual evapotranspiration from potential evapotranspiration. Nord Hydrol 6:70–80

Patil KH (2008) A model for allocation of water resources. M. Tech. thesis (Unpublished), MPKV, Rahuri, Ahmednagar, Maharashtra

Refsgaard J, Seth C, Barthurst SM, Erlich JC, Jorgensen B (1992) Application of SHE to catchment in India – Part II: General results. J Hydrol 140:23–59

Singh R, Subramanian K (1999) Hydrological modelling of a small watershed using MIKE SHE for irrigation planning. Agric Water Manag 41(3):149–166

Impact of Climate Change on Water Resources

Mamta Gautam and Anil Kumar Singh

Abstracts

In recent times, several studies around the globe have indicated that climate change is likely to significantly impact freshwater resources availability. In India, demand for water has already increased manifold over the years due to urbanization, agricultural use, increasing population, rapid industrialization, and economic development. At present, the change in cropping and land-use pattern, overexploitation of water, storage, and change in irrigation and drainage are modifying the hydrological cycle in many climate regions and river basins of India. An assessment of the availability of water resources in the context of future national requirements and expected impacts of climate change and its variability is critical for relevant national and regional long-term development strategies and sustainable development. This article examines the potential for sustainable development of surface water and groundwater resources within the constraints imposed by climate change and envisages future research needs of relevance to India.

Keywords

Climate change • Groundwater • Water resources

Introduction

Globally, one third of the population depends on groundwater for its drinking requirements, in urban as well as rural areas. Groundwater plays an even more critical role in agriculture, and an increasing portion of groundwater extracted is used for irrigation. It is estimated that at least 40 % of the world's food is produced by groundwater-irrigated farming, both in low-income and high-income countries. In arid and semiarid areas, the dependency on groundwater for water supply is between 60 and 100 %. Therefore, the aim of halving the number of people below the poverty line without sustainable access to safe drinking water and basic sanitation (Millennium Development Goal, MDG 7) depends very much on the development and management of groundwater resources. Since

M. Gautam (✉) • A.K. Singh
Natural Resources Management, Indian Council of Agricultural Research, KAB II, IARI Campus, New Delhi 110012, India
e-mail: gmamta007@gmail.com

A.K. Singh et al. (eds.), *Climate Change Modelling, Planning and Policy for Agriculture*,
DOI 10.1007/978-81-322-2157-9_21, © Springer India 2015

groundwater resources and their replenishment are influenced by long-term climatic conditions, climate change will have a great impact on its quality and quantity.

Climate change is likely to increase pressure on groundwater resources by reducing recharge capacities in some regions and decrease surface water availability due to increased uncertainty and variability in rainfall. Groundwater contamination is also expected to increase in coastal zones due to sea-level rise. In many vulnerable areas, such impacts on groundwater resources may render the only available freshwater reserve unavailable or unsuitable for use in the near future (IPCC 2007). Obviously there is a need to improve the understanding and modeling of climate changes with specific reference to the hydrological cycle at scales relevant for decision making. Currently, information about water-related impacts of climate change is inadequate – especially with respect to water quality, aquatic ecosystems, and groundwater – including socio-economic dimensions (IPCC 2008).

It is estimated that groundwater levels have already declined in about 0.34 million km^2 area. Although efforts are made for improved water management practices including water conservation, artificial recharge and watershed management, utilization of nonconventional energy, and integrated water development, the projected water demand of a minimum of 980 BCM during 2050 will require intensive development of groundwater resources, exploiting both dynamic and in-storage potential. It is apparent that the projected climate change will disturb the hydrological balance in different parts of India and quality of groundwater along the coastal regions in particular.

The purpose of this paper is to highlight the impact of climate change with special reference to groundwater resources so that they are considered a "water bank" of which the "principal amount" should never be touched and all activities should be managed judiciously through the "accrued interest" only.

Changing Temperature and Rainfall

The 4th Assessment Report (2007) of IPCC has projected that by 2100, the Earth's mean temperature will rise by 1.4–5.8 °C, precipitation will decrease in the subtropical areas, and frequency of extreme events will increase significantly. As of now, in reality, in the past 100 years, the global mean temperature has increased by 0.74 °C.

It was also reported in IPCC (2007) that 11 of the last 12 years (between 1995 and 2009) rank among the 12 warmest years in the instrumental record of global surface temperature (since 1850) (Fig. 1). Melting of glaciers and rising sea level are some of the most important manifestations of it. IPCC (2007) has further reported change in their frequency and/or intensity of extreme events over the last 50 years. The report has made the following significant observations.

However, climate change impacts are already being felt, as the last 60 years were the warmest in the last 1,000 years. Since the beginning of the twenty-first century, India has experienced droughts in quick succession, of which the 2009 one was the most recent one that significantly affected *kharif* crop production. It was the 2nd largest all India monsoon rainfall deficit since 1972 (23 % below normal). Incidentally, 2009 also achieved the distinction of being the warmest year in past several centuries across the world. However, after 2009, 2010 proved to the warmest year on record since 1850. 2011 is now the 11th warmest year on record since 1850. Apart from that, 1998 was one of the warmest years, 2003 experienced unprecedented heat and cold waves across the globe, and occurrence of high temperature in March 2004 adversely affected crops like wheat, apple, potato, etc. across northern India; 2005 witnessed destructive hurricanes/cyclones across the globe. Again, 2007 was as warm as 1998 in the entire northern hemisphere, and unusual summer rains and floods were experienced in many parts

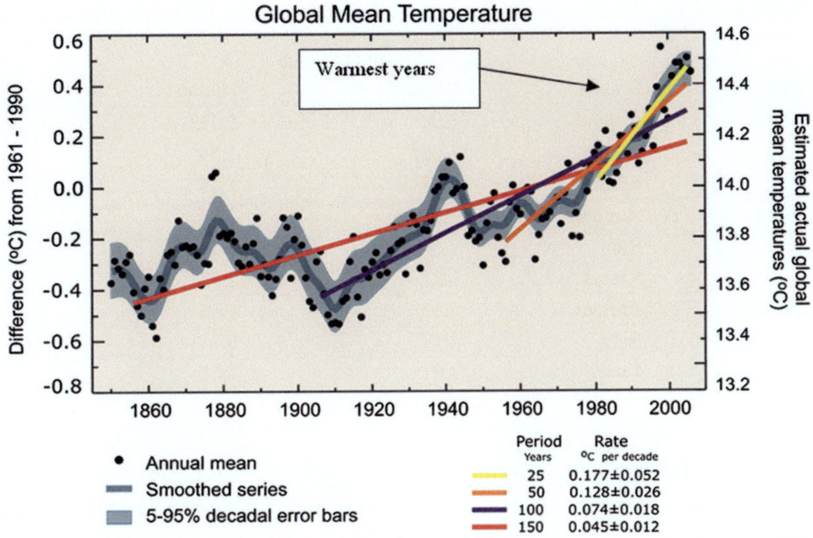

Fig. 1 Increase of global mean temperatures during the last 100 years (Source: IPCC 2007)

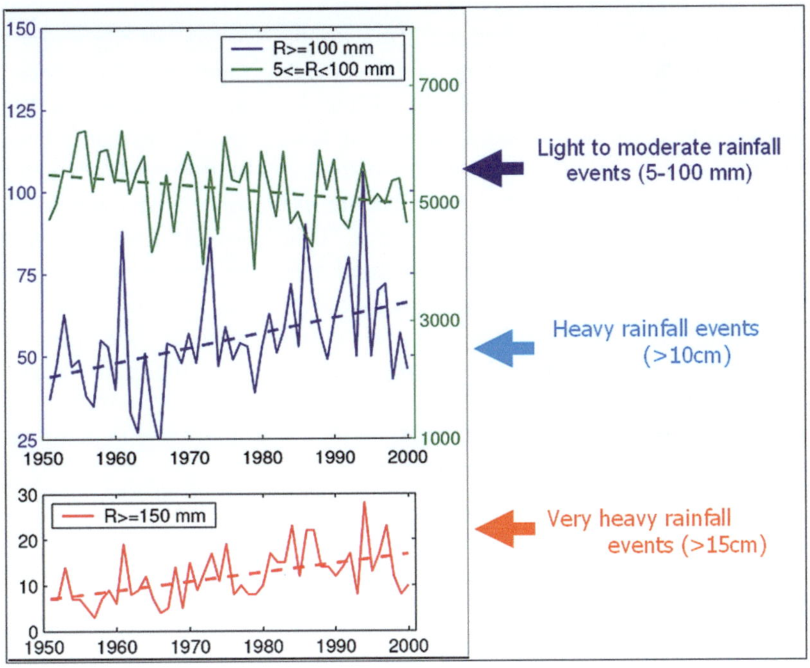

Fig. 2 Change in intensity of rainfall over Central India in the last 50 years (Source: Goswami 2006)

of India. Besides that, the amount and distribution of rainfall is becoming more and more erratic which is causing greater incidences of droughts and floods globally. The increase in frequency of heavy rainfall events in the last 50 years over Central India (Fig. 2) points toward a significant change in climate pattern in India (Goswami 2006).

In 2005, the conference (Defra 2005) "Avoiding Dangerous Climate Change" raised concerns that climate change is occurring more quickly than previously anticipated. Large-scale, irreversible system disruption and the destabilization of the Antarctic ice sheets are serious risks: changes to polar ice, glaciers, and rainfall regimes have already occurred.

Table 1 Projected mean temperature and rainfall changes over the Indian subcontinent

		Temperature change (°C)		Rainfall change (%)	
Year	Season	Lowest	Highest	Lowest	Highest
2020s	Annual	1.00	1.41	2.16	5.97
	Rabi	1.08	1.54	−1.95	4.36
	Kharif	0.87	1.17	1.81	5.10
2050s	Annual	2.23	2.87	5.36	9.34
	Rabi	2.54	3.18	−9.22	3.82
	Kharif	1.81	2.37	7.18	10.52
2080s	Annual	3.53	5.55	7.48	9.90
	Rabi	4.14	6.31	−24.83	−4.50
	Kharif	2.91	4.62	10.10	15.18

Lal (2001)

The rainfall analysis of India (Rao et al. 2009) showed that significant negative trends of rainfall were observed in the eastern parts of Madhya Pradesh, Chhattisgarh, and parts of Jharkhand, U.P., and northeast India. Lal (2001) reported that annual mean area-averaged surface warming over the Indian subcontinent is likely to range between 3.5 and 5.5 °C by 2080s (Table 1). These projections showed more warming in winter season over summer. The spatial distribution of surface warming suggests a mean annual rise in surface temperatures in North India by 3 °C or more by 2050. The study also indicated that during winter, the surface mean air temperature could rise by 3 °C in the northern and central parts, while it would rise by 2 °C in the southern parts by the same period. In case of rainfall, a marginal increase of 7–10 % in annual rainfall is projected over the subcontinent by 2080. Nevertheless, the study suggests a fall in rainfall by 5–25 % in winter, while it would increase by 10–15 % in summer. Marked variability is seen even in the onset and withdrawal of monsoon over the period. The glaciers and the snowfields in the Himalayas are on the decline as a result of climate variability. The rate of retreat of the snow of Gangotri glacier demonstrated a sharp rise in the first half of the twentieth century. Glacial melt would lead to increased summer river flow and floods over the next few decades, followed by a serious reduction in flows thereafter.

Greenhouse Gases

The increasing levels of greenhouse gases (GHGs) in the atmosphere have been attributed as one of the major driving forces behind the rapid climate change phenomenon. Though the increase in the level of CO_2 (Fig. 3) is expected to produce some beneficial effects on crop dry matter production, it may soon be nullified by associated water and thermal stresses leading to overall deterioration of agro-climatic conditions for food production systems. At the global scale, the historical temperature-yield relationships indicate that warming from 1981 to 2002 is very likely to offset some of the yield gains from technological advance, rising CO_2, and other non-climatic factors (Lobell and Field 2007).

Role of Carbon Dioxide

Research suggests that atmospheric carbon dioxide levels may affect water availability through its influence on vegetation. Controlled experiments indicate that elevated CO_2 concentrations would increase the resistance of plant "pores," that is, stomata, to water vapor transport. Experiments suggest that a doubling of CO_2 would increase stomatal resistance and reduce the rate of transpiration vis-à-vis the

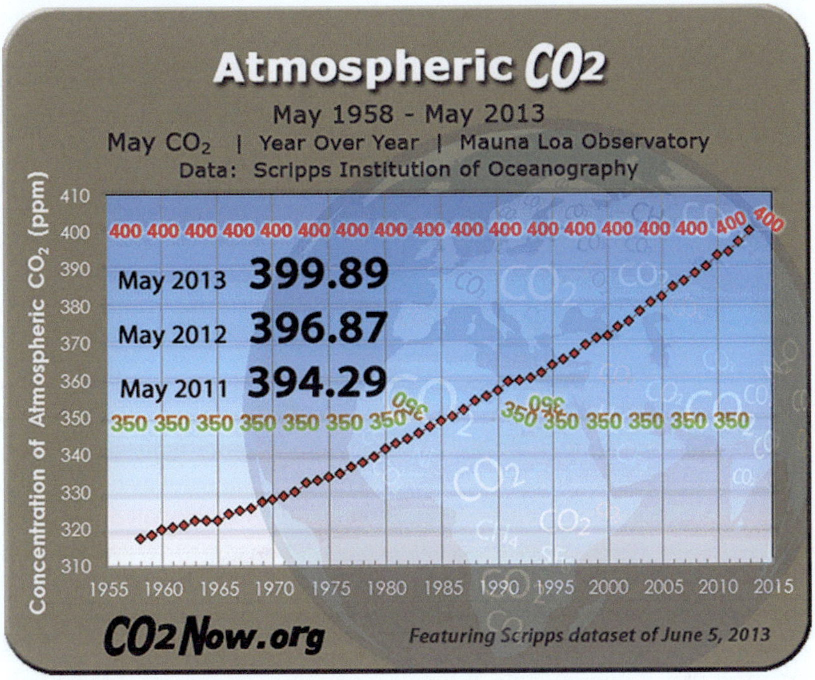

Fig. 3 Current global CO_2 concentration levels (Source: NOAA-ESRL)

passage of water vapor from plants by about 50 % on average. The resulting decrease in transpiration would tend to increase runoff. On the other hand, CO_2 also has been demonstrated to increase plant growth, leading to a larger area of transpiring tissue and a corresponding increase in transpiration. Other factors that might offset increases in plant water-use efficiency associated with a CO_2-enriched atmosphere are a potential increase in leaf temperatures caused by reduced transpiration rates and species changes in vegetation communities. The net effect of opposing influences on water supplies would depend on the type of vegetation and other interacting factors, such as soil type and climate.

The management and labor costs of farm production will rise to a great level owing to increased incidences of pests and diseases as well as weeds. Additionally, extreme events like hails and frosts will also negatively impact crop production. As of now, most of the developing countries, including India, are not fully prepared to deal with the adverse impacts expected as a consequence of climate change and are therefore relatively vulnerable.

Water Resources

Of the total water present on the earth, only 2.7 % is freshwater and the rest 97.3 % is saline and brackish water. Of the total freshwater, 68.7 % is locked up in icecaps and glaciers; 30.1 % is the groundwater, and only 0.3 % is the surface water present in lakes, swamps, and rivers (FAO 1990) that may be readily available. This surface water resource is insufficient to meet the gregarious demand of the growing population, industrialization, urbanization, and intensive agriculture. In India, about 60 % of the total irrigated area depends on groundwater (CWC 2000) and about 60 % of irrigated food production depends on irrigation from groundwater wells (Shah et al. 2000).

Overexploitation of groundwater has resulted in polluting aquifers through toxic and undesirable chemical constituents coming from underlying minerals containing aquifers. Along the coastal areas this has resulted in seawater intrusion causing deterioration in the quality of adjoining groundwater.

Global warming and its impact on the hydrological cycle and nature of hydrological events have posed an additional threat to this mountainous region of the Indian subcontinent. Extreme precipitation events have geomorphologic significance in the Himalayas, where they may cause widespread landslides (Ives and Messerli 1989). The response of hydrological systems, erosion processes, and sedimentation in this region could alter significantly due to climate change. It is estimated that the Himalayan mountains cover a surface area of permanent snow and ice in the region which is about 97,020 km^2 with 12,930 km^2 volume. In these mountains, 10–20 % of the total surface area is covered by glaciers, while an area ranging from 30 to 40 % has seasonal snow cover (Upadhyay 1995; Bahadur et al. 1999). These glaciers provide snow and the glacial meltwaters keep the Himalayan rivers perennial. Bahadur et al. (1999) reported that a conservative estimate gives at least 500 km^3/year as snow and ice melt water contributions to Himalayan streams, while Alford (1992) reports about 515 km^3/year from the upper Himalayan mountains. The most useful facet of glacial runoff is the fact that glaciers release more water in a drought year and less water in a flood year and thus ensure water supply even during the lean years. The snow line and glacier boundaries are sensitive to changes in climatic conditions. Almost 67 % of the glaciers in the Himalayan mountain ranges have retreated in the past decade (Ageta and Kadota 1992; Fushimi 2000). Available records suggest that the Gangotri glacier is retreating about 28 m per year. A warming is likely to increase the melting more rapidly than the accumulation. Glacial melt is expected to increase under changed climate conditions, which would lead to increased summer flows in some river systems for a few decades, followed by a reduction in flow as the glaciers disappear (IPCC 1998).

Impact on Water Resources

A warmer climate will modify the hydrologic cycle, altering rainfall, magnitude, and timing of runoff. Warm air holds more moisture and increases evaporation of surface moisture. With more moisture in the atmosphere, rainfall and snowfall events tend to be more intense, increasing the potential for floods. However, if there is little or no moisture in the soil to evaporate, the incident solar radiation will lead to increase in the temperature, contributing to longer and more severe droughts (Trenberth 1999). Therefore, change in climate will affect the soil moisture, groundwater recharge and frequency of flood or drought episodes, and finally groundwater level in different areas.

Table 2 Annual requirement of freshwater (billion cubic meters – BCM)

User sector	2008–2009	2025	2050	Ratio
Irrigation	501	910	1,072	2.1
Domestic	30	73	102	3.4
Industries	20	22	63	3.1
Thermal power	20	31	130	6.5
Others	24	72	80	3.3

Source: National Commission for Integrated Water Resources Development Planning, GOI

At present, available statistics on water demand show that the agriculture sector is the largest consumer of water in India using more than 80 % of the available water. The quantity of water used for agriculture has increased progressively through the years as more and more areas were brought under irrigation. Since 1950–1951, the net irrigated area in India rose from 28.85 to 62.29 million ha up to 2007–2008. Contribution of surface water and groundwater resources for irrigation has played a significant role in India attaining self-sufficiency in food production during the past three decades, but water is likely to become more scarce and critical input in the future. By judicious utilization, the demand for water from farm sector can be pegged at 68 % by the year 2050, but agriculture will still remain the largest consumer (Table 2). In order to meet this demand, augmentation of the existing water resources by development of additional sources of water or conservation of the existing resources through impounding more water in the existing water bodies and their conjunctive use will be needed (Mall et al. 2006). Table 3 (adapted from

Table 3 Impact of climate change on water resources during the next century over India

Region/location	Impact
Indian subcontinent	Increase in monsoonal and annual runoff in the central plains
	No substantial change in winter runoff
	Increase in evaporation and soil wetness during monsoon and on an annual basis
Orissa and West Bengal	One-meter sea-level rise would inundate 1,700 km^2 of prime agricultural land
Indian coastline	One-meter sea-level rise on the Indian coastline is likely to affect a total area of 5,763 km^2 and put 7.1 million people at risk
All-India	Increases in potential evaporation across India
Central India	Basin located in a comparatively drier region is more sensitive to climatic changes
Kosi Basin	Decrease in discharge on the Kosi River. Decrease in runoff by 2–8 %
Southern and Central India	Soil moisture increases marginally by 15–20 % during monsoon months
Chenab river	Increase in discharge in the Chenab river
River basins of India	General reduction in the quantity of the available runoff, increase in Mahanadi and Brahmani basins
Damodar Basin	Decreased river flow
Rajasthan	Increase in evapotranspiration

Adapted from Mall et al. (2006)

Mall et al. 2006) summarizes the possible impacts of climate change on water resources during the next century over India. The enhanced surface warming over the Indian subcontinent by the end of the next century would result in an increase in pre-monsoonal and monsoonal rainfall and no substantial change in winter rainfall over the central plains. This would result in an increase in the monsoonal and annual runoff in the central plains, with no substantial change in winter runoff and increase in evaporation and soil wetness during the monsoon on an annual basis (Fig. 4).

In another study made by IITM, Pune, in collaboration with DEFRA, the UK reported that there will be an increase in annual rainfall and annual flow in Godavari and Ganga basins by 2071–2100 period.

Goswami and Rao (2003) used a combination of a hydrological model (SWAT: Soil and Water Assessment Tool) and weather data generated by the Hadley Centre for Climate Prediction (HadRM2) at a resolution of $0.44^0 \times 0.44^0$ on a GIS framework for quantifying the impact of climate change on the water balance components in 12 river basins of India. The results of their study have been presented in Fig. 5. The projected impact is likely to vary from basin to basin. There would be basins which will have an increase in rainfall, runoff, and evapotranspiration (ET), e.g., Mahanadi, while in some basins, there would be a decrease in all the three components, e.g., Sabarmati.

In the recent report submitted by IPCC (2013), it has been brought out explicitly that each of the last three decades was warmer than the previous one but the rate of warming per decade during the period 1998–2012 was less than the period 1951–2012. The report has also stated that the contributions made by anthropogenic sources are responsible for global warming.

Chaturvedi et al. (2012) used the representative concentration pathways (RCPs) along with the Coupled Model Intercomparison Project 5 (CMIP5) described by IPCC (2013) to generate scenarios of temperature and precipitation for India. They concluded that the ensemble model's output was closer to the observed climate than any individual model. They reported that under the "business as usual scenario," the mean warming is likely to be between 1.7–2 °C by 2030s and 3.3–4.8 °C by 2080s compared to preindustrial era. The precipitation is likely to increase by 4–5 % by 2030s and 6–14 % by 2080s. Temperature projections are more robust

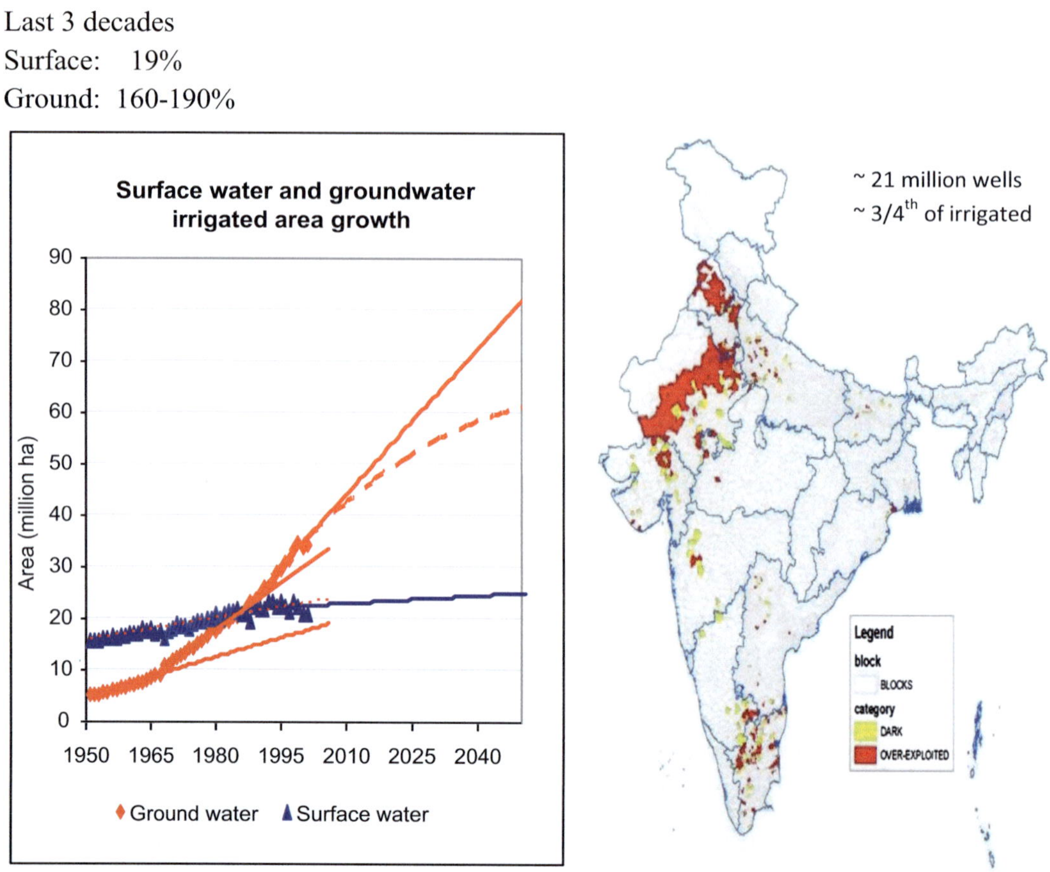

Fig. 4 Overexploitation and underutilization of groundwater (Source: GOI 2004)

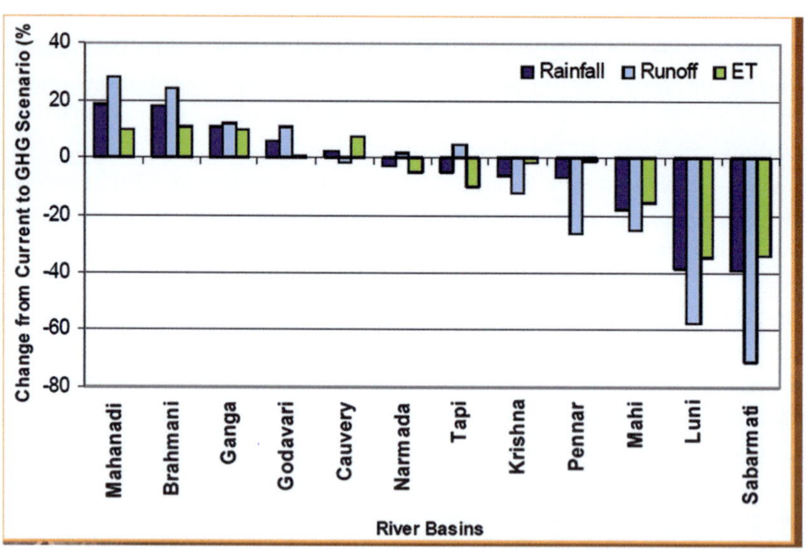

Fig. 5 Percent change in mean annual water balance for control and GHG climate scenarios

Table 4 Zone-wise groundwater resources availability, utilization, and stage of development

States/ union territories	Net annual groundwater availability	Annual groundwater draft			Projected demand for domestic and industrial uses up to 2025	Groundwater availability for future irrigation	Stage of groundwater development (%)
		Irrigation	Domestic and industrial uses	Total			
North	105.45	86.55	5.19	91.71	8.01	11.43	87
South	75.71	42.34	4.01	46.34	6.43	30.58	61
East	112.12	28.87	4.31	32.99	6.52	76.66	29
West	105.93	54.77	4.79	59.58	8.21	43.62	56
Total	399.20	212.53	18.29	230.62	29.16	162.28	58

Source: Central Ground Water Board, Annual Report, 2005–2006

than precipitation projections. Another important projected impact is the likely increase in extreme rainfall events of >40 mm/day for 2060s and beyond.

Groundwater Resources Availability, Quantity, and Quality

Rainfall is the major source of groundwater recharge in India, which is supplemented by other sources such as recharge from canals, irrigated fields, and surface water bodies. A major part of the groundwater withdrawal takes place from the upper unconfined aquifers, which are also the active recharge zones and hold the replenishable groundwater resource. The replenishable groundwater resource in the active recharge zone in the country has been assessed by Central Ground Water Board jointly with the concerned State Government authorities. The assessment was carried out with Block/Mandal/Taluka/Watershed as the unit and as per norms recommended by the Ground Water Estimation Committee (GEC)-1997. As per the latest assessment, the annual replenishable groundwater resource in this zone has been estimated as 432 billion cubic meter (bcm), out of which 399 bcm is considered to be available for development for various uses after keeping 34 bcm for natural discharge during non-monsoon period for maintaining flows in springs, rivers, and streams (Table 4) (Central Ground Water Board 2006).

Groundwater extraction for various uses and evapotranspiration from shallow water table areas constitute the major components of groundwater draft. In general, the irrigation sector remains the main consumer of groundwater. The groundwater draft for the country as a whole has been estimated as 231 bcm (Central Ground Water Board 2006), about 92 % of which is utilized for irrigation and the remaining 8 % for domestic and industrial uses. Hence, the stage of groundwater development, computed as the ratio of groundwater draft to total replenishable resource, works out as about 58 % for the country as a whole. However, the development of groundwater in the country is highly uneven and shows considerable variations from place to place.

As a part of the resource estimation following the GEC norms, the assessment units have been categorized based on the stage of groundwater development and long-term declining trend of groundwater levels. As per the assessment, out of the total of 5,723 assessment units in the country, groundwater development was found to exceed more than 100 % of the natural replenishment in 839 units (14.7 %) which have been categorized as "overexploited." Groundwater development was found to be the extent of 90–100 % of the utilizable resources in 226 assessment units (3.9 %), which have been categorized as "critical." 550 assessment units with stage of groundwater development in the range of 70–100 % and long-term decline of water levels either during pre- or post-monsoon period have been categorized as "semi-critical," and 4,078 assessment units with stage of groundwater development below 70 % have been categorized as "safe." 30 assessment units have

Table 5 Groundwater resources availability and status of its utilization in India

S. no.	Regions	Annual discharge groundwater resource (bcm)	Natural discharge during non-monsoon season (bcm)	Net annual water available (bcm)	Annual groundwater draft	Stage groundwater development (%)	Categorization of assessment units (Blocks/Mandals) Total assessment units	Overexploited nos./%	Critical/%
1	2	3	4	5	6	7	8	9	10
1	Northern Himalayan states	5.4	0.48	4.92	1.84	37	30	2/6.67	0
2	Northeastern hilly states	33.99	3.02	30.98	5.63	18	118	0/0	0
3	Eastern plain states	111.63	9.03	102.5	43.97	43	1,895	1/.05	2/.11
4	Northwestern plain states	80.78	6.92	73.85	72.17	98	277	201/72.56	28/10.11
5	Western arid region	27.38	1.97	25.4	24.48	96	462	172/37.23	62/13.42
6	Central Plateau states	90.723	5.19	85.53	36.11	42	985	31/3.15	6/.61
7	Southern peninsular states	82.78	7.14	75.65	46.4	61	1,946	432/22.2	128/6.58
8	Islands	0.34	0.01	0.32	0.01	4	10	0	0
	Country total	433.02	33.73	399.26	230.63	58	5,723	839	226

Note: Southern peninsular states include Andhra Pradesh, Karnataka, Kerala, Tamil Nadu, and Pondicherry; northeastern hilly states include Arunachal Pradesh, Assam, Manipur, Meghalaya, Mizoram, Nagaland, Sikkim, and Tripura; eastern plain states include Bihar, Orissa (part), Eastern Uttar Pradesh, and West Bengal; Central Plateau states include Chhattisgarh, Jharkhand, Madhya Pradesh, Maharashtra, Dadra, and Nagar Haveli; North Western plain states include Delhi, Haryana, Punjab, Western Uttar Pradesh, and Chandigarh; Western arid states include Gujarat, Rajasthan, Daman, and Diu; Northern Himalayan states include Himachal Pradesh, Jammu and Kashmir, and Uttarakhand; and islands Andaman and Nicobar and Lakshadweep

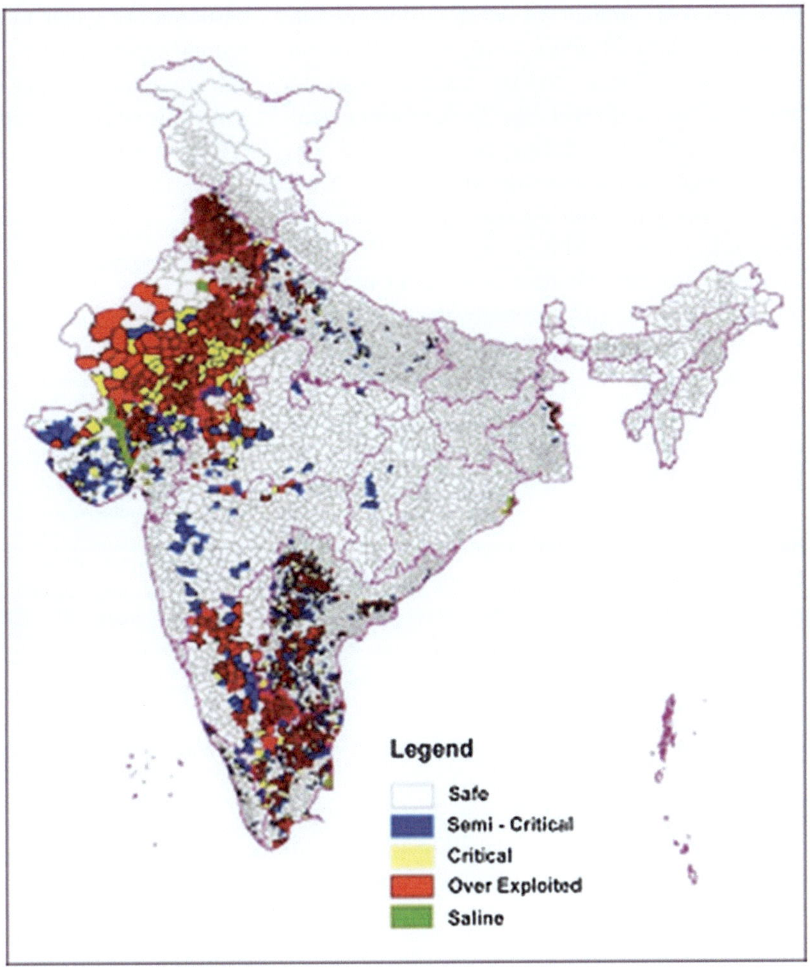

Fig. 6 Geographical distribution of various categories of assessment units in India

been excluded from the assessment due to the salinity of groundwater in the aquifers in the replenishable zone. Salient details of groundwater resource availability, utilization, stage of development, and categorization of assessment units for the above regions of the country are given in Table 5, and geographic distribution of various categories of assessment units is shown in Fig. 6.

In addition to the resources available in the zone of water-level fluctuation, extensive groundwater resources have been proven to occur in the deeper confined aquifers in the country, a major chunk of which is in the Ganga–Brahmaputra alluvial plains (Romani 2006). Such resources are also available in the deltaic and coastal aquifers to a lesser extent. These aquifers have their recharge zones in the upper reaches of the basins. The resources in these deep-seated aquifers are termed "in-storage groundwater resources." The quantum of these resources has been tentatively estimated as ~10,800 bcm. Though the groundwater resources in these aquifers are being exploited to a limited extent in parts of Punjab, Haryana, and western Uttar Pradesh, detailed studies are to be taken up to fully understand the yield potentials and characteristics of these aquifers.

The climate is expected not only to affect input (recharge) and output (discharge) but also to influence the quantity and quality of the groundwater, through drying up of millions of wells/bore wells, poor quality of groundwater, salinity ingress, drying up of rivers like

Swarnamukhi, impact of waste effluents into river (river–aquifer interaction), overexploitation of freshwater from coastal aquifer, seawater intrusion, and sea-level rise. Groundwater quality water recharged during an arid period may have a higher concentration of salts and hence higher TDS, while during a wet period the converse may occur (Sukhija et al. 1988). However, to appreciate such changes, long-term monitoring of rainfall and groundwater quality is required. It is also possible to link the occurrence of certain ions in groundwater to particular water–rock processes that occurred during specific past climatic periods:

Fluoride (>1.5 ppm) – 14 states affecting a total of 69 districts

Salinity – 73 districts and three blocks of Delhi

Iron (>0.3 ppm) – 23 districts from 4 states

Arsenic (>50 ppb) – alluvial plains of Ganges covering six districts of West Bengal

Presence of heavy metals – 40 districts from 13 states

Nonpoint pollution caused by fertilizers and pesticides

Nitrate (>45 ppm) – 95 districts from 11 states (Source: India Assessment 2002 Water Supply and Sanitation, Planning Commission, Govt. of India).

Conclusion

On the basis of a bird's-eye view on the above study, it could be suggested there is an urgent need to improve understanding and modeling of climate changes related to the hydrological cycle at scales relevant to decision making. "Information about the water-related impacts of climate change is inadequate – especially with respect to water quality, aquatic ecosystems and groundwater – including their socio-economic dimensions (IPCC 2008). Agricultural water crisis can be solved through identification of input use efficient crops and cropping systems, integrated use of rain, surface and groundwaters, judicious use of polluted and poor quality waters, awareness among farmers about the value and scarcity of water and negative fallouts of improper use, participatory water management, recharging the groundwater aquifers, multiple uses of water, enhancing income per unit quantity of water consumed, speedy development and transfer of cost-effective and eco-friendly technologies.

References

Ageta Y, Kadota T (1992) Predictions of changes of glacier mass of air temperature increase for three glaciers. Ann Glaciol 16:89–94

Alford D (1992) Hydrological aspects of the Himalayan region. ICIMOD Occasional paper no. 18, Kathmandu

Bahadur J, Dash SK, Bahadur J (1999) The Himalayan environment. New Age International (P) Ltd., New Delhi

Central Ground Water Board (2006) Dynamic ground water resources of India. Ministry of Water Resources, Government of India, New Delhi

Chaturvedi RK, Joshi J, Jayaraman M, Bala G, Ravindranath NH (2012) Multi – model climate change projections for India under representative concentration pathways. Curr Sci 103:791–802

CWC (2000) Water and related statistics. Information Systems Organisation, Water Planning & Project Wing, Central Water Commission, New Delhi

Defra (2005) http://www.defra.gov.uk/environment/climatechange/internat/dangerous-cc.htm

FAO (1990) Production of year book. Food and Agricultural Organization, Rome

Fushimi H (2000) In: Domoto A et al (eds) Tsukiji-Shokan Pub Co Ltd., Japan and the World Conservation Union, Gland, Switzerland

GOI (2004) Central ground water board ministry of water resources. http://cgwb.gov.in/documents/DGWR2004.pdf

Goswami BN (2006) Increasing trend of extreme rain events and possibility of extremes of seasonal mean Indian monsoon in a warming world (http://saarc-sdmc.nic.in/pdf/workshops/kathmandu/pres16.pdf)

Goswami AK, Rao S (2003) Impact of climate change on water sector. In: Shukla PR, Sharma SK, Ravindranath NH, Garg A, Bhattacharya S (eds) Climate change and India – vulnerability, assessment and adaptation. University Press, Hyderabad, pp 159–192

IPCC (Intergovernmental Panel for Climate Change) (1998) In: Watson RT, Zinyowera MC, Moss RH, WMO/UNEP (eds) The Regional impacts of climate change: an assessment of vulnerability: a special report of IPCC Working Group II. Cambridge University Press, Cambridge

IPCC (Intergovernmental Panel for Climate Change) (2007) Inter-Governmental Panel on Climate Change, 4th assessment report-synthesis report. https://www.ipcc-wg2.gov/AR4/website/intro.pdf

IPCC (Intergovernmental Panel for Climate Change) (2008) Climate change and water. IPCC technical paper VI. URL: http://www.ipcc.ch/ pdf/technical-papers/climate-change-water-en.pdf. Accessed 10 Dec 2009

IPCC (Intergovernmental Panel for Climate Change) (2013) Summary for policymakers. In: Stocker TF, Qing D, Plattner GK, Tignor M, Allen SK, Boschung J, Nauels A, Xia Y, Bex V, Midgley PM (eds) Climate change 2013: the physical science basis: contribution of Working Group I to the fifth assessment report of the Intergovernmental Panel on Climate Changes. Cambridge University Press, Cambridge/New York

Ives JD, Messerli B (1989) The Himalayan dilemma: reconciling development and conservation. Routledge, London

Lal M (2001) Future climate change: implications for Indian summer monsoon and its variability. Curr Sci 81(9):1205

Lobell DB, Field CB (2007) Global scale climate-crop yield relationships and the impacts of recent past. Environ Res Lett 2:7

Mall RK, Gupta A, Singh R, Singh RS, Rathore LS (2006) Water resources and climate change: an Indian perspective. Curr Sci 90(12):1610–1626

Rao GGSN, Rao AVMS, Rao VUM (2009) Trends in rainfall and temperature in rainfed India in previous century. In: Aggarwal PK (ed) Global climate change and Indian agriculture – case studies from ICAR network project. ICAR Publication, New Delhi

Romani S (2006) Groundwater management – emerging challenges: groundwater governance – ownership of groundwater and its pricing. In: Proceedings of the 12th national symposium on hydrology, New Delhi, 14–15 November 2006

Shah T, Molden D, Sakthivadivel R, Seckler D (2000) The global ground water situation: overview of opportunity and challenges. International Water Management Institute, Colombo

Sukhija BS, Reddy DV, Nagabhushanam P, Chand R (1988) Validity of the environmental chloride method for recharge evaluation of coastal aquifers, India. J Hydrol 99:349–366

Trenberth KE (1999) Conceptual framework for changes of extremes of the hydrological cycle with climate change. Clim Chang 42:327–339

Upadhyay DS (1995) Cold climate hydrometeorology. New Age International (P) Ltd, New Delhi

Building Climate Resilient Agriculture and Enhancing Grassroot-Level Adaptive Capacity in the Semiarid Tropics of India: Indicative Policies for Action

Naveen P. Singh, K. Byjesh, and Cynthia Bantilan

Abstract

Climate change has been recognized as a potential threat to livelihood of the poor farmers in the marginal agricultural productive environment especially in the semiarid tropics of India. The impacts may vary spatially, and the rural poor are more challenged of its impacts. Initiatives at national level are underway to address the consequences especially rural and agriculture. The research initiative coordinated by the International Crops Research Institute for the semiarid tropics (ICRISAT) tracked the climate change impacts, adaptation strategies, and constraints at the households' level through a rigorous quantitative and qualitative analysis in the semiarid tropics of India. This explorative exercise identified challenges and opportunities towards climate resilience through recommendations and policy directive for action. This chapter comprehends the policy needs that emerged from the regional study in identifying the impacts and constraints to effective adaptation by climate change. This evidence-based indicative policy stresses the need to channelize resources effectively in enhancing the grassroot-level resilience to climate change.

Keywords

Climate-resilient agriculture • Policies • Local-level adaptation • Resource management

Introduction

The 4th Assessment Report of the Intergovernmental Panel on Climate Change (IPCC) provides the latest situation and outlook on the science, impacts, and potential measures to address climate change (IPCC 2007). However, the state of knowledge available at the global level is far from being comprehensive and holistic. There has been a strong focus on the different scales of understanding beyond regional and subregional levels (INCCA 2010). In response to the growing awareness of governments worldwide, including those from developing countries, "early action" plans

N.P. Singh (✉) • K. Byjesh • C. Bantilan
Research Program – Markets, Institutions and Policies, International Crops Research Institute for Semi-Arid Tropics (ICRISAT), Hyderabad 502324, India
e-mail: naveenpsingh@gmail.com

A.K. Singh et al. (eds.), *Climate Change Modelling, Planning and Policy for Agriculture*,
DOI 10.1007/978-81-322-2157-9_22, © Springer India 2015

are expected to strengthen confidence, capacity, knowledge, and experience to focus on enhancing resilience of the system particularly local level agricultural production systems against climatic risks. The research agenda on enhancing climate resilience in agriculture with a considerable focus on the microlevel understanding of impacts, opportunities, and constraints are crucial.

The climate resilient agriculture should evolve from science-based solutions and pro-poor approaches that enable agricultural systems to effectively deal with the climate-related challenges so that the poor and the most vulnerable farmers in the semiarid regions are benefitted. Contributing to this global effort, the International Crops Research Institute for the Semi-Arid Tropics (ICRISAT) together with five countries in Asia implemented a project entitled "Vulnerability to Climate Change: Adaptation Strategies and Layers of Resilience" supported by the Asian Development Bank (ADB). This research identified and prioritized regions most at risk and agricultural adaptation and mitigation strategies at microlevel as an integral part of agricultural development in the most vulnerable areas. Research activities were designed and undertaken with a goal to improve innovations in agricultural institutions, crop and resource management, social capital, and social networks (Adger 2003) in the target countries. The research activities generated valuable output in the form of useful information repository to inform policy decisions on critical issues affecting the future of agriculture and livelihoods in the target domain and valid information with policy and livelihood impacts. To achieve these impacts, the research activities should implement a robust approach with reliable and in-depth understanding with an optimal set of key information that should be undertaken. This will pitch for enhanced information from different components of analysis, argumentation, and advocacy (Adger and Vincent 2005; Kelly and Adger 2000).

Ensuring sufficient food for the ever-increasing global population through improved productivity and increased resource use efficiency continues to be a key challenge throughout this century. Since competition for natural resources like water and land is increasing, catapulted by the risks of climate-related impacts on agriculture, the challenge appears to be even more daunting (Shiferaw and Bantilan 2004). With the pressure to produce more from shrinking natural resources under climate uncertainties, the agricultural production systems are also to be environment friendly and climate neutral by minimizing carbon emissions. Indeed this is a daunting task. To achieve this task of paving the way for a "climate smart agriculture," several measures must be taken, including enabling policies, institutions and infrastructure in place, and farm communities better informed and empowered with necessary resources. As a response to impacts of climatic extremes and the initiatives to tackle the expected impacts, countries including India have come up with strategies and plans, e.g., India (NATCOM 2009; NAPCC 2008). However, these strategies and plans are not properly oriented to cater the regional or local specific needs that are critical to the agriculture and rural sector. In India, there are timely initiatives from the government understanding the magnitude of impact by climate change, for example, national communications to the UNFCCC, National Action Plan on Climate Change (NAPCC), and other related projects (Table 1).

The Government of India has several policy decisions on the anvil and enabled that reduce risks and enhance the adaptive capacity of the most vulnerable sectors and groups including that of farming community (Fig. 1). These efforts are primarily driven by the objective of sustainable livelihoods and poverty alleviation. According to the latest information on the governmental initiatives undertaken and/or already existing, those are primarily aimed to mainstream as adaptation components in the country. There exist seven major initiatives that include crop improvement and research, poverty alleviation and livelihood preservation, health improvement and prevention of diseases, forest conservation, disaster management, risk financing, and drought proofing and flood control. These initiatives have

Table 1 National initiatives on climate change resilience

Submitted first national communication to UNFCCC in 2004 and second national communication in 2012
National Action Plan on Climate Change (NAPCC) was released in 2008. It identifies eight missions in the area of solar energy, enhanced energy efficiency, sustainable agriculture, sustainable habitat, water, Himalayan ecosystem, increasing forest cover, and strategic knowledge on climate change
ICAR has launched a major project entitled National Initiative on Climate Resilient Agriculture (NICRA) during 2010–2011 in the 11th national plan in conjunction with the proposed NAPCC
To achieve coherence between strategies and action at national and state level. State-level action plan on climate change (SAPCC) was drafted to enable to address existing and future climate risks and vulnerability. 14 out of 28 states have drafted SAPCC, and further planning is underway towards implementation

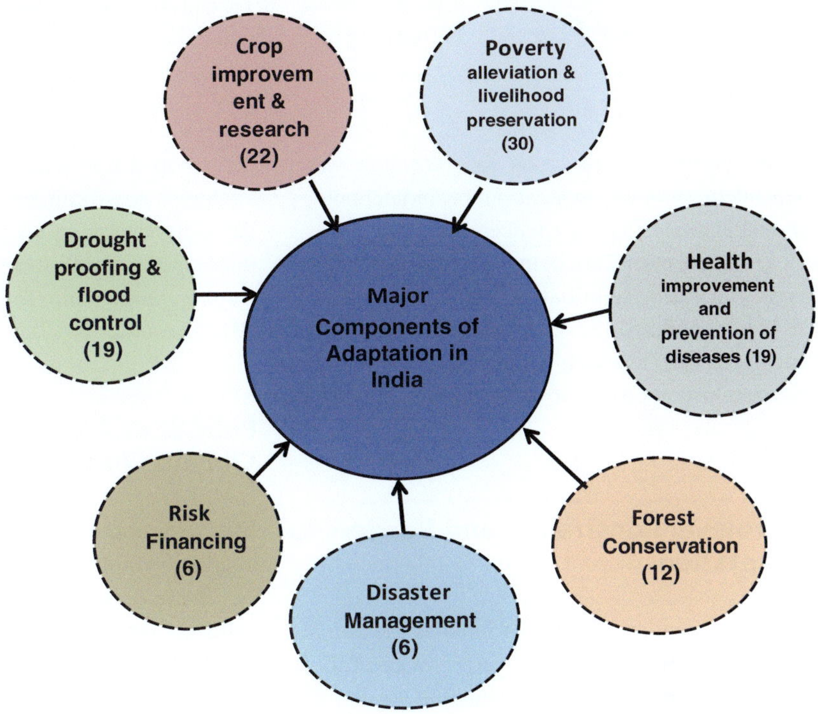

Fig. 1 Major components of adaptation planning in India (Source: GOI (2013). *Parenthesis indicates number of schemes identified under each category)

components of natural resource management, socioeconomic support program, agricultural development program, and other allied programs aimed for strengthening the adaptive capacity of farmers. In total, the schemes are around 100, and most of the schemes are in crop improvement and research and poverty alleviation and livelihood preservation.

The government schemes that currently exist target region/groups that are in need (Table 2). For example, for the agriculture sector, the currently existing schemes include drought proofing measures; promoting zero tillage practices; developing drought-resistant varieties; promoting crop diversification (Walker and Ryan 1990); promoting on-farm water-efficient technologies, farmer credit, and loan system to improve at local level; promoting national agricultural insurance scheme; and encouraging resource-conserving technologies (RCTs) for enhancing input use efficiency and crop productivity. Similarly, in the water resource sector, it includes integrated water resource management strategy, the national water policy, schemes for revival of diverse and community-based irrigation systems and soil and water conservation, managing drought through early warning, flood mapping, appropriate drought protection measures, and schemes for reducing the

Table 2 Major government initiatives and programs in agriculture and water resources sector in India

Agriculture	Water resources
Drought proofing measures	Integrated water resources management strategy
Promoting zero tillage practices	The National Water Policy (2002)
Developing drought-resistant varieties	Revival of diverse and community-based irrigation systems, soil and water conservation, etc.
Promoting crop diversification	Technological management of drought through early warning, flood mapping, etc.
Promoting on-farm water-efficient technologies	Appropriate drought protection measures
Farmer credit and loan system	Reducing the water requirement of crops and developing crops that are less dependent on water through application of biotechnology
Promoting the national agricultural insurance scheme	Encouraging resource-conserving technologies for crop production

Source: Ray (2010)

water requirement of crops that are less water dependent through application of biotechnology and other sophisticated sciences.

These programs are being implemented; however, how far these initiatives have real impact is always unrealized. Several studies including the one from our research stressed the need for a downstream approach to have maximum impact with greater efficiency.

Microlevel Constraints and Barriers to Adaptation

Studies on semiarid tropics of India as the domain conducted by ICRISAT were successful in augmenting improved understanding of the climate challenge influencing the changes in cropping patterns, crop yields, structures of income and employment, and adaptation-coping strategies of the rural poor in semiarid tropic (SAT) villages; the best practices and institutional innovations for mitigating the effects of climate change and other related shocks; and strategies to address socioeconomic problems relating to changing weather patterns and availability of a range of initiatives for their alleviation. Attempt was made to pull together the entire aspects for action. The following list is indicative and is neither exhaustive nor specific. The idea is to suggest policies/strategies to create an enabling environment for the farmers in SAT India to address climate variability-related socioeconomic problems. The illustrative policies are grouped into three subheads: (a) policies and strategies to minimize climate change impacts; (b) tools, technologies, and infrastructure for climate smart agriculture; and (c) financing and forging partnership for transformational change (Bantilan et al. 2012).

Policies and Strategies to Minimize Climate Change Impacts

It is very important that all initiatives to address adaptation and mitigation to climate change must be integrated with government policies that address agriculture, food production (Klein et al. 2005), and livelihood. This will ensure effective mainstreaming. The measures identified should be sustainable based on location specificity and adaptation gains:

- *Integration of climate change initiatives (such as NAPAs,[1] NAPCC,[2] NICRA,[3] NDMA,[4]etc.) with the national agricultural policies/ programs (food security, disaster management, natural resource conservation technology adoption, livelihood enhancement, etc.) to*

[1] National Adaptation Programs of Action identifies priority activities that respond to their urgent and immediate needs to climate change by which further delay would increase vulnerability and/or costs in the future.

[2] National Action Plan on Climate Change.

[3] National Initiatives for Climate Resilient Agriculture.

[4] National Disaster Management Authority.

encourage rural communities to engage and adopt the proposed adaptation measures to address climate change impacts.

Response to climatic shock may not be efficient, but identifying these and work towards improving their capacity to adopt these during the time of climatic extremes (Eriksen et al. 2011). There is a need to implement measures that will enable farmers to invest in mitigation and adaptation measures (short duration varieties, soil and water conservation technologies, crop management practices, replenishing the feed, and fodder management). An example can be to encourage farmers by giving subsidies on interest on loans for implementing adaptation measures. Subsidized weather-based crop insurance crop insurance are a measure to tackle the climate risks associated with extreme weather events such as drought in the region. Development of strong collective initiatives such as cooperative movements will improve economic status and help in facing climate shocks.

- *Prioritizing regions of climate change vulnerability in arid and semiarid tropics; preparation and implementation of comprehensive district-wise (local level) agriculture and livelihood contingency plans*[5] *of actions for effectively managing the climate risk.*

From meso-level data analysis, the regions vulnerable to long-term climate change need to be identified. Regional crop-contingency plans, i.e., district-wise, which will be a response to anticipated climate change developed on an annual basis, with sufficient flexibility. In all study countries, regional-level plans exist, and identifying vulnerable sectors and regions is a prerequisite.

- *Encourage crop and livelihood diversification*[6] *and ensure rural income flow; managing the common property resources (ponds, wells, tanks, grazing land, etc.) judiciously by community participation enabling long-term sustainability.*

Increasing dry spells in the wet seasons, delayed monsoons, and other climate change-related effects require tailoring a location-specific cropping calendar and developing suitable crop management techniques through research and interaction with the farmers. The farming income is not considered sufficient to cover the increasing risks due to uncertainty and variability in rainfall and occurrences of extreme events like droughts, floods, etc. Farmers are increasingly looking for diversification to high-value crops and other income-generating enterprises from traditional agriculture to cushion the risk associated with agricultural production and income loss. Farmers need an enabling environment that creates or assists in innovation by the farmers to diversify their income sources. This could be achieved through rural developmental agencies. Hence, revamping of rural developmental agencies such as SFDA[7] and DRDA[8] focused towards small farmers in India, policies on sustainable development, livestock production, irrigation and fisheries, and aqua cultural development. In India, evoking the focus of these rural development agencies to farm and nonfarm evenly is a must.

- *Support to implement pasture conservation and better feed and fodder management*[9] *approaches for improved productivity of livestock, fisheries, poultry, and other enterprises.*

There is a need to improve feed and fodder management approaches to enhance fodder quality and availability for improved livestock productivity. Cereal-based systems particularly coarse cereals are slowly being replaced with other cash crops in villages of SAT India. As a result the availability of dry fodder to feed the

[5] Includes state-/district-level contingency plans, disaster management plans, and other reliefs.

[6] Enable opportunities to diversify more into high-value crops, livestock and other nonfarm income sources.

[7] Small Farmer Development Agency.

[8] District level Rural Development Agency.

[9] Programs/schemes on dairy development, development of small ruminants, fodder and feed development, livestock entrepreneur programs, etc. In India, the concerns of demand for fodder and pasture are on the 12th plan call for rehabilitation of pasture and fodder resource in the country.

livestock population has become an issue. Options for improved fodder management and availability will ensure a healthy development of livestock sector in the villages which will further help the farmers to diversify their income options. There are several livestock/poultry/fisheries programs at state/district/national level officially being implemented in the regions; however, the time has come to relook the impacts of these programs and policies on livelihood and ensuring better effectiveness and efficiency.

- *Ensure equitability in access of government support/relief programs such as Antyodaya*[10] *program, food security programs in India, etc. These programs focus on food security, agricultural and enterprise subsidies, rural finances, poverty reduction programs, technology adoption support, etc.*

All groups of farmers must be able to get loans under easy conditions. This will enable small and disadvantaged farmers to implement adaptation measures to address climate change. This is true only if there is no recurrence of drought in this period. In reality droughts recur in the time span, and many of these farmers fall into perpetual debt traps. Access to finance on easy terms and highly subsidized interest will help them come out of the debts.

Support in terms of subsidies must be given for choosing adaptation measures and innovative technologies to address climate change impacts as well as productivity-improving measures of watersheds, integrated water, and nutrient management options for efficient use of resources like land, water, etc., as well as any other inputs. This is mainly because farmers in vulnerable areas do not have any social safety nets and require support to sustain and continue crop production. Easy access to support mechanisms like government interventions in terms of knowledge flow and/or financing options might help.

- *Strengthen and empower the final beneficiary, i.e., farmers, to make them meaningful partners. Supplement their traditional/experiential knowledge with valid scientific knowhow and technology options, engage them more meaningfully in climate information management systems, provide incentives to farmers to adopt natural resource conservation measures, and support to improve the existing indigenous technologies that are eco-friendly.*

Although the farmers had a wealth of information and experience in dealing with climate change variability and the harsh realities of moisture stress, they were still lacking in knowledge on accessing information and taking optimum use of services provided by the governments. Often they are unaware of their entitlements, reliefs on offer, and other government support programs and thus fall prey to ignorance and consequences of extreme climate conditions. In vulnerable areas, farmers also lacked social capital and the organizational capabilities. They are often passive suppliers of information to the state and research establishments, but not integrated as valuable and active stakeholders in the climate change debates or intervention programs. The concept of "climate change schools" could have sufficient potential for sharing information and knowledge (indigenous knowledge). The weather data collected at local levels once synthesized centrally must go back to the farmers as useful outputs, so that they can and are assisted to make effective use of inferences drawn. The study also calls for strengthening extension program and institutionalizing effective mechanism of information dissemination through Agricultural Technology Centre (KVK,[11] ATMAs[12] in India) in every block/mandal/sub-country level.

- *Prioritize investment in training officials, extension, and local development workers to make them more effective change agents in assisting farmers and strengthening*

[10] Schemes under the program included land allotment, agriculture and land development, animal husbandry, village and cottage industries, wage employment, old age pension, housing subsidy, etc.

[11] Krishi Vigyan Kendra is a district-level institution engaged in transfer of latest agricultural technologies to the end users for bridging the gap between production and productivity.

[12] Agricultural Technology Management Agencies addresses the constraints faced by extension system.

institutions to improve climate adaptation capacity at local levels.

Officials responsible for the farmers' socio-economic well-being may be educated on climate change and mitigation through a series of awareness programs. Such programs may be conducted at the village level, and required incentives need to be provided. To illustrate, it is observed that the common property resources (CPRs) like grazing lands have degraded over the last several decades due to lack of collective action in managing them. It will be appropriate for extension officials to educate the farmers on low moisture availability in their ecosystem and ways to mitigate. It is essential to emphasize on capacity building for the government employees dealing with farmers problems in particular and agriculture in general. The lack of needed competitiveness in understanding climate change-related implications and experience is highly recommended for the region. Moreover, various stakeholders involved in nation building through agriculture development are not well aware about global policies, decisions, and other related information.

Tools, Technologies, and Infrastructure for "Climate Smart Agriculture"

- *Increasing the density of weather observatories, establishing rain gauges at village level, and enabling access and efficient management of weather-related information (remote sensing and GIS) and repository.*

Weather especially rainfall is variable across the regions. Analysis of single-station data may not represent the accurate climate resources. Microlevel weather data analysis showed a decreasing trend in the rainfall compared to the positive trends at a district level. This feature was observed in two selective project locations in India. Village-level rainfall observations are important in characterizing the environment at microlevel. Therefore, there is a need to increase the density of network of weather stations for better interpretation of variability of weather parameters and for accurate planning for improved and sustainable agricultural production. Droughts and flash floods are common in the Asian countries. In the event of increased frequencies of extreme weather events, agricultural production gets affected considerably. The best way to reduce the impact is to prepare the farmer well in advance to manage the situation in order to minimize the losses. Weather-based agro-advisories benefit the farming community in ensuring effective agricultural operations. In spite of best efforts to alert the farmers, extreme weather events often cause huge losses subjecting the farmer to extreme hardships. To save the farmer from the weather hazards, weather insurance is quite beneficial. To cope with disasters such as typhoons, flash foods, or droughts, identification of geographical boundaries for such events followed by preparation of regional crop-contingency plans must be put in place. These will form ready-reckoners to meet out any eventuality. They should be prepared to meet out the year-to-year variability. Modern tools such as "remote sensing and GIS" should provide an excellent opportunity to analyze spatial land use and land cover changes to climate change. There have been initiatives[13] from the government on this front in the study countries to improve the infrastructure and database on climate information and to use advanced methods.

- *Institutionalize continuous mechanism to collect and collate microlevel information (climate, crops, socioeconomic, natural resources, etc.) and efficiently transmit to be used as an input in formalizing macro-level policies.*

Most of the macro-level policies are formulated with inputs from aggregated level. The aggregated information and existing microlevel information could be highly diverse. There is a pressing need of having microlevel information on climate, crops, socioeconomics, natural resources, governance, trends, efficiencies, etc., especially in the context of

[13] In India, the Indian Meteorological Department (IMD) and the Allied Department are greatly involved in enhancing weather information by improving weather station density across the country.

climate change issues. Microlevel information needs to be collected and collated to be accessed and used by various national/regional, governmental/nongovernmental, and other developmental agencies for efficient planning.

- *Blending of farmers' traditional/indigenous knowledge on resource conservation, coping strategies, etc., with advanced technological interventions (varieties, crop management, community resource conservation, rainwater harvesting, storage, etc.) for coping against climate change and associated stress.*

Farmers have inherited the knowledge of managing and understanding the climate through their ancestors. Hence, there is a need to utilize this ancient wisdom[14] along with modern knowhow. For effective utilization of the modern technologies, combining traditional knowledge may improve the reliability and acceptability.

- *Encourage investment in research and development of locally adaptable crops, management practices, input sources, etc., decision support systems (DSS), and models for analyzing the impacts of climate change and mitigation strategies in the semiarid tropics in view of future climate scenarios.*

With the changes in and increasing variability of weather patterns, and introduction of new crops and varieties, the pest and disease behavior is likely to be altered in any given location. There is a need to identify such location and crop-specific pest and disease incidence and approaches to manage such situations that developed. For example, in Maharashtra (India), introduction of sugarcane in Shirapur and soybean in Kanzara and improved and short duration pigeon pea in Kalman villages brought in new diseases and pests that needed different management practices from the normal. Improved on-farm water harvesting and water conservation measures are useful in rainfed agriculture; similarly, improved technologies like drip irrigation and precision timing of irrigation will reduce the risk associated with the variability of rainfall (Rockstrom et al. 2010; Lundqvist and Falkenmark 2010; Barron et al. 2010). Incorporating organic matter or mulching to increase the water holding capacity of soils, in situ water storage using different devices is a time-tested measure adopted by farmers in cover cropping, mulching, composting, etc. The different techniques adopted by farmers in the region provide an array of options for field validation in other countries and subsequent adoption.

- *Encourage adoption of location-specific conservation techniques (cover cropping, in situ moisture conservation, rainwater harvesting, groundwater recharge techniques, locally adapted cropping mixture, etc.) for water-efficient agriculture and demonstration of these available technologies[15] in the farmers' field.*

Incentives or support must be given for choosing adaptation measures and innovative technologies to address climate change impacts as well as productivity-improving measures for efficient use of resources like land, water, etc. as well as any other inputs. For example, modern technology and external support by government in India helped the farmers in many villages to harvest groundwater through agro wells and tube wells. In the recent decades, there has been a rapid, uncontrolled expansion in the number of tube wells in many villages resulting in the receding of the groundwater table. Such "tragedy of the commons" should be avoided through collective action, regulation by external agencies, or systems of incentives and disincentives. The ground situation is sometimes aggravated due to low levels of education of farmers. Thus improving the knowledge of farmers may be a first step before adopting other measures. This calls for sensitivity to local socioeconomic contexts when addressing mitigatory measures.

- *Managing climate risks effectively through weather-based agro-advisories and*

[14] On weather prediction, water conservation and storage, and cultivation practices, viz., organic farming, natural pesticides, etc.

[15] Support in soil and water conservation, soil health, irrigation, fertilizer, etc.

developing equally accessible innovative weather insurance products.[16]

In the event of increased frequencies of extreme weather events, agricultural production gets affected considerably. The best way to reduce the impact is to prepare the farmer well in advance to manage the situation in order to minimize the losses. Weather-based agro-advisories come in a big way to benefit the farming community on timely agricultural operations. In spite of the best efforts to alert the farmers, extreme weather events often cause huge losses subjecting the farmer to extreme hardships, and weather insurance can be an effective strategy to offset the losses. In order to prepare the weather insurance products for different agro-climatic regions, research efforts on crop-weather relations need to be strengthened.

- *Harnessing nonconventional energy*[17] *sources in agriculture and other allied sectors.*

The use of nonconventional sources of energy, such as biofuels, solar energy, and wind power in agricultural operations, is very limited where there are more effective state interventions, with high levels of adoption in the rural area. In order to reduce the GHG emissions from the different sources, more research is required to estimate the emission levels and on measures to restore the balance.

Financing and Partnerships for Transformational Change

- *Enabling environment to attract public and private finances to invest in "climate smart agriculture"*

Increasing the level of state financing for promoting climate smart agriculture is a priority, considering the long-term goals of minimizing food insecurity, reducing carbon emissions, and mitigating climate change effects. Public investment in the field of agriculture research and development must be increased. The focus should be to invest in tools and technologies as well as policies. For example, the National Initiative for Climate Resilient Agriculture (NICRA) is a major research and capacity-building national project launched by the Government of India and ICAR to develop location-specific tools and technologies and capacity building.

- *Encouraging the role of the nongovernmental organizations and public and philanthropic organizations for enhancing adaptation preparedness among the local community*

Along with the government efforts, NGOs are also important for the development of the rural community. There is a need to generate partnerships between public funding and financing from foundations and charitable private institutions for investment into smart agriculture promotion (Vermeulen et al. 2012; Vogel et al. 2007). Many NGO and other research organizations funded by various societies and trusts have been doing commendable job in various sectors. Their involvement in conducting research to manage climate change threats may be encouraged, and enabling environment must be created.

- *Forging international/regional partnerships for developing tools and technologies adaptable to suit local requirement through pooling finance and intellectual resources*

International partnerships among neighboring countries that share similar ecosystems as well as similar agricultural practices might be useful in sharing financial and intellectual resources to develop appropriate adaptation tools and technologies.[18] The technologies generated at

[16] Weather-based insurance schemes, government support through subsidies on premium. When weather indices differ from the guaranteed indices of major crops, a payment equal to the deviation/shortfall is payable to all insured farmers.

[17] The progressing demand and initiatives by the respective governments are existing. These are well highlighted in the related policies and strategies, national action plans, etc.

[18] Drought, flood and salt tolerance varieties, robust methodologies to predict climate change impacts, resource conservation technologies, innovative safety nets, etc.

various locations in the world may be collected and identified for its suitability to other regions. The SAARC[19] is a potential platform for cooperation and exchange of tools, technologies, skills, finance, and other related resources to combat climate change and enhance resilience in India.

Conclusion

Future adaptation and/or agriculture policies should explicitly draw on the evidence-based grassroots level insights and should create enabling environment overcoming the constraints at the grassroots. In order to design efficient and effective adaptation and mitigation strategies, there is a prior need to know the impacts of climate change and locate existing knowledge on climate change, including local practices and indigenous knowledge. The overarching recommendation is to diagnose and understand farmers' adaptation strategies to climate variability and change with special focus on the dynamics of adaptations, implying search for and promoting approaches and options to harness the opportunities in the changing economic, technological, and institutional opportunities at microlevel, which should even exceed what farmers have been practicing in the subsistence-oriented, locally focused contexts in the present. Dynamism, diversity, and flexibility are essential for enhancement and reorientation of the capacities of the farmers and fostering enabling environment among the rural communities, though institutional arrangements and innovations are in dire need. Future policies and actions should be devised from evidence-based information on climate change impacts and implication at the household and community level.

[19] South Asian Association for Regional Cooperation (SAARC) with an objective of providing promotion of economic and social progress and cultural development within the South Asia region.

References

Adger WN (2003) Social capital, collective action, and adaptation to climate change. Econ Geogr 79 (4):387–404

Adger WN, Vincent K (2005) Uncertainty in adaptive capacity. Geosciences 337:399–410

Bantilan C, Singh N P, Byjesh K, Padmaja R, Jayatilaka W (2012) Helping communities adapt: climate change perceptions and policy in Asia (Policy brief no. 23). Research Program – Markets, Institutions and Policies. International Crops Research Institute for the semi-arid tropics, Patancheru, p 12

Barron J, Enfors E, Cambridge H, Moustapha A (2010) Coping with rainfall variability in semi-arid agroecosystems: implications on catchment scale water balances by dry spell mitigation strategies among small-scale farmers in Niger. Int J Water Resour Dev 26(4):543–559

Eriksen S, Aldunce P, Bahinipati CS, D'almeida T, Molefe JI, Nhemachena C, O'brien K, Olorunfemi F, Park J, Sygna L, Ulsrud K (2011) When not every response to climate change is a good one: identifying principles for sustainable adaptation. Clim Dev 3:7–20

GOI (2013) Department of Agriculture & Cooperation; Ministry of Agriculture, Government of India. http://agricoop.nic.in/

INCCA (2010) Climate change and India: a 4 × 4 assessment – a sectoral and regional analysis for 2030s (Indian Network for Climate Change Assessment, report no. 2). Ministry of Environment and Forest, Government of India, New Delhi, p 164, November 2010

IPCC (2007) Climate change 2007: synthesis report. Contribution of Working Groups I, II and III to the fourth assessment report of the Intergovernmental Panel on Climate Change. IPCC, Geneva, Switzerland, 104 pp

Kelly PM, Adger WN (2000) Theory and practice in assessing vulnerability to climate change and facilitating adaptation. Clim Chang 47:325–352

Klein RJT, Schipper LEF, Dessai S (2005) Integrating mitigation and adaptation into climate and development policy: three research questions. Environ Sci Policy 8:579–588

Lundqvist J, Falkenmark M (2010) Adaptation to rainfall variability and unpredictability: new dimensions of old challenges and opportunities. Water Resour Dev 26(4):595–612

NAPCC (2008) National action plan on climate change. Government of India, Prime minister's council on climate change, p 59. http://pmindia.nic.in/pg01-52.pdf

NATCOM (2009). India's National Communication to UNFCCC. Data Extraction tool for Regional Climate Scenario (PRECIS) for India. Ministry of Environment and Forests, Government of India

Ray R (2010) India: adaptation approaches and strategies. Ministry of Environment & Forests, Government of India, New Delhi

Rockstrom J, Karlberg O, Wani SP, Barron J, Hatibu N, Oweis T, Bruggeman A, Farahani J, Qiang Z (2010) Managing water in rainfed agriculture – the need for a paradigm shift. Agric Water Manag 97:543–550

Shiferaw B, Bantilan MCS (2004) Agriculture, rural poverty and natural resource management in less favored environments: revisiting challenges and conceptual issues. Food Agric Environ 2(1):328–339

Vermeulen S, Zougmore R, Wollenberg E, Thornton P, Nelson G, Kristjanson P, Kinyangi J, Jarvis A, Hansen J, Challinor A, Campbell B, Aggarwal PK (2012) Climate change, agriculture and food security: a global partnership to link research and action for low-income agricultural producers and consumers. Curr Opin Environ Sustain 4(1):128–133

Vogel C, Moser SC, Kasperson RE, Dabelko GD (2007) Linking vulnerability, adaptation, and resilience science to practice: pathways, players, and partnerships. Glob Environ Chang 17:349–364

Walker T, Ryan J (1990) Village and household economies in India's semi-arid tropics. Johns Hopkins University Press, Baltimore, p 394

Epilogue

Information embodied in the papers pertains to the wide variety of location-specific characteristics across India's landscapes. Adaptation and mitigation options, therefore, appear to be location and crop specific. The initiative of the Indian Council for Agricultural Research has been discussed in great detail including the recent National Initiative on Climate Resilient Agriculture. Strategies for conservation and management of crops, soil, water, nutrients, and carbon in particular have been discussed with implications for sustaining transitions. Biomass energy options and waste recycling also feature on the menu of options. Interestingly investigations from the state of Odisha, parts of central India, Punjab, Tamil Nadu, Rajasthan, Maharashtra, Himalayas, and the North-East have provided significant empirical insights guiding modeling efforts. Crop performance evaluation and contingency planning, linkages with synoptic disturbances, drought analysis, and their best fit with several models are discussed. Remote sensing and other geospatial tools have also been mainstreamed to justify model assessments. The use of several integrated ICT strategies has also featured along with interventions to assess the impacts. A chapter on farmer's perceptions and preparedness to adopt climate-resilient measures provides valuable insights regarding the community interface. Valuable lesions for development of integrated policies from a wide range of initiatives from across the world strengthen the mosaic of inputs, for easy learning.

The present compilation is therefore an initiative to help consolidate learning and stimulate convergence of ideas. We wish the readers will find the effort useful.

A.K. Singh et al. (eds.), *Climate Change Modelling, Planning and Policy for Agriculture*,
DOI 10.1007/978-81-322-2157-9, © Springer India 2015

MIX
Papier aus verantwortungsvollen Quellen
Paper from responsible sources
FSC® C105338

If you have any concerns about our products,
you can contact us on
ProductSafety@springernature.com

In case Publisher is established outside the EU,
the EU authorized representative is:
Springer Nature Customer Service Center GmbH
Europaplatz 3, 69115 Heidelberg, Germany

Printed by Libri Plureos GmbH
in Hamburg, Germany